ALGEBRA
for College Students

ALGEBRA
for College Students

Nancy Myers

Bunker Hill Community College

D. VAN NOSTRAND COMPANY

New York   Cincinnati   Toronto   London   Melbourne

Cover photograph by Fred Charles.

D. Van Nostrand Company Regional Offices:
New York   Cincinnati

D. Van Nostrand Company International Offices:
London   Toronto   Melbourne

Library of Congress Catalog Card Number: 78-069759
ISBN: 0-442-25625-6

Published by D. Van Nostrand Company
135 West 50th Street, New York, N.Y. 10020

10 9 8 7 6 5 4 3 2

Preface

ALGEBRA FOR COLLEGE STUDENTS is intermediate in level in the sense that it assumes the student has some background in the basic concepts of algebra. We have not called the book Intermediate Algebra, however, because it is different in many ways from standard books of that title.

In particular, we do not begin with a general review, but rather we review each concept as it is encountered, so that each is reviewed at the time it is required. We have included a uniquely broad range of topics, each including many worked examples and detailed explanations. Complicated proofs are left for the ends of units where they do not interrupt the flow of the text. The book contains topics in the form of applications which are not generally found in standard books, including conic sections, linear programming, and exponential growth and decay.

The unique order in which the topics are presented is essential to the unique scope of this book. We introduce quadratic equations early, so that they are available for use throughout the book. Applications are introduced early and then presented whenever possible throughout the book. Functions are introduced early through the specific examples of the linear and quadratic functions, and then other basic types of functions are introduced throughout the book.

We have chosen the unit form of organization because it allows the student to master a single concept in each section and a limited number of related concepts in each unit. Specific Objectives are stated at the beginning of each unit. Each section number and topic within the unit correspond to an objective number. There is a Self-Test at the end of each unit. The self-test items may be in random order, but the answers also are keyed to the objective numbers. Thus a student can refer by number from a self-test item to the relevant objective and corresponding section. This organization within each unit, along with Cumulative Review units spaced strategically throughout the book, make this book suitable for self-paced courses and other types of self-instruction as well as traditional courses.

The unit form of organization also allows the instructor flexibility in choosing topics. The book falls generally into three parts. The first part, Units 1 to 7, introduces basic techniques of equation solving, applications, and functions, for both linear and quadratic forms. These units provide background and motivation to explore other algebraic forms, such as rational expressions, exponents, and radicals, in the second part consisting of Units 8 to 16. The third part, Units 17 to 24, then uses these forms to explore more sophisticated equation solving, applications, and relations and functions. Units 1 to 16 may be considered a core course at the intermediate level. In Units 17 to 24, every effort has been made to allow choices among units, or parts of units, to fit a variety of needs and time limitations.

The answers to all Exercises and Self-Tests are given at the back of the book. An Instructor's Manual is available which contains four quiz forms for each unit, as well as a pretest and two forms of a final exam.

As the text developed, detailed comments were provided on several occasions by Richard Semmler, of Northern Virginia Community College, and Ara Sullenberger, of Tarrant County Junior College. Other reviewers included Arnold Knuppel, of the University of Houston, Albert Liberi, of Westchester Community College, Ronald Schryer, of Orange Coast College, and Charlotte Ann Webster, of Appalachian State University.

At Bunker Hill Community College, the mathematics department with Joan McGowan, chairperson, provided help and encouragement. In particular, Virginia Campiola used the manuscript in her sections of our course, and Robert (Ted) Carlson checked all the worked examples and exercise answers. Maria Trenga, a former student, helped with typing. And scores of students helped with their comments and questions.

Nancy Myers

Contents

Unit 7 Quadratic Functions   94

Unit 8 Factoring   117

Unit 9 Rational Expressions   131

Unit 10 Common Denominators   145

Unit 11 Cumulative Review   158

Unit 12 Integral Exponents   161

Unit 13 Rational Exponents   177

Unit 14 Radical Expressions   189

Unit 15 Imaginary Numbers   208

Unit 16 Cumulative Review   224

Unit 17 Equations with Rational Expressions   228

Unit 18 Equations with Radical Expressions   248

Unit 19 Systems of Equations   269

Unit 20 Inequalities   304

Unit 21 Absolute Value   330

Unit 22 Cumulative Review   345

Unit 23 Logarithms   350

Unit 24 Exponential Equations   369

Answers   381

Appendix   436

Index   443

Simplifying Expressions

INTRODUCTION

In this unit you will review some basic techniques for simplifying algebraic expressions that you will need to use in the following units. All these techniques are related to one very important property, the distributive property. The use of the distributive property will be emphasized in developing each technique.

OBJECTIVES

When you have finished this unit you should be able to:

1. Multiply a polynomial by a common factor and remove a common factor from a polynomial.
2. Combine like terms of an algebraic expression.
3. Divide an algebraic expression by a common factor.
4. Multiply two binomials.
5. Factor a trinomial into two binomial factors.

Section

1.1

The Distributive Property

Algebra is the study of algebraic expressions. An **algebraic expression** is a representation of a number. It may include numbers, letters, and any of the operations of addition, subtraction, multiplication, division, exponentiation, and root extraction. If numbers are substituted for all of the letters in the expression and the operations are performed, an algebraic expression will simply be a number. In algebra, we call the numbers within an algebraic expression **constants**, and the letters **variables**.

    A **monomial** is an algebraic expression consisting of constants, or variables, or both, joined only by multiplication. Some examples of monomials are:

$$2, \quad x, \quad x^2, \quad x^2y, \quad 2x^2y^2z, \quad \sqrt{3}\,xy, \quad \frac{1}{2}xy.$$

A monomial may include exponentiation since exponentiation can be thought of as multiplication. For example,

$$x^2 = x \cdot x,$$

which consists of variables joined by multiplication. A constant may appear as a radical; for example, $\sqrt{3}$ is a constant. There could not be a variable in the radical, however. A constant may also be a fraction, as in $\frac{1}{2}xy$, but a variable may not be in the denominator of a fraction.

A **polynomial** is any number of monomials added together. For example,

$$2x^2y + 2xy^2$$

is a polynomial. We may also consider

$$2x^2y - 2xy^2$$

to be a polynomial, since we may think of this expression as $-2xy^2$ added to $2x^2y$. Another example of a polynomial is

$$x^2 + 3x - 2,$$

where $3x$ is added to x^2, and then -2 is added to the result. Each monomial addend of a polynomial is called a **term** of the polynomial. In the example above, the terms are x^2, $3x$, and -2.

A polynomial consisting of exactly two terms is called a **binomial**. Some examples of binomials are:

$$x - 2, \quad xy + 4, \quad x^2 - z^2.$$

Suppose we wish to multiply a binomial represented by $B + C$, where B and C are the two terms of the binomial, by a monomial A. We may do this by multiplying B by A and C by A. This statement describes the **distributive property for multiplication over addition**. Written symbolically, the distributive property for multiplication over addition is

$$A(B + C) = AB + AC.$$

We can easily show examples of the distributive property using numbers:

$$2(4 + 7) = 2(4) + 2(7)$$

since

$$2(4 + 7) = 2(11) = 22$$

and

$$2(4) + 2(7) = 8 + 14 = 22.$$

EXAMPLE 1.1.   Multiply $2(x + y)$.

Solution.   Using the distributive property, we multiply x and y each by 2:

$$2(x + y) = 2x + 2y.$$

To multiply a polynomial of more than two terms by a monomial, we simply multiply each term, however many, by the monomial. This property is called the **extended distributive property**, although we will usually refer to it also as the distributive property.

EXAMPLE 1.2.   Multiply $3x(x + y - 2)$.

Solution.
$$3x(x + y - 2) = 3x(x) + 3x(y) + 3x(-2)$$

$$= 3x^2 + 3xy - 6x.$$

Multiplication of a polynomial by a negative monomial reverses all the signs of the polynomial.

EXAMPLE 1.3.   Multiply $-5(x + 2y - 1)$.

Solution.

$$-5(x + 2y - 1) = -5(x) + (-5)(2y) + (-5)(-1)$$
$$= -5x - 10y + 5.$$

A minus sign alone in front of a polynomial simply changes all the signs of the polynomial. We may think of the minus sign as -1.

EXAMPLE 1.4.   Multiply $-(3x - 2y - 1)$.

Solution.

$$-(3x - 2y - 1) = -1(3x - 2y - 1)$$
$$= -1(3x) + (-1)(-2y) + (-1)(-1)$$
$$= -3x + 2y + 1.$$

It is usual to skip the two middle steps and simply write

$$-(3x - 2y - 1) = -3x + 2y + 1.$$

When each term of a polynomial is multiplied by a monomial A, we call A a **common factor**. Just as a common factor may be multiplied through a polynomial using the distributive property, a common factor may be removed from a polynomial using the distributive property.

Let us think of the distributive property with the left- and right-hand sides reversed:

$$AB + AC = A(B + C).$$

A is a factor of AB and also a factor of AC. Therefore, A is a common factor of AB and AC. Whenever a polynomial contains a factor common to every term, the factor may be taken outside the polynomial using the distributive property.

EXAMPLE 1.5.   Factor $2x + 2y$.

Solution.   Observe that $2x$ and $2y$ have a common factor 2. Therefore,

$$2x + 2y = 2(x + y).$$

You should check such examples by multiplication:

$$2(x + y) = 2x + 2y,$$

and so the factoring is correct.

EXAMPLE 1.6.   Factor $4x^2 + 12x$.

Solution.   Since $12 = 4 \cdot 3$, 4 is a common factor. Thus,

$$4x^2 + 12x = 4(x^2 + 3x).$$

However, there is still a common factor x, so the factoring is not yet complete:

$$4x^2 + 12x = 4(x^2 + 3x)$$
$$= 4x(x + 3).$$

To check,

$$4x(x + 3) = 4x(x) + 4x(3)$$

$$= 4x^2 + 12x.$$

You must always be sure to factor completely; that is, be sure there are no common factors left in the polynomial. Then check your factoring by multiplication.

Exercise 1.1

Multiply:

1. $4(x - y)$
2. $-3(x - y)$
3. $-(x - y)$
4. $2x(x + 3)$
5. $-4(x^2 - 2x - 1)$
6. $5x(2x + 4)$
7. $2x(2x^2 + 3x + 2)$
8. $-(x^2 - 2x - 3)$
9. $-2x(x - 10)$
10. $-x(x + 6)$

Factor completely:

11. $3x + 3y$
12. $2x - 4$
13. $6x + 3$
14. $2x^2 - 6x + 2$
15. $x^2 - 2x$
16. $2x^2 + 4x$
17. $10x^2 - 10x - 5$
18. $12x^2 + 8x - 16$
19. $x^3 + 2x^2 + 2x$
20. $3x^3 - 6x^2 - 3x$

Section 1.2 Combining Like Terms

At this time, we look at two other important properties besides the distributive property. These properties are the associative property and the commutative property.

The **associative property** states that we may add three numbers by adding the first two and then the third to the result, or by adding the second two and then the first to the result. For example, to add $2 + 3 + 4$ we may write

$$(2 + 3) + 4 = 5 + 4 = 9,$$

or

$$2 + (3 + 4) = 2 + 7 = 9.$$

More generally, if A, B, and C are terms of a polynomial, then

$$(A + B) + C = A + (B + C).$$

The associative property is also true for multiplication:

$$(A \cdot B) \cdot C = A \cdot (B \cdot C).$$

The **commutative property** states that we may add two numbers in either order: the first to the second, or the second to the first. For example,

$$2 + 3 = 5,$$

and also

$$3 + 2 = 5.$$

More generally, if A and B are terms of a binomial, then

$$A + B = B + A.$$

The commutative property is also true for multiplication:

$$A \cdot B = B \cdot A.$$

If we have an expression such as $(2 + x) + 3$, we may write:

$$(2 + x) + 3 = (x + 2) + 3 \qquad \text{commutative property}$$

$$= x + (2 + 3) \qquad \text{associative property}$$

$$= x + 5.$$

This process is called **combining like terms**. In combining like terms, we will not generally specify uses of the commutative and associative properties. We will concentrate on the distributive property and how it may be used to combine like terms.

EXAMPLE 1.7.   Combine $2 - (x - 1)$.

Solution.   We use the distributive property to remove the parentheses. Once the application of the distributive property has been completed, the other properties are easily applied:

$$2 - (x - 1) = 2 - x + 1$$

$$= 2 + 1 - x$$

$$= 3 - x.$$

Two or more terms are called **like terms** if they are constants, or if all the variable parts are identical. $2x$ and $3x$ are like terms. $2x$ and $3x^2$ are not like terms because the first has an x and the second an x^2. $2x$ and $3y$ are not like terms because the first has an x and the second a y. $2x^2y$ and $3x^2y$ are like terms.

We may use the distributive property to combine like terms of a polynomial. For example, to combine

$$2x + 3x,$$

we factor out the common factor x:

$$2x + 3x = (2 + 3)x$$

$$= 5x.$$

To combine

$$5x^2y - 2x^2y$$

we factor out the common factor x^2y:

$$5x^2y - 2x^2y = (5 - 2)x^2y$$

$$= 3x^2y.$$

Normally we do not write the first step showing the factoring. Thus, we simply write:

$$2x + 3x = 5x$$

and

$$5x^2y - 2x^2y = 3x^2y.$$

EXAMPLE 1.8. Combine like terms of $(2x^2 + x - 3) - (x^2 + 4x - 2)$.

Solution.

$$
\begin{aligned}
(2x^2 + x - 3) - (x^2 + 4x - 2) &= 2x^2 + x - 3 - x^2 - 4x + 2 \\
&= 2x^2 - x^2 + x - 4x - 3 + 2 \\
&= x^2 - 3x - 1.
\end{aligned}
$$

EXAMPLE 1.9. Combine like terms of $-3x(x + 2) + 4(x^2 - 2x - 1)$.

Solution.

$$
\begin{aligned}
-3x(x + 2) + 4(x^2 - 2x - 1) &= -3x^2 - 6x + 4x^2 - 8x - 4 \\
&= -3x^2 + 4x^2 - 6x - 8x - 4 \\
&= x^2 - 14x - 4.
\end{aligned}
$$

Exercise 1.2

Combine like terms:

1. $(2 + x) - 3$

2. $6 - (x - 3)$

3. $4x + 6x$

4. $12x - 5x$

5. $(2x - 3) + (x + 2)$

6. $(3x + 4) - (2x - 3)$

7. $(x - 3) - (x - 4)$

8. $2(x + 1) + 3(x - 5)$

9. $6(2x - 3) - 3(5x - 2)$

10. $4(x - 2) - 2(3x - 4)$

11. $x(x + 3) - 2(x + 3)$

12. $x(x - 2) - 2x(x - 1)$

13. $-(2x^2 + x - 5) - 2(x^2 - x + 4)$

14. $3x(x - 2) - (x^2 + 2x - 1)$

15. $-4x(x - 1) + 6x(2x - 3)$

16. $-3x(x^2 - 2x - 1) - 4x(x^2 + x - 2)$

<table>
<tr><td>Section
1.3</td><td><h1 style="text-align:center">Division by a Common Factor</h1></td></tr>
</table>

We often encounter the problem of reducing fractions. In arithmetic, you learn to reduce fractions by dividing out a common factor. For example,

$$\frac{2}{4} = \frac{1}{2} \quad \text{since} \quad \frac{2}{4} = \frac{2 \cdot 1}{2 \cdot 2} = \frac{2}{2} \cdot \frac{1}{2} = 1 \cdot \frac{1}{2} = \frac{1}{2}.$$

Observe, however, that we cannot divide out *addends*, that is, terms that are added to other terms. We can only divide out factors.

If a fraction with a polynomial numerator has a common factor in the numerator, it may be possible to reduce the fraction.

EXAMPLE 1.10.   Reduce $\dfrac{2 + 2x}{4}$.

Solution.   Observe that the numerator has a common factor 2. We factor the numerator:

$$2 + 2x = 2(1 + x).$$

Then,

$$\frac{2 + 2x}{4} = \frac{2(1 + x)}{4}$$

$$= \frac{1 + x}{2},$$

dividing a factor of 2 from the numerator and the denominator.

EXAMPLE 1.11.   Reduce $\dfrac{x^2 + 3x + xy}{x}$.

Solution.

$$\frac{x^2 + 3x + xy}{x} = \frac{x(x + 3 + y)}{x}$$

$$= x + 3 + y.$$

EXAMPLE 1.12.   Reduce $\dfrac{8 - 16\sqrt{3}}{12}$.

Solution.   We observe that, like 2 and $2x$, the terms 8 and $16\sqrt{3}$ are *not* like terms and cannot be combined. Therefore, we reduce the fraction by factoring out the common factor 8 in the numerator:

$$\frac{8 - 16\sqrt{3}}{12} = \frac{8(1 - 2\sqrt{3}\,)}{12}$$

$$= \frac{2(1 - 2\sqrt{3}\,)}{3}$$

$$= \frac{2 - 4\sqrt{3}}{3}.$$

Exercise 1.3

Reduce:

1. $\dfrac{3x + 6}{6}$ 2. $\dfrac{5 - 10x}{5}$ 3. $\dfrac{15 + 20x}{30}$ 4. $\dfrac{9x^2 - 12x}{6x}$

5. $\dfrac{x^2 - 4x - xz}{x}$ 6. $\dfrac{2y + xy + y^2}{y}$ 7. $\dfrac{2xy - 2xz + 4x}{2x}$ 8. $\dfrac{5x^3 - 5x^2 + 10x}{10x^2}$

9. $\dfrac{4 - 4\sqrt{2}}{8}$ 10. $\dfrac{-12 + 18\sqrt{5}}{15}$

<table>
<tr><td>Section
1.4</td><td><h1 style="text-align:center">Multiplying Binomials</h1></td></tr>
</table>

Using the distributive property twice, we can derive a method for multiplying two binomials. Suppose, for example, we wish to multiply $2x - 3$ times $x + 2$. We use A to represent $x + 2$:

$$(2x - 3)(x + 2) = (2x - 3)A$$
$$= 2x(A) - 3(A).$$

Now we substitute $x + 2$ back for A, and use the distributive property a second time:

$$2x(A) - 3(A) = 2x(x + 2) - 3(x + 2)$$
$$= 2x^2 + 4x - 3x - 6$$
$$= 2x^2 + x - 6.$$

Thus,

$$(2x - 3)(x + 2) = 2x^2 + x - 6.$$

Leaving out the use of A, we have

$$(2x - 3)(x + 2) = 2x(x + 2) - 3(x + 2)$$
$$= 2x^2 + 4x - 3x - 6$$
$$= 2x^2 + x - 6.$$

Observe that the middle step,

$$(2x - 3)(x + 2) = 2x^2 + 4x - 3x - 6,$$

is the result of multiplying

the two first terms, $2x$ and x,
the two outside terms, $2x$ and 2,
the two inside terms, -3 and x,
the two last terms, -3 and 2.

This observation leads to a shorter method for multiplying two binomials. Mentally, we use the diagram

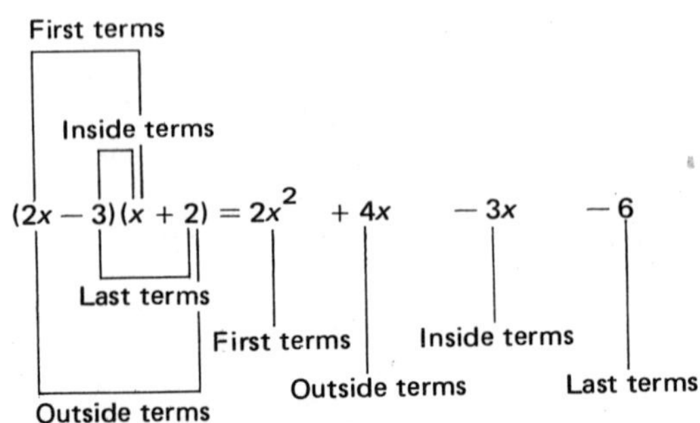

Then, combining the results of the outside and inside terms,

$$(2x - 3)(x + 2) = 2x^2 + 4x - 3x - 6$$

$$= 2x^2 + x - 6.$$

EXAMPLE 1.13.   Multiply $(5x - 3)(2x - 4)$.

Solution.
$$(5x - 3)(2x - 4) = 10x^2 - 20x - 6x + 12$$

$$= 10x^2 - 26x + 12.$$

EXAMPLE 1.14.   Multiply $(x - 2y)(x + 2y)$.

Solution.
$$(x - 2y)(x + 2y) = x^2 + 2xy - 2xy - 4y^2$$

$$= x^2 - 4y^2.$$

Since $2xy - 2xy = 0$, the result of the outside and inside terms does not appear in the answer.

Exercise 1.4

Multiply:

1. $(x + 1)(x + 2)$ 2. $(x - 2)(x + 3)$ 3. $(2x - 1)(5x - 8)$ 4. $(3x + 4)(6x - 3)$

5. $(x + 3y)(3x + y)$ 6. $(2x - y)(5x + y)$ 7. $(3x + 5)(3x + 5)$ 8. $(2x - 1)(2x - 1)$

9. $(x - 5)^2$ 10. $(4x + 3)^2$ 11. $(4x - 2)(4x + 2)$ 12. $(6x + 3y)(6x - 3y)$

Section
1.5

Factoring Trinomials

When two binomials are multiplied, the result is usually a three term expression called a **trinomial**. For example,

$$(2x - 3)(x + 2) = 2x^2 + x - 6,$$

where $2x^2 + x - 6$ is a trinomial. (There is an exception when the outside and inside terms add to zero and so the result is a binomial.)

Many trinomials can be factored into two binomial factors where the constants are integers. Recall from arithmetic that **integers** are positive or negative whole numbers. However, not all trinomials can be so factored. Consider the trinomial $x^2 + x + 1$.
We write

$$x^2 + x + 1 = (x \quad 1)(x \quad 1).$$

The first terms must both be x because $x^2 = x \cdot x$. The last terms must both be 1 or -1 because

$1 = 1 \cdot 1$ or $1 = (-1)(-1)$. However, neither 1 nor -1 is possible since

$$(x + 1)(x + 1) = x^2 + 2x + 1$$

and

$$(x - 1)(x - 1) = x^2 - 2x + 1.$$

Since neither product is the same as $x^2 + x + 1$, this trinomial cannot be factored into two binomial factors involving only integers.

For the rest of this unit, we will consider only trinomials which can be factored into two binomial factors where the constants are integers.

EXAMPLE 1.15.   Factor $x^2 + 5x + 6$.

Solution.   The first terms must each be x since $x^2 = x \cdot x$. Thus we write

$$x^2 + 5x + 6 = (x \quad)(x \quad).$$

The second terms of the binomial factors may be 1 and 6, -1 and -6, 2 and 3, or -2 and -3. We now consider the sum of the outside and inside terms of the binomial factors, which we will call the **middle term** of the trinomial, $5x$ in our example. The middle term is positive, therefore the second terms of the binomial factors cannot both be negative. We now consider 1 and 6, or 2 and 3 as the second terms.

$$(x + 1)(x + 6) = x^2 + 7x + 6$$

and

$$(x + 2)(x + 3) = x^2 + 5x + 6.$$

Therefore,

$$x^2 + 5x + 6 = (x + 2)(x + 3).$$

Observe that in the trinomial, if the middle term is negative but the third term is positive, the two second terms of the binomial factors will be negative. For example,

$$x^2 - 5x + 6 = (x - 2)(x - 3).$$

If the third term of the trinomial is negative, one of the second terms of the binomial factors is positive and the other is negative, regardless of the sign of the middle term.

EXAMPLE 1.16.   Factor $x^2 + x - 6$.

Solution.   We write

$$x^2 + x - 6 = (x \quad)(x \quad).$$

Since the third term of the trinomial is negative, the two second terms of the binomial factors can be 1 and -6, -1 and 6, 2 and -3, or -2 and 3. We rule out the cases where the larger number has the negative sign, because in these cases the middle term will be negative. Now we can go right to the answer. Since $-1 + 6 = 5$ and $-2 + 3 = 1$, the answer must be

$$x^2 + x - 6 = (x - 2)(x + 3).$$

Observe also that

$$x^2 - x - 6 = (x + 2)(x - 3).$$

EXAMPLE 1.17.   Factor $2x^2 - 3x - 2$.

Solution.   Since $2x^2 = 2x \cdot x$, we write

$$2x^2 - 3x - 2 = (2x \quad)(x \quad).$$

The third term of the trinomial is negative, so the second terms of the binomial factors will be -1 and 2 or 1 and -2. Suppose we try

$$(2x - 1)(x + 2) = 2x^2 + 3x - 2.$$

The factorization is not correct. Since the first terms of the binomial factors are different, we must also try

$$(2x + 2)(x - 1) = 2x^2 - 2.$$

These are the two possibilities for -1 and 2. We now try 1 and -2:

$$(2x + 1)(x - 2) = 2x^2 - 3x - 2.$$

This factorization gives the original trinomial. Therefore,

$$2x^2 - 3x - 2 = (2x + 1)(x - 2).$$

Had this combination not worked, the fourth possibility is to try 1 and -2 reversed. When the trinomial has several possible combinations, you will have to do some searching by trial and error. It helps to be able to find the middle term without writing out the entire multiplication of the possible factors you are considering.

EXAMPLE 1.18.   Factor $4x^2 - 19x + 12$.

Solution.   The first terms may be $2x$ and $2x$ or $4x$ and x. The second terms must both be negative, and can be -1 and -12, -2 and -6, or -3 and -4. We might try

$$4x^2 - 19x + 12 = (2x \quad)(2x \quad).$$

Trying each of the possible pairs of second terms, we find

$(2x - 1)(2x - 12)$ has middle term $-26x$,
$(2x - 2)(2x - 6)$ has middle term $-16x$,
$(2x - 3)(2x - 4)$ has middle term $-14x$.

Since none of these possibilities gives the correct middle term, we try

$$4x^2 - 19x + 12 = (4x \quad)(x \quad).$$

Now, since the first terms are different, we must try each possible pair of second terms with each first term. Thus,

$(4x - 1)(x - 12)$ has middle term $-49x$,
$(4x - 12)(x - 1)$ has middle term $-16x$,

so the last terms are not -1 and -12.

$(4x - 2)(x - 6)$ has middle term $-26x$,
$(4x - 6)(x - 2)$ has middle term $-14x$,

so the last terms are not -2 and -6.

$(4x - 3)(x - 4)$ has middle term $-19x$, thus

$$4x^2 - 19x + 12 = (4x - 3)(x - 4).$$

EXAMPLE 1.19.   Factor $6x^2 + x - 15$.

Solution.   The first terms may be $2x$ and $3x$ or $6x$ and x. The second terms must have opposite signs, and may be 1 and -15, -1 and 15, 3 and -5, or -3 and 5.
We might try

$$6x^2 + x - 15 = (2x \quad)(3x \quad).$$

Trying each of the possible pairs of second terms, we find 1 and -15, -1 and 15, and 3 and -5, all give the wrong middle term. However, for -3 and 5,

$$(2x - 3)(3x + 5) \text{ has middle term } x,$$

and so

$$6x^2 + x - 15 = (2x - 3)(3x + 5).$$

The trinomial is factored, and we need not try $6x$ and x as first terms.

EXAMPLE 1.20.   Factor $4x^2 - 9$.

Solution.   The expression $4x^2 - 9$ is really a binomial. However, we may think of it as

$$4x^2 + 0x - 9.$$

Therefore, the middle term must be zero. In this case, the two first terms will always be identical.
We write

$$4x^2 - 9 = (2x \quad)(2x \quad).$$

The second terms can be 3 and -3, 1 and -9, or -1 and 9. Since the middle term is zero, and

$$(2x + 3)(2x - 3) \text{ has middle term } 0x,$$

we have

$$4x^2 - 9 = (2x + 3)(2x - 3).$$

You should always be sure to factor out any common factors before factoring a trinomial.

EXAMPLE 1.21.   Factor $4x^2 - 16x + 12$.

Solution.   We factor out the common factor 4:

$$4x^2 - 16x + 12 = 4(x^2 - 4x + 3).$$

The remaining trinomial expression is easily factored. Since the first terms are both x, and the second terms must be -1 and -3,

$$x^2 - 4x + 3 = (x - 1)(x - 3).$$

Thus,

$$4x^2 - 16x + 12 = 4(x^2 - 4x + 3)$$
$$= 4(x - 1)(x - 3).$$

Exercise 1.5

Factor:

1. $x^2 + 5x + 4$
2. $x^2 + 9x + 18$
3. $x^2 - 7x + 12$
4. $x^2 - 10x + 16$

5. $x^2 - 2x - 15$
6. $x^2 + 2x - 48$
7. $2x^2 + x - 6$
8. $3x^2 - 16x + 5$

9. $x^2 + 12x + 36$
10. $4x^2 - 12x + 9$
11. $4x^2 - 1$
12. $9x^2 - 25$

13. $6x^2 + 25x + 24$
14. $8x^2 - 75x - 50$
15. $3x^2 + 12x + 9$
16. $10x^2 - 25x - 60$

Self-test

1. Reduce $\dfrac{3 - 15x}{3}$.

1. _______________

2. Factor $2x^2 - 7x - 30$.

2. _______________

3. Factor $2x^2 - 8x$.

3. _______________

4. Combine like terms of $2(x - 3) - 3(x + 2)$.

4. _______________

5. Multiply $(3x - 5)(4x + 1)$.

5. _______________

Solving Equations

INTRODUCTION

Historically, the growth of the study of algebra, from prehistoric times to about the sixteenth century, may be attributed to attempts to solve different kinds of equations. In this unit you will learn the basic techniques for solving equations and will apply these techniques to solving a few elementary types of equations.

OBJECTIVES

When you have finished this unit you should be able to:

1. Apply addition, subtraction, multiplication, and division rules to solve linear equations.
2. Solve linear equations with constant denominators.
3. Solve linear equations with constant common factors.
4. Apply the product rule to solve quadratic equations with variable common factors.
5. Solve quadratic equations where the quadratic expression can be factored.

Linear Equations

An **equation** consists of two equal algebraic expressions; that is, two algebraic expressions with an equal sign between them. In a **linear equation** the algebraic expressions contain only variables to the first power. There are no variables with exponents greater than 1 and no variables in denominators or under radical signs in linear equations. These are some examples of linear equations:

$$x + 3 = 4, \quad \frac{x}{5} = 2, \quad 4x - 10 = 6, \quad 3 - 2x = x + 7, \quad x\sqrt{3} = \sqrt{6}.$$

Observe that a linear equation can have constants in denominators or under radical signs, but not variables.

An example of a very simple linear equation in one variable is

$$x = 2.$$

Here, the variable x is clearly equal to the constant 2. In solving linear equations in one variable, the goal is to find a constant to which the variable is equal. To accomplish this goal, there are four basic operations:

1. Add the same expression to both sides of the equation.
2. Subtract the same expression from both sides of the equation.
3. Multiply both sides of the equation by the same nonzero expression.
4. Divide both sides of the equation by the same nonzero expression.

In later sections of this unit, and in other units, it will be important that we cannot multiply or divide both sides of an equation by an expression that is equal to zero.

Given the equation $x + 3 = 4$, we may subtract the expression 3 from both sides of the equation:

$$x + 3 = 4$$
$$x + 3 - 3 = 4 - 3$$
$$x + 0 = 1$$
$$x = 1.$$

Thus the equation is solved, and the solution is 1.

Given the equation $x - 4 = 2$, we add 4 to both sides:

$$x - 4 = 2$$
$$x - 4 + 4 = 2 + 4$$
$$x + 0 = 6$$
$$x = 6.$$

The solution to this equation is 6.

In multiplying and dividing, we must be careful not to multiply or divide by zero. Given the equation $2x = 8$, since 2 is not zero, we may divide both sides by 2:

$$2x = 8$$
$$\frac{2x}{2} = \frac{8}{2}$$
$$1x = 4$$
$$x = 4.$$

The solution to this equation is 4.

Given the equation $\frac{x}{5} = 2$, since 5 is not zero, we may multiply both sides by 5:

$$\frac{x}{5} = 2$$
$$5\left(\frac{x}{5}\right) = 5(2)$$
$$1x = 10$$
$$x = 10.$$

The solution to this equation is 10.

In general, it is easier to do addition and subtraction before multiplication and division.

EXAMPLE 2.1. Solve $4x - 10 = 6$.

Solution. First, adding 10 to both sides,

$$4x - 10 = 6$$
$$4x - 10 + 10 = 6 + 10$$
$$4x = 16.$$

Now, dividing both sides by 4,

$$\frac{4x}{4} = \frac{16}{4}$$
$$x = 4.$$

The solution is 4. Observe we have skipped the steps involving $x + 0$ and $1x$. The first should be understood always to be involved in addition and subtraction, and the second always in multiplication and division.

EXAMPLE 2.2. Solve $3 - 2x = x + 7$.

Solution. There are several ways to solve this equation. We may start by collecting the terms involving x. Adding $2x$ to both sides,

$$3 - 2x = x + 7$$
$$3 - 2x + 2x = x + 7 + 2x$$
$$3 = 3x + 7.$$

Now we isolate the term involving x. Subtracting 7 from both sides,

$$3 - 7 = 3x + 7 - 7$$
$$-4 = 3x.$$

Finally, dividing both sides by 3,

$$\frac{-4}{3} = \frac{3x}{3}$$
$$-\frac{4}{3} = x$$

or

$$x = -\frac{4}{3}.$$

The solution is $-\frac{4}{3}$.

When one or both of the expressions in an equation contain like terms, we usually combine the like terms before applying the techniques for solving equations.

EXAMPLE 2.3. Solve $2x - 3 + x = 4 - x + 2$.

Solution. First, combining like terms,

$$2x - 3 + x = 4 - x + 2$$
$$3x - 3 = 6 - x.$$

Now, adding x and then 3 to both sides,

$$3x - 3 + x = 6 - x + x$$

$$4x - 3 = 6$$

$$4x - 3 + 3 = 6 + 3$$

$$4x = 9.$$

Finally, dividing by 4,

$$\frac{4x}{4} = \frac{9}{4}$$

$$x = \frac{9}{4}.$$

Thus the solution is $\frac{9}{4}$.

EXAMPLE 2.4.   Solve $2(x + 3) - (x - 2) = 5$.

Solution.   We use the distributive property to simplify the expression on the left-hand side, and then combine like terms:

$$2(x + 3) - (x - 2) = 5$$

$$2x + 6 - x + 2 = 5$$

$$x + 8 = 5.$$

Subtracting 8 from both sides,

$$x + 8 - 8 = 5 - 8$$

$$x = -3.$$

The solution is -3.

You may check the solution to an equation by substituting the solution for the variable, wherever the variable appears in the original equation. To check the solution $x = -3$ in the previous example, we write:

$$2(x + 3) - (x - 2) = 5$$

$$2(-3 + 3) - (-3 - 2) \overset{?}{=} 5$$

$$2(0) - (-5) \overset{?}{=} 5$$

$$0 + 5 \overset{?}{=} 5$$

$$5 = 5.$$

Observe that we did not use any of the rules for solving equations, but only ordinary addition and subtraction. Since it is true that $5 = 5$, our solution $x = -3$ is correct.

Exercise 2.1

Solve for x and check:

1. $x - 3 = 5$ 2. $x + 10 = 4$ 3. $x + 1 = 3$ 4. $x - 12 = -15$

5. $\dfrac{x}{4} = 2$ 6. $3x = 15$ 7. $-5x = 20$ 8. $-\dfrac{x}{3} = 3$

9. $3x + 2 = 10$ 10. $10x - 3 = 6$ 11. $5 - 4x = -1$ 12. $4 - 2x = -4$

13. $2 - x + 5 = 3x - 4 + x$ 14. $4x - 2 - x = 3 + 2x - 1$

15. $2(x + 7) - 3(x + 6) = 4$ 16. $4 - (3 - x) = 2(x - 6)$

Section 2.2 — Constant Denominators

An equation may have constant denominators and still be considered a linear equation if the variable is linear. The equation

$$\frac{x}{5} = 2$$

is such an equation. Recall that we solved equations of this type by multiplying both sides by the nonzero number in the denominator. In the equation above, we multiplied both sides by 5.

If there is more than one term in one or both of the expressions of an equation, we will need to use the distributive property in applying the multiplication rule.

EXAMPLE 2.5.   Solve $\dfrac{x}{5} + 9 = 2x$.

Solution.   We multiply both sides of the equation by 5, and use the distributive property on the left-hand side:

$$\frac{x}{5} + 9 = 2x$$

$$5\left(\frac{x}{5} + 9\right) = 5(2x)$$

$$5\left(\frac{x}{5}\right) + 5(9) = 5(2x)$$

$$x + 45 = 10x.$$

Then, subtracting x from both sides,

$$x - x + 45 = 10x - x$$

$$45 = 9x.$$

Finally, we divide both sides by 9:

$$\frac{45}{9} = \frac{9x}{9}$$

$$5 = x$$

or

$$x = 5.$$

The solution is 5. To check this solution, we write

$$\frac{x}{5} + 9 = 2x$$

$$\frac{5}{5} + 9 \stackrel{?}{=} 2(5)$$

$$1 + 9 \stackrel{?}{=} 10$$

$$10 = 10.$$

The first step in the preceding example may be skipped if we observe that this use of the distributive property is the same as multiplying each term of each expression in the equation by the same number.

EXAMPLE 2.6.   Solve $\dfrac{2x}{3} + 4 = \dfrac{x}{3}$.

Solution.   We multiply both sides by 3, which is the same as multiplying each term by 3:

$$\frac{2x}{3} + 4 = \frac{x}{3}$$

$$3\left(\frac{2x}{3}\right) + 3(4) = 3\left(\frac{x}{3}\right)$$

$$2x + 12 = x.$$

Subtracting x from both sides,

$$2x - x + 12 = x - x$$

$$x + 12 = 0.$$

Finally, subtracting 12 from both sides,

$$x + 12 - 12 = 0 - 12$$

$$x = -12.$$

The solution is -12. To check this solution, we write

$$\frac{2x}{3} + 4 = \frac{x}{3}$$

$$\frac{2(-12)}{3} + 4 \stackrel{?}{=} \frac{-12}{3}$$

$$\frac{-24}{3} + 4 \stackrel{?}{=} -4$$

$$-8 + 4 \stackrel{?}{=} -4$$

$$-4 = -4.$$

If there are two or more different constant denominators in an equation, we multiply each term by the **least common denominator**. Recall from arithmetic that the least common denominator is the smallest number which is divisible by each of the separate denominators. Often we will be able to find the least common denominator by inspection. If the least common denominator is not easily found, any number which is divisible by each of the separate denominators may be used.

EXAMPLE 2.7. Solve $\dfrac{x}{3} - 2 = \dfrac{x}{5} + 1$.

Solution.   By inspection, the least common denominator is 15:

$$\frac{x}{3} - 2 = \frac{x}{5} + 1$$

$$15\left(\frac{x}{3}\right) - 15(2) = 15\left(\frac{x}{5}\right) + 15(1)$$

$$5x - 30 = 3x + 15$$

$$5x - 3x - 30 = 3x - 3x + 15$$

$$2x - 30 = 15$$

$$2x - 30 + 30 = 15 + 30$$

$$2x = 45$$

$$\frac{2x}{2} = \frac{45}{2}$$

$$x = \frac{45}{2}.$$

The solution is $\frac{45}{2}$. To check this solution, we write

$$\frac{x}{3} - 2 = \frac{x}{5} + 1$$

$$\frac{\frac{45}{2}}{3} - 2 \overset{?}{=} \frac{\frac{45}{2}}{5} + 1$$

$$\frac{\frac{45}{2}}{\frac{3}{1}} - 2 \overset{?}{=} \frac{\frac{45}{2}}{\frac{5}{1}} + 1$$

$$\frac{45}{2} \cdot \frac{1}{3} - 2 \overset{?}{=} \frac{45}{2} \cdot \frac{1}{5} + 1$$

$$\frac{15}{2} - 2 \overset{?}{=} \frac{9}{2} + 1$$

$$\frac{15}{2} - \frac{4}{2} \overset{?}{=} \frac{9}{2} + \frac{2}{2}$$

$$\frac{15 - 4}{2} \overset{?}{=} \frac{9 + 2}{2}$$

$$\frac{11}{2} = \frac{11}{2}.$$

EXAMPLE 2.8.   Solve $\dfrac{2x}{5} - \dfrac{3}{10} = \dfrac{x}{4} - 3$.

Solution.   The least common denominator is 20:

$$\frac{2x}{5} - \frac{3}{10} = \frac{x}{4} - 3$$

$$20\left(\frac{2x}{5}\right) - 20\left(\frac{3}{10}\right) = 20\left(\frac{x}{4}\right) - 20(3)$$

$$4(2x) - 2(3) = 5x - 60$$

$$8x - 6 = 5x - 60$$

$$8x - 5x - 6 = 5x - 5x - 60$$

$$3x - 6 = -60$$

$$3x - 6 + 6 = -60 + 6$$

$$3x = -54$$

$$\frac{3x}{3} = \frac{-54}{3}$$

$$x = -18.$$

The solution is -18. You should check this solution in the original equation.

We will also encounter equations in which a constant denominator applies to an entire expression.

EXAMPLE 2.9.   Solve $\frac{1}{3}(3x - 2) = \frac{5}{6}$.

Solution.   The constant denominator 3 applies to the entire expression $3x - 2$. However, we need not simplify the expression $\frac{1}{3}(3x - 2)$. We simply multiply both sides of the equation by the least common denominator, which is 6:

$$\frac{1}{3}(3x - 2) = \frac{5}{6}$$

$$(6)\left(\frac{1}{3}\right)(3x - 2) = 6\left(\frac{5}{6}\right)$$

$$2(3x - 2) = 5.$$

Now we simplify the resulting expression and solve the equation:

$$6x - 4 = 5$$

$$6x - 4 + 4 = 5 + 4$$

$$6x = 9$$

$$\frac{6x}{6} = \frac{9}{6}$$

$$x = \frac{3}{2}.$$

The solution is $\frac{3}{2}$. To check, we write

$$\frac{1}{3}(3x - 2) = \frac{5}{6}$$

$$\frac{1}{3}\left[3\left(\frac{3}{2}\right) - 2\right] \stackrel{?}{=} \frac{5}{6}$$

$$\frac{1}{3}\left(\frac{9}{2} - 2\right) \stackrel{?}{=} \frac{5}{6}$$

$$\frac{1}{3}\left(\frac{9}{2} - \frac{4}{2}\right) \stackrel{?}{=} \frac{5}{6}$$

$$\frac{1}{3}\left(\frac{9 - 4}{2}\right) \stackrel{?}{=} \frac{5}{6}$$

$$\frac{1}{3}\left(\frac{5}{2}\right) \stackrel{?}{=} \frac{5}{6}$$

$$\frac{5}{6} = \frac{5}{6}.$$

Exercise 2.2

Solve for x and check:

1. $\dfrac{x}{3} = 5$ 2. $\dfrac{x}{-2} = 6$ 3. $\dfrac{2x}{3} - 3 = 2$ 4. $\dfrac{3x}{5} + 5 = \dfrac{-2}{5}$

5. $\dfrac{x}{2} + 5 = \dfrac{1}{3}$ 6. $\dfrac{3x}{4} = \dfrac{2x}{5} + 1$ 7. $\dfrac{5x}{6} - \dfrac{2}{3} = \dfrac{1}{2}$ 8. $\dfrac{3x}{10} + \dfrac{1}{2} = \dfrac{x}{5}$

9. $\dfrac{1}{9} = \dfrac{2}{3} - \dfrac{x}{6}$ 10. $\dfrac{5x}{6} - \dfrac{1}{12} = \dfrac{3x}{4} - \dfrac{1}{8}$ 11. $\dfrac{1}{2}(x - 5) = \dfrac{1}{2}$ 12. $\dfrac{1}{4}(4x + 9) = \dfrac{3}{2}$

13. $\dfrac{1}{5}(3x + 2) = \dfrac{7}{10}$ 14. $\dfrac{5}{8}(2x - 3) = \dfrac{5}{6}$

Section 2.3

Constant Factors

Just as we can multiply both sides of an equation by the same nonzero number, we can divide both sides by the same nonzero number. The process is actually the same as dividing a polynomial by a constant factor.

EXAMPLE 2.10.   Solve $2x + 8 = 6$.

Solution.   We may divide both sides of the equation by 2:

$$2x + 8 = 6$$

$$\frac{2x + 8}{2} = \frac{6}{2}$$

$$\frac{2x + 8}{2} = 3.$$

The left-hand side is now a division of a binomial by a constant factor:

$$\frac{2(x + 4)}{2} = 3$$

$$x + 4 = 3.$$

We complete the solution by subtracting 4 from both sides:

$$x + 4 - 4 = 3 - 4$$

$$x = -1.$$

The solution is -1. You should check this solution in the original equation.

Again, observe that we may skip some steps by simply dividing each term of the equation by the constant factor. Thus, we could write

$$\frac{2x}{2} + \frac{8}{2} = \frac{6}{2}$$

$$x + 4 = 3$$

$$x + 4 - 4 = 3 - 4$$

$$x = -1.$$

Of course, we could also have solved the equation by using just subtraction and division, without first dividing out the constant factor. When we solve more complicated equations, however, it is useful to be able to simplify the equation by dividing out any constant factors.

EXAMPLE 2.11.   Solve $25 - 10x = 10x - 35$.

Solution.   Dividing each term by 5,

$$25 - 10x = 10x - 35$$

$$\frac{25}{5} - \frac{10x}{5} = \frac{10x}{5} - \frac{35}{5}$$

$$5 - 2x = 2x - 7.$$

Now we use addition and division to solve the simplified equation:

$$5 - 2x + 2x = 2x + 2x - 7$$

$$5 = 4x - 7$$

$$5 + 7 = 4x - 7 + 7$$

$$12 = 4x$$

$$\frac{12}{4} = \frac{4x}{4}$$

$$3 = x$$

or

$$x = 3.$$

The solution is 3. You should check this solution in the original equation.

Dividing out a constant factor is useful in simplifying an equation where we have a constant factor applied to an entire expression.

EXAMPLE 2.12. Solve $3(4x + 3) = 12$.

Solution. The constant factor 3 applies to the entire expression $4x + 3$. On the right-hand side, 3 also is a factor of 12. We divide both sides of the equation by 3:

$$3(4x + 3) = 12$$

$$\frac{3(4x + 3)}{3} = \frac{12}{3}$$

$$4x + 3 = 4$$

$$4x + 3 - 3 = 4 - 3$$

$$4x = 1$$

$$\frac{4x}{4} = \frac{1}{4}$$

$$x = \frac{1}{4}.$$

The solution is $\frac{1}{4}$. To check, we write

$$3(4x + 3) = 12$$

$$3\left[4\left(\frac{1}{4}\right) + 3\right] \stackrel{?}{=} 12$$

$$3(1 + 3) \stackrel{?}{=} 12$$

$$3(4) \stackrel{?}{=} 12$$

$$12 = 12.$$

Exercise 2.3

Solve for x and check:

1. $4x = 8$

2. $-3x = 6$

3. $3x - 6 = 12$

4. $10x + 5 = 20$

5. $12x + 16 = 8$

6. $9 - 12x = 18$

7. $2x - 4 = 6x - 2$

8. $7x - 14 = 28 - 7x$

9. $12 - 18x = 24 - 12x$

10. $32x - 16 = 48x - 64$

11. $2(2x + 5) = 6$

12. $5(3x + 3) = 10$

13. $4(6x - 3) = 20$

14. $3(5x - \frac{1}{2}) = 21$

<table>
<tr><td>Section
2.4</td><td><h1>Variable Factors</h1></td></tr>
</table>

Although an equation may be simplified by dividing each term of the equation by a *constant* factor, we cannot divide each term of an equation by a *variable* factor. For example, consider the equation

$$x^2 = x.$$

It is tempting to write

$$\frac{x^2}{x} = \frac{x}{x}$$

$$x = 1.$$

It is possible, however, that the variable x is equal to zero. If this is the case, we would have divided by zero, which is not allowed. Indeed, since substituting $x = 0$ into the original equation gives

$$0^2 = 0$$

$$0 = 0$$

which is a true equality, $x = 0$ is another solution to the equation. Thus, in dividing by x we have divided by zero, and in doing so we have lost the solution $x = 0$.

Since we cannot divide an equation by a variable factor, we must use another method for solving equations like the one above. We use a theorem which we will call the product rule.

> **Product rule:** If $ab = 0$, then $a = 0$ or $b = 0$ or both.

The proof of the product rule is given at the end of this unit.

To use the product rule to solve an equation such as $x^2 = x$, we collect the nonzero terms on one side with zero on the other side, and then factor out the common factor.

EXAMPLE 2.13. Solve $x^2 = x$.

Solution. Collecting the terms on the left-hand side, we have

$$x^2 = x$$

$$x^2 - x = 0.$$

Now we factor out the common factor x:

$$x(x - 1) = 0.$$

Then, using the product rule,

$$x = 0 \text{ and } x - 1 = 0$$

$$x = 0 \text{ and } x = 1.$$

Since $x = 0$ and $x = 1$ each can be checked in the original equation, both 0 and 1 are solutions of the equation.

EXAMPLE 2.14.   Solve $2x^2 + 4x = 0$.

Solution.   We may remove the constant factor 2 by dividing every term of the equation by 2:

$$\frac{2x^2}{2} + \frac{4x}{2} = \frac{0}{2}$$

$$x^2 + 2x = 0.$$

However, we must not divide by x, so we continue by factoring and using the product rule:

$$x(x + 2) = 0$$

$$x = 0 \text{ and } x + 2 = 0$$

$$x = 0 \text{ and } x = -2.$$

The solutions are 0 and -2. You should check these and all other solutions to equations.

By now you should be familiar enough with the addition, subtraction, multiplication, and division rules so that their use need not be shown in detail.

EXAMPLE 2.15.   Solve $3x^2 - 4x = x^2 + x$.

Solution.

$$3x^2 - 4x = x^2 + x$$

$$2x^2 - 5x = 0$$

$$x(2x - 5) = 0$$

$$x = 0 \text{ and } 2x - 5 = 0$$

$$x = 0 \text{ and } x = \frac{5}{2}.$$

The solutions are 0 and $\frac{5}{2}$.

Exercise 2.4

Solve for x and check:

1. $x^2 = 5x$ 2. $2x^2 = x$ 3. $2x^2 = 6x$ 4. $3x^2 + 9x = 0$

5. $4x^2 + 6x = 0$ 6. $5x = 15x^2$ 7. $4x^2 + 3x = x^2 - 2x$ 8. $x^2 - 6x = 3x^2 + 2x$

Section
2.5
Quadratic Equations

The algebraic expressions making up a **quadratic equation** contain variables to the second power, or **second-degree terms**. They may also contain variables to the first power, or **first-degree terms**, and nonzero constants. The equations in Section 2.4 are quadratic equations. Some other examples of quadratic equations are:

$$x^2 - 3x + 2 = 0, \quad 2x^2 - 8 = 0, \quad x^2 = 0.$$

Observe that, although there must be a second-degree term in a quadratic equation, there need not be a first-degree term or a term consisting of a nonzero constant.

To solve the equations in Section 2.4, we collected all the nonzero terms on one side, factored, and used the product rule. Since these equations contained no nonzero constant, factoring was a matter of taking out a common factor x. The product rule can be used for quadratic equations which do have a nonzero constant, if the algebraic expression consisting of all the nonzero terms can be factored by the methods in Section 1.5.

EXAMPLE 2.16.   Solve $x^2 - 3x + 2 = 0$.

Solution.   Since all the nonzero terms are on the left-hand side, we try to factor the trinomial into two binomial factors. This trinomial factors easily, and we have

$$x^2 - 3x + 2 = 0$$

$$(x - 1)(x - 2) = 0.$$

Now, using the product rule,

$$x - 1 = 0 \text{ and } x - 2 = 0$$

$$x = 1 \text{ and } x = 2.$$

Both 1 and 2 are solutions, as you can easily verify by checking in the original equation.

If the nonzero terms are not collected on one side of the equation, we must collect them before we factor the quadratic expression. This step is necessary in order to have a zero on one side of the equation. If one side of the equation is not zero, the product rule does not hold. For example, if $ab = 1$, then it is not necessarily true that $a = 1$ or $b = 1$ or both. We could have $a = -1$ and $b = -1$, or $a = 2$ and $b = \frac{1}{2}$, and so on. Thus we must always write the equation with all the nonzero terms on one side and zero on the other.

EXAMPLE 2.17.   Solve $2x^2 - x = 1$.

Solution.   First we collect the nonzero terms:

$$2x^2 - x = 1$$

$$2x^2 - x - 1 = 0.$$

Now we may factor the trinomial and use the product rule to complete the solution:

$$(x - 1)(2x + 1) = 0$$

$$x - 1 = 0 \text{ and } 2x + 1 = 0$$

$$x = 1 \text{ and } x = -\frac{1}{2}.$$

The solutions are 1 and $-\frac{1}{2}$.

EXAMPLE 2.18.   Solve $x + 1 = 6x^2$.

Solution.   When it is more convenient, the nonzero terms can be collected on the right-hand side of the equation, with zero on the left. Collecting on the right, we have

$$x + 1 = 6x^2$$

$$0 = 6x^2 - x - 1$$

$$0 = (2x - 1)(3x + 1)$$

$$2x - 1 = 0 \text{ and } 3x + 1 = 0$$

$$x = \frac{1}{2} \text{ and } x = -\frac{1}{3}.$$

The solutions are $\frac{1}{2}$ and $-\frac{1}{3}$.

The expression made up of the nonzero terms of a quadratic equation need not be a trinomial. We also know how to factor expressions where the middle term is $0x$.

EXAMPLE 2.19.   Solve $x^2 = 9$.

Solution.   First we collect the nonzero terms:

$$x^2 = 9$$

$$x^2 - 9 = 0.$$

Factoring and using the product rule, we have

$$(x + 3)(x - 3) = 0$$

$$x + 3 = 0 \text{ and } x - 3 = 0$$

$$x = -3 \text{ and } x = 3.$$

The solutions are -3 and 3.

Although you must never divide out a term involving a variable, you should always divide out any constant factors.

EXAMPLE 2.20.   Solve $2x^2 - 8 = 0$.

Solution.   Since 2 is a constant factor, we may divide by 2:

$$2x^2 - 8 = 0$$

$$\frac{2x^2}{2} - \frac{8}{2} = \frac{0}{2}$$

$$x^2 - 4 = 0.$$

Factoring and using the product rule,

$$(x + 2)(x - 2) = 0$$

$$x + 2 = 0 \text{ and } x - 2 = 0$$

$$x = -2 \text{ and } x = 2.$$

The solutions are -2 and 2.

Finally, we have a special case where the factored expression has identical factors.

EXAMPLE 2.21.   Solve $4x = x^2 + 4$.

Solution.   Collecting on the right, we have

$$4x = x^2 + 4$$

$$0 = x^2 - 4x + 4$$

$$0 = (x - 2)(x - 2)$$

$$x - 2 = 0$$

$$x = 2.$$

Since this equation has identical factors, we need only write the repeated factor once, and we only write the solution derived from the factor once. However, all quadratic equations appear to have two solutions. It is sometimes convenient to think of an equation with two identical factors as having two identical solutions. For example, the equation above has two identical solutions each of which is 2. This situation is often called a "double root." In the case of a double root, you need only write the repeated solution once.

Exercise 2.5

Solve for x and check:

1. $x^2 - 1 = 0$ 2. $x^2 = 25$ 3. $4x^2 = 1$ 4. $9x^2 - 4 = 0$

5. $2x^2 = x$ 6. $x^2 + 3x = 0$ 7. $x^2 - 5x + 6 = 0$ 8. $x^2 + 4x = 60$

9. $3x^2 - x = 2$ 10. $6 = 4x^2 + 5x$ 11. $x^2 + 6x + 9 = 0$ 12. $2x^2 + 2 = 4x$

13. $3x^2 = 6 - 3x$ 14. $26x - 5 = 5x^2$ 15. $14x - 4x^2 = 8 - x^2$ 16. $6x^2 + 7x = 4x + 3$

Proof of the Product Rule

Suppose $ab = 0$. Consider the factor a. Either $a = 0$ or $a \neq 0$. Thus we have two cases: (1) $a = 0$ and (2) $a \neq 0$.

Case 1.   If $a = 0$, then $a = 0$ or $b = 0$ or both, and the theorem is proved for this case.

Case 2.   If $a \neq 0$, then we can divide by a since we will not be dividing by zero:

$$\frac{ab}{a} = \frac{0}{a}$$

$$b = 0.$$

Thus $a = 0$ or $b = 0$ or both, and the theorem is proved for this case.

Self-test

Solve for x and check:

1. $3x^2 = 2x$

2. $6x - 6 = 12x - 9$

3. $\dfrac{x}{4} - \dfrac{3x}{5} = \dfrac{1}{10}$

4. $3(x - 1) = 4 - (x - 1)$

5. $5x - 2 = 2x^2$

1. ________________

2. ________________

3. ________________

4. ________________

5. ________________

Unit 3

Applications

INTRODUCTION

The main reason to be able to solve equations is in order to be able to solve applied problems. An applied problem is a verbal description of a situation which can be described in mathematical terms. In this unit you will see a few sample types of applied problems which can be described in terms of algebraic equations. You will learn how to translate a problem into an equation, and then solve the equation to find the solution of the problem.

OBJECTIVES

When you have finished this unit you should be able to:

1. Translate simple verbal phrases into algebraic expressions.
2. Solve problems describing numbers.
3. Solve problems involving mixtures and solutions.
4. Solve problems involving rate, time, and distance.
5. Solve problems involving areas of rectangles, parallelograms, triangles, and trapezoids, and volumes of rectangular solids.

Section 3.1 — Translating Phrases

Solving an applied problem is very much like translating a paragraph from one language to another. We must translate a paragraph written in English into another language called "Mathematics." This process involves learning the meaning of English words in terms of mathematical symbols, and then translating English phrases into mathematical phrases. We start with three basic translations.

"More" in English means "plus," the addition symbol, in mathematics. Of course, we must be careful where we put the plus. For example, if one number is four more than another number, we must add 4 to the smaller number in order to represent the larger number. If x represents the smaller number, then $x + 4$ represents the larger number.

"Less" in English means "minus," the subtraction symbol, in mathematics. Again, we must be careful where we put the minus. Also, we must be careful of the order in which we subtract. For example, if one number is six less than another number, we must subtract 6 from the larger number in order to represent the smaller number. If x represents the larger number, then $x - 6$ represents the smaller number.

"Of" in English means "times," the multiplication symbol, in mathematics. Suppose one number is three-fourths of another. If the first number is x, three-fourths of it is $\frac{3}{4}x$.

EXAMPLE 3.1.   Translate into mathematical expressions:

    a.   Ten more than a number.
    b.   Fifteen less than a number.
    c.   One-fifth of a number.

Solutions.   Suppose in each case the number is x. Using the translations described above, we have:

$$\text{a. } x + 10 \qquad \text{b. } x - 15 \qquad \text{c. } \frac{1}{5}x$$

In translating phrases such as "fifteen less than a number" in Example 3.1b, we must subtract in the correct order. There is a tendency to want to write $15 - x$ because fifteen was mentioned first. Recalling, however, the given translations, we see that $15 - x$ means "some amount less than fifteen"; that is, start with fifteen and decrease it by some amount x. Correctly, "fifteen less than a number" means start with the number x and decrease it by fifteen, or $x - 15$.

EXAMPLE 3.2.   Translate into mathematical expressions:

    a.   Eight less than some amount.
    b.   Some amount less than eight.

Solutions.   Suppose in each case the amount is x.

    a.   "Eight less than some amount" means the amount decreased by eight, or $x - 8$.
    b.   "Some amount less than eight" means eight decreased by the amount, or $8 - x$.

We will often find "of" combined with "more" or "less" in one phrase. In this case we need another symbol of mathematics, the parentheses. Parentheses tell us which comes first, the "of" or the "more" or "less." For example, consider the phrases "one-half of two more than a number" and "two more than one-half of a number." For the first, we must write the translation of "two more than a number," and then one-half of that entire phrase. We write $x + 2$, and then $\frac{1}{2}(x + 2)$, using parentheses to indicate one-half of the entire phrase. For the second, we must write the translation of "one-half of a number," and then two more than that phrase. We write $\frac{1}{2}x$, and then $\frac{1}{2}x + 2$. In this case there are no parentheses because we have one-half only of x, not of an entire phrase.

EXAMPLE 3.3.   Translate into mathematical expressions:

    a.   Three-halves of six less than a number.
    b.   Six less than three-halves of a number.

Solutions.   Suppose in each case the number is x.

    a.   We have six less than x, or $x - 6$. Then we have three-halves of the entire phrase. The translation is $\frac{3}{2}(x - 6)$.
    b.   We have three-halves of x only, or $\frac{3}{2}x$. Then we have six less than that. The translation is $\frac{3}{2}x - 6$.

Exercise 3.1

Translate into mathematical expressions:

1.   Fourteen more than a number.

2.   Some amount more than three.

3.   Five less than some amount.

4.   Some amount less than thirty.

5.   Some amount less seven.

6.   Nine less some amount.

7.   One-third of a number.

8.   Five-halves of a number.

9.   Two-fifths of eight more than a number.

10.   Ten less than one-fifth of a number.

11.   Eighteen more than two-thirds of a number.

12.   Three-fourths of twelve less than a number.

<table><tr><td>Section
3.2</td><td><h1 style="text-align:center">Describing Numbers</h1></td></tr></table>

A common type of problem simply describes a number. This type of problem uses the phrases in Section 3.1, or similar phrases. The problem is to find the number described. To do this, we write an equation made up of the expressions which are the mathematical translations of the phrases in the problem. We will also need to know that "is," or any form of "is," in English translates to "equals," the equal sign, in mathematics.

EXAMPLE 3.4.   One-third of a number is five-halves. Find the number.

Solution.   Suppose the number is x. The phrase "one-third of a number" translates into the mathematical expression $\frac{1}{3}x$. Since "is" means "equals," we have the equation

$$\frac{1}{3}x = \frac{5}{2}.$$

To solve the equation, we multiply both sides by 3:

$$3\left(\frac{1}{3}x\right) = 3\left(\frac{5}{2}\right)$$

$$x = \frac{15}{2}.$$

The number is $\frac{15}{2}$. To check, we see that one-third of fifteen-halves is

$$\frac{1}{3}\left(\frac{15}{2}\right) = \frac{5}{2}$$

as required.

EXAMPLE 3.5.   Fifteen less than a number is thirty. Find the number.

Solution.   If the number is x, "fifteen less than a number" translates to $x - 15$. The equation is

$$x - 15 = 30.$$

Adding 15 to both sides,

$$x = 45.$$

The number is 45. To check, we observe that fifteen less than 45 is $45 - 15 = 30$.

EXAMPLE 3.6.   One-third of five more than a number is four-ninths. Find the number.

Solution.   If the number is x, we have $\frac{1}{3}(x + 5)$, and so

$$\frac{1}{3}(x + 5) = \frac{4}{9}.$$

We multiply each side by the least common denominator, which is 9:

$$9\left(\frac{1}{3}\right)(x + 5) = 9\left(\frac{4}{9}\right)$$

$$3(x + 5) = 4$$

$$3x + 15 = 4.$$

Now, solving the resulting equation,

$$3x = -11$$

$$x = -\frac{11}{3}.$$

The number is $-\frac{11}{3}$. To check, five more than $-\frac{11}{3}$ is

$$-\frac{11}{3} + 5 = -\frac{11}{3} + \frac{15}{3}$$

$$= \frac{-11 + 15}{3}$$

$$= \frac{4}{3}$$

and one-third of this number is

$$\frac{1}{3}\left(\frac{4}{3}\right) = \frac{4}{9}$$

as required. You should always check using the original wording of the problem rather than your equation, since the equation could be an incorrect translation of the words.

EXAMPLE 3.7.   Six less than one-quarter of a number is seven-halves. Find the number.

Solution.   If the number is x, we have $\frac{1}{4}x - 6$, and so

$$\frac{1}{4}x - 6 = \frac{7}{2}.$$

We may multiply by the least common denominator 4, but since the expression on the left-hand side is not in parentheses, we must remember to multiply each term:

$$4\left(\frac{1}{4}x\right) - 4(6) = 4\left(\frac{7}{2}\right)$$

$$x - 24 = 14$$

$$x = 38.$$

The number is 38. You should check this solution using the original wording of the problem.

A number problem might involve translated phrases on each side of the equal sign.

EXAMPLE 3.8. Two times five more than a number is twenty-five less than the number. Find the number.

Solution.   If the number is x, we have the two phrases $2(x + 5)$ and $x - 25$. Thus the equation is

$$2(x + 5) = x - 25.$$

Removing the parentheses and solving, we have

$$2x + 10 = x - 25$$

$$x + 10 = -25$$

$$x = -35.$$

The number is -35. Five more than the number is $-35 + 5 = -30$, and two times that is -60. Twenty-five less than -35 is $-35 - 25 = -60$. Since we get the same number using the wording of the phrases on each side of the equal sign, the solution -35 checks.

Exercise 3.2

1. One-eighth of a number is one-fourth. Find the number.

2. Three-fifths of a number is three-fourths. Find the number.

3. One-half is nine-tenths of a number. Find the number.

4. Twelve is three-eighths of a number. Find the number.

5. Thirty-three less than a number is fifty-one. Find the number.

6. Twenty-four more than a number is eighteen. Find the number.

7. One-fourth of five less than a number is six. Find the number.

8. One-fifth of seven more than a number is twelve. Find the number.

9. Two-thirds of fifteen more than a number is ten-thirds. Find the number.

10. Four-fifths of one-fourth less than a number is three-halves. Find the number.

11. Five less than one-half of a number is one-fourth. Find the number.

12. Six more than two-thirds of a number is five-thirds. Find the number.

13. Twenty more than five times a number is ten. Find the number.

14. Thirty less than four times a number is twenty. Find the number.

15. One more than one-half a number is three-halves less than the number. Find the number.

16. Six more than three times a number is one-third of the number. Find the number.

Mixtures and Solutions

Mixture and solution problems offer an example of a type of problem which may look complicated, but which can be resolved by means of a chart. The chart displays the pieces of information in the problem, and their relationships to one another. For mixture problems, we need a chart with spaces for each component, the unit price of each, the number of units of each, and the total prices:

	Unit price	Number of units	Total price
First component			
Second component			
Mixture			

EXAMPLE 3.9.   A 10-pound bag of potatoes contains Idahos which sell for 24¢ a pound and all-purpose Maines which sell for 16¢ a pound. The bag costs $1.84. How many pounds of each kind of potato does it contain?

Solution.   We make a chart of the price per pound and number of pounds of each type of potato. If there are x pounds of Idahos, and a total of 10 pounds in the mixture, then there are $10 - x$ pounds of Maine potatoes:

	Unit price	Number of units	Total price
Idaho	24	x	
Maine	16	$10-x$	
Mixture		10	

The total cost of the Idahos is 24¢ a pound times x pounds, or $24x$. The total cost of the Maines is 16¢ a pound times $10 - x$ pounds, or $16(10 - x)$. We are not given the unit price of the mixture, but the total price of the bag is given as $1.84, or 184¢. Now we can fill in the total prices on our chart:

	Unit price	Number of units	Total price
Idaho	24	x	$24x$
Maine	16	$10 - x$	$16(10 - x)$
Mixture		10	184

The sum of the total prices of each type of potato is equal to the total price of the bag. Thus the equation is

$$24x + 16(10 - x) = 184.$$

Solving this equation,

$$24x + 160 - 16x = 184$$

$$8x + 160 = 184$$

$$8x = 24$$

$$x = 3.$$

The bag contains 3 pounds of Idaho potatoes and 7 pounds of Maine potatoes. To check, we must use the wording of the problem and not the equation. Three pounds of Idaho potatoes at 24¢ a pound cost 72¢. Seven pounds of Maine potatoes at 16¢ a pound cost 112¢. The total cost is 72¢ + 112¢ = 184¢, or $1.84 for the bag.

EXAMPLE 3.10.   Creme-filled chocolates worth $2.50 a pound are mixed with plain chocolates worth $1.30 a pound to make 20 pounds of a mixture which sells for $1.60 a pound. How many pounds of each type of chocolate are in the mixture?

Solution.   We make a chart for the price per pound and number of pounds of each type of chocolate. Since the total price of the mixture is not given, we use the final line to give the price per unit and number of units for the mixture:

	Unit price	Number of units	Total price
Creme	2.50	x	2.50x
Plain	1.30	$20 - x$	1.30(20 − x)
Mixture	1.60	20	32.00

Since the sum of the total price of each type of chocolate is equal to the total price for the mixture, the equation is:

$$2.50x + 1.30(20 - x) = 32.00.$$

It is often easiest to write the prices in cents before solving the equation:

$$250x + 130(20 - x) = 3200$$

$$250x + 2600 - 130x = 3200$$

$$120x = 600$$

$$x = 5.$$

There are 5 pounds of creme-filled chocolates and 15 pounds of plain chocolates in the mixture. You should check these answers, using the original wording of the problem.

Solution problems work in essentially the same way as mixture problems. The amount of a substance in a solution is given as the percent of the substance in the solution. We generally write percents as decimals for computation.

EXAMPLE 3.11.   A solution which is 10% alcohol is mixed with a solution which is 4% alcohol to make 5 quarts of a solution which is 6% alcohol. How many quarts of each of the original parts are in the solution?

Solution.   The chart for this problem is:

	Percent	Number of units	Total amount
First part	.10	x	.10x
Second part	.04	$5 - x$	.04$(5 - x)$
Solution	.06	5	.30

Since the sum of the total amounts of alcohol in each of the two parts is equal to the total amount of alcohol in the solution, we have the equation

$$.10x + .04(5 - x) = .30,$$

or, multiplying each term by 100,

$$10x + 4(5 - x) = 30.$$

Solving this equation,

$$10x + 20 - 4x = 30$$

$$6x = 10$$

$$x = \frac{10}{6}$$

$$x = \frac{5}{3} \text{ or } 1\frac{2}{3}.$$

There are $1\frac{2}{3}$ quarts of the 10% solution and $3\frac{1}{3}$ quarts of the 4% solution. We will check using fractions since thirds do not convert to exact decimals. To write the percents as fractions, $10\% = \frac{10}{100} = \frac{1}{10}$, and $4\% = \frac{4}{100} = \frac{1}{25}$. Thus 10% of $1\frac{2}{3}$ is

$$\frac{1}{10}\left(\frac{5}{3}\right) = \frac{1}{6}$$

and 4% of $3\frac{1}{3}$ is

$$\frac{1}{25}\left(\frac{10}{3}\right) = \frac{2}{15}.$$

The total is

$$\frac{1}{6} + \frac{2}{15} = \frac{5}{30} + \frac{4}{30}$$

$$= \frac{9}{30}$$

$$= \frac{3}{10}$$

$$= .3.$$

Also, five quarts of 6% solution is 5(.06), or .3 units of alcohol.

If one part of a solution contains none of a substance at all, then it has 0% of that substance.

EXAMPLE 3.12.   Water in a pool contains 5% pollutants. Clear water is added to reduce the pollutants to 1% of the solution. If the total amount of solution in the pool is then 100 gallons, how much pure water was added?

Solution.   We let x be the amount of pure water. The pure water contains no pollutants, so its percent is 0%. The chart is:

	Percent	Number of units	Total amount
Water	0	x	0
Pool	.05	$100 - x$	$.05(100 - x)$
Solution	.01	100	1

The equation is

$$0 + .05(100 - x) = 1$$

or

$$5(100 - x) = 100.$$

Solving this equation,

$$500 - 5x = 100$$

$$400 = 5x$$

$$80 = x.$$

80 gallons of pure water were added. You should check this answer.

Possibly we will know the amount of one of the parts of a solution, rather than the total amount of the solution.

EXAMPLE 3.13.   An insecticide spray which is 66% poison is thinned by adding 4 quarts of water. The resulting spray is 2% poison. How much of the original spray was used?

Solution.   We let x be the amount of original spray. The amount of water is 4, and so the amount of solution is $x + 4$. The chart is:

	Percent	Number of units	Total amount
Spray	.66	x	$.66x$
Water	0	4	0
Solution	.02	$x+4$	$.02(x+4)$

The equation is

$$.66x + 0 = .02(x + 4)$$

or

$$66x = 2(x + 4)$$

$$66x = 2x + 8$$

$$64x = 8$$

$$x = \frac{8}{64}$$

$$x = \frac{1}{8}.$$

The original amount of spray was $\frac{1}{8}$ of a quart. You should check this answer.

Exercise 3.3

1. Two varieties of pears are mixed in a 5-pound package. The package contains Bosc pears, which sell for 59¢ a pound, and Bartlett pears, which sell for 49¢ a pound. The package sells for $2.65. How many pounds of each variety of pear does it contain?

2. A 50-pound bag of sand sells for $1.70. The bag contains a mixture of fine sand worth 5¢ a pound and coarse sand worth 3¢ a pound. How many pounds of each is in the bag?

3. An assortment of 30 pens contains two types. There are ball-point pens, which sell for 35¢ each, and felt-tip pens, which sell for 65¢ each. The assortment sells for $13.50. How many pens of each type does it contain?

4. A mixed case of 24 cans includes light tuna worth 90¢ a can and white tuna worth $1.30 a can. The case sells for $25.20. How many cans of each type does it contain?

5. Ten pounds of bean salad are made by mixing kidney beans worth 40¢ a pound and cut green beans worth 60¢ a pound. The salad sells for 52¢ a pound. How many pounds of each type of bean does it contain?

6. Beef flavor and liver flavor dog food are mixed to make an 80-ounce bag. The beef flavor costs 20¢ an ounce and the liver flavor costs 30¢ an ounce. The bag is worth $22\frac{1}{2}$¢ an ounce. How many ounces of each flavor are in it?

7. A mixture of 25 pecks of soil is made to sell for $1.26 a peck. It contains peat moss worth $1.40 a peck and potting soil worth $1.20 a peck. How many pecks of each does it contain?

8. Olive oil and vegetable oil are mixed to make 50 quarts of salad oil worth $1.54 a quart. The olive oil is worth $2.20 a quart and the vegetable oil is worth $1.00 a quart. How many quarts of each does the salad oil contain?

9. A solution which is 8% salt and another which is 2% salt are combined to make 9 quarts of a solution which is 3% salt. How many quarts of each original part does the solution contain?

10. A solution which is 20% detergent is reduced to $15\frac{1}{2}$% by adding a solution which is 5% detergent. There are 20 gallons of the resulting solution. How many gallons of each of the original solutions are used?

11. An all-purpose cleaner is 8% ammonia. It is diluted with water to make 4 quarts of a cleaner with 2% ammonia. How many quarts of water are added?

12. An aquarium contains water with 2% pollutants. Pure water is added to reduce the amount of pollutants to .5%. If the aquarium holds 10 gallons, how much pure water was added?

13. A tank contains water with 10% pollutants. The amount of pollutants is reduced to 1% by adding 54 gallons of pure water. How many gallons of polluted water were in the tank originally?

14. A solution which is 25% corn starch is thinned by adding 1 cup of water. The resulting solution is 10% corn starch. How many cups of the original solution were there?

15. Twelve ounces of a chowder contain 50% non-milk products. How much milk should be added to dilute the chowder to 30% non-milk products?

16. Water is added to $\frac{1}{4}$ of a gallon of fertilizer which is 20% nitrogen to reduce the strength to .2% nitrogen. How many gallons of water are added?

Section 3.4

Rate, Time, and Distance

An important relationship in mathematics and science is rate times time equals distance:

$$R \cdot T = D.$$

For example, if you walk at 3 miles per hour for 4 hours, you can walk $3 \cdot 4 = 12$ miles. From the basic $R \cdot T = D$ formula, we also have

$$T = \frac{D}{R}$$

and

$$R = \frac{D}{T}.$$

If you travel 200 miles at an average rate of 40 miles per hour, it will take you $\frac{200}{40} = 5$ hours. If you wish to travel 60 miles in 2 hours, you must average $\frac{60}{2} = 30$ miles per hour. "Miles per hour" means literally miles divided by hours, and is often written $\frac{m}{h}$.

Problems involving rates, times, and distances related by $R \cdot T = D$ or its variations can often be solved using a chart similar to the one used in the preceding section.

EXAMPLE 3.14.   Campers leave their camp in a valley to climb a hill. They can climb at an average rate of 2 miles per hour, and return downhill at 3 miles per hour. If it takes 1 hour longer for them to climb the hill than to return, how much time must they allow for the trip each way?

Solution.   If we use t for the time of the return trip, the time needed to climb is $t + 1$. Now, we can make a chart using $R \cdot T = D$:

	Rate	Time	Distance
Climb	2	$t+1$	$2(t+1)$
Return	3	t	$3t$

Since the distance to climb the hill and the distance to return are the same, we can set these two distances equal to obtain the equation

$$2(t + 1) = 3t.$$

Solving this equation, we have

$$2t + 2 = 3t$$

$$2 = t.$$

Thus the campers must allow 3 hours to climb and 2 hours to return. To check, observe that the distance is 6 miles each way, and so the two distances are indeed equal.

EXAMPLE 3.15.   A swimmer goes across a lake, and then returns by boat. His average rate swimming across the lake is 3 miles per hour less than the average rate of the boat returning. It takes him 2 hours to swim across the lake, and $\frac{4}{5}$ of an hour to return in the boat. What is his average rate swimming and what is the average rate of the boat?

Solution.   If we use r for the rate of the boat, the rate swimming is $r - 3$. The chart is:

	Rate	Time	Distance
Swimming	$r-3$	2	$2(r-3)$
Boat	r	$\frac{4}{5}$	$\frac{4}{5}r$

Again the distances are the same, so we have the equation

$$2(r - 3) = \frac{4}{5}r$$
$$5(2)(r - 3) = 5(\frac{4}{5}r)$$
$$10(r - 3) = 4r$$
$$10r - 30 = 4r$$
$$6r = 30$$
$$r = 5.$$

The average rate of the boat is 5 miles per hour, and the average rate swimming is 2 miles per hour. You should check these answers by seeing that the distances are the same.

There are many types of problems involving equal distances. The following example is of another type.

EXAMPLE 3.16.   A cyclist starts out at an average rate of 12 miles per hour. Fifteen minutes later, another cyclist starts at an average rate of 15 miles per hour. How long will it take the second cyclist to overtake the first?

Solution.   We want the time for the second cyclist, so we call this time t. Fifteen minutes is $\frac{1}{4}$ of an hour, so the first cyclist travels for $t + \frac{1}{4}$ hours. The chart is:

	Rate	Time	Distance
First	12	$t+\frac{1}{4}$	$12(t+\frac{1}{4})$
Second	15	t	$15t$

Since the second cyclist overtakes the first, they cover the same distance. The equation is

$$12\left(t + \frac{1}{4}\right) = 15t$$
$$12t + 3 = 15t$$
$$3 = 3t$$
$$1 = t.$$

It will take 1 hour for the second cyclist to overtake the first. To check, observe that the first cyclist will have traveled $1\frac{1}{4}$ hours. Each will cover 15 miles.

We can use the same chart for problems where the distances are not the same. In the following example, we know the sum of the distances. An interesting variation, where we know the sum of the times, is in Unit 17.

EXAMPLE 3.17.   You drive from your home towards a friend's home at an average rate of 30 miles per hour. One-half hour later, your friend leaves her home driving towards your home, also at an average rate of 30 miles per hour. If your homes are 60 miles apart, how long will you drive until you meet your friend?

Solution.   If your time is t, your friend's time is $\frac{1}{2}$ hour less, or $t - \frac{1}{2}$. The chart is:

	Rate	Time	Distance
You	30	t	$30t$
Friend	30	$t-\frac{1}{2}$	$30(t-\frac{1}{2})$

Since the sum of the two distances traveled is 60 miles, the equation is

$$30t + 30\left(t - \frac{1}{2}\right) = 60.$$

Solving this equation, we have

$$30t + 30t - 15 = 60$$
$$60t = 75$$
$$t = \frac{75}{60}$$
$$t = 1\frac{1}{4}.$$

You will drive for $1\frac{1}{4}$ hours. You should check this answer by finding the distance each of you will travel, and seeing that the sum of the distances is 60 miles.

Exercise 3.4

1. A cyclist goes from one town to another, mostly downhill, at an average rate of 10 miles per hour, but returns at an average rate of only 4 miles per hour. It takes him 3 hours longer to return. How long does it take him to go each way?

2. Suppose you walk into town at an average rate of 3 miles per hour, where you pick up a bicycle and return at an average rate of 6 miles per hour. It takes you $\frac{1}{2}$ hour longer to get to town than to return. How long does it take you to go each way?

3. A jogger goes from her home to a park, but walks back at an average rate of 4 miles per hour less. If it takes her $\frac{1}{2}$ hour to get to the park and 1 hour to return, what are her average rate jogging and her average rate walking?

4. An airplane flies from one town to another against the wind but returns with a tail wind, which makes its average rate 50 miles per hour more. If the plane takes $1\frac{3}{4}$ hours to go against the wind, and $1\frac{1}{4}$ hours to return, what are its average rates each way?

5. A car starts out from a town, and goes at an average rate of 40 miles per hour. One-half hour later, another car starts from the same point along the same route at an average rate of 50 miles per hour. How long will it take the second car to overtake the first?

6. A local train starts from a terminal and runs at an average rate of 20 miles per hour. Twenty minutes later, an express train starts from the same terminal along the same track at an average rate of 40 miles per hour. How long will it take for the express to overtake the local?

7. The Boston Marathon includes the National Wheelchair Marathon which starts 15 minutes ahead of the Marathon. Suppose the winner of the Wheelchair Marathon goes at an average rate of 10 miles per hour and the Marathon winner goes at an average rate of $11\frac{1}{2}$ miles per hour. How long will it take the Marathon winner to overtake the Wheelchair winner?*

8. A car goes by a billboard averaging 60 miles per hour (one mile per minute). Two minutes later, the motorcycle policeman behind the billboard gives chase at an average rate of 80 miles per hour ($\frac{4}{3}$ miles per minute). How long, in minutes, will it take the policeman to catch the car?

9. Two cities are 200 miles apart. A train leaves the first city for the second at an average rate of 60 miles per hour. One hour later, a train leaves the second city for the first at an average rate of 80 miles per hour. How long after the first train leaves will they meet one another?

10. Two airports are 125 miles apart. Two airplanes leave at the same time, each headed for the other airport. The second plane flies at an average rate of 30 miles per hour less than the first. If they pass each other after flying for half an hour, what is the average rate of each plane?

<table>
<tr><td>Section
3.5</td><td><h1 style="text-align:center">Areas</h1></td></tr>
</table>

Areas of geometric figures provide an application of the quadratic equations in Section 2.5. You should recall from geometry, or learn now, the names of these geometric figures and the formula for the area of each.

A **rectangle** has four sides, where each pair of opposite sides is parallel and equal in length, and each angle is a right angle.

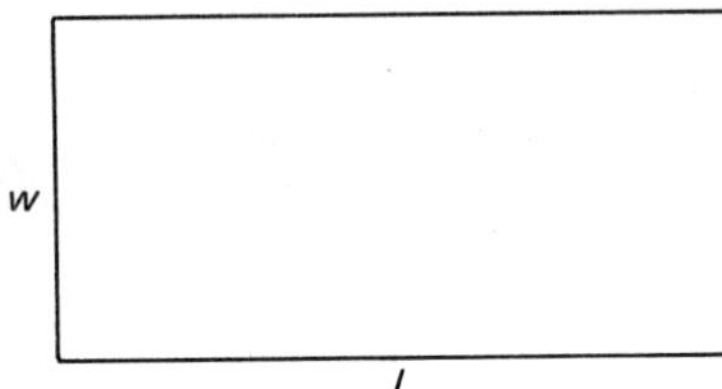

If w is the width of a rectangle and l is the length, the area is

$$A = w \cdot l,$$

the width times the length.

*During the time this book was being prepared, the 1978 Boston Marathon was held. The Wheelchair Marathon started 20 minutes ahead. The winner of the Marathon covered the 26 miles, 385 yards in a near-record 2 hours, 10 minutes, 13 seconds. However, he never overtook the Wheelchair Marathon winner, who finished in an astonishing 2 hours, 26 minutes, 5 seconds!

A **parallelogram** has four sides, where each pair of opposite sides is parallel and equal in length. The angles are not necessarily right angles. If the angles are right angles, then the parallelogram is a rectangle. If the parallelogram is not a rectangle, we pick one side to be the base b. Then the height h is perpendicular to b.

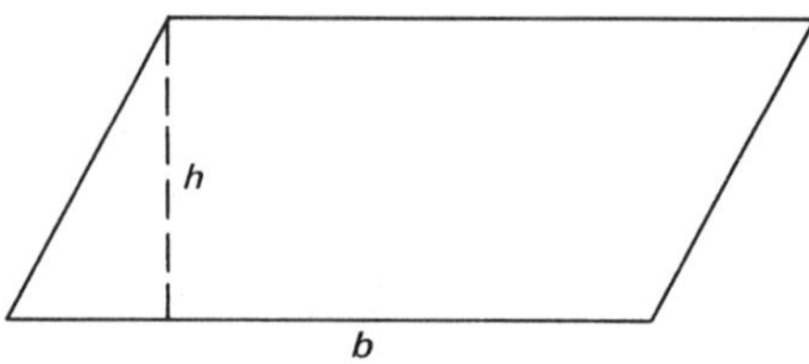

The area is given by

$$A = b \cdot h,$$

the base times the height.

A **triangle** has three sides and three angles. If we pick one side to be the base b, the height h is drawn from the angle opposite the base and perpendicular to the base.

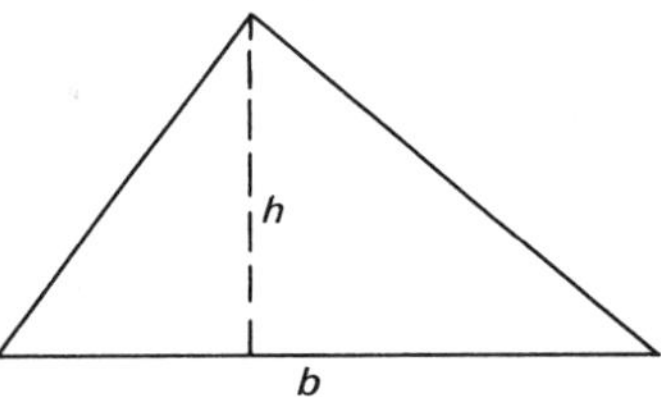

The area of a triangle is given by

$$A = \frac{1}{2}(b \cdot h),$$

one-half of the base times the height.

A **trapezoid** has four sides, where one pair of opposite sides is parallel. The two parallel sides are the bases, b and B. The height h is the perpendicular distance between the bases.

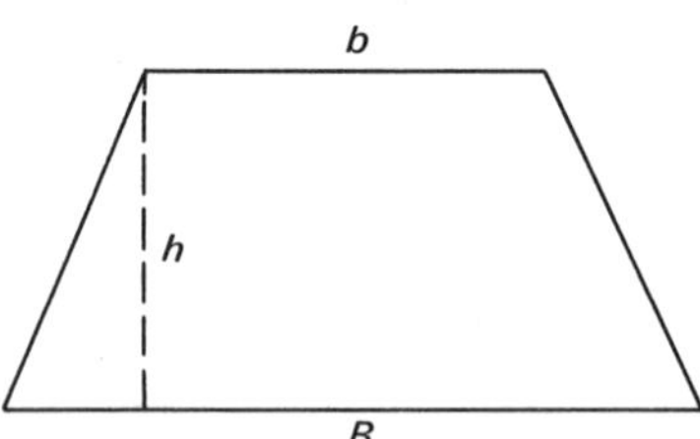

The area of a trapezoid is given by

$$A = \frac{1}{2}(b + B)h,$$

one-half of the sum of the bases times the height.

To solve area problems, we make a diagram of the figure and label the given parts. The formula for the area of the figure should give us our equation.

EXAMPLE 3.18.   A rectangle has a length of 8 inches more than its width. Its area is 33 square inches. Find the width and the length.

Solution.   Since the length is 8 inches more than the width, if the width is w, the length is $w + 8$. The diagram is

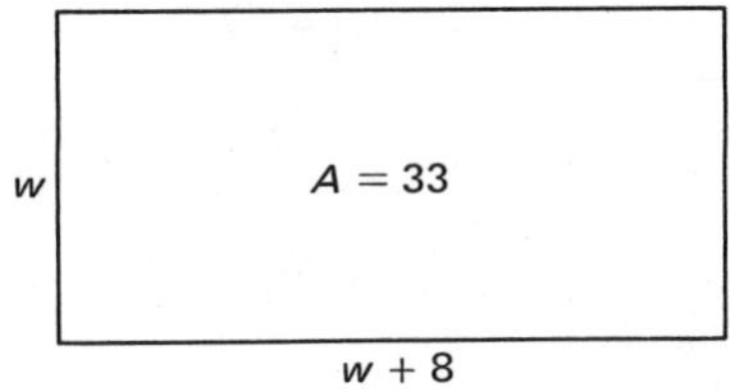

The area is given by the width times the length, so we write

$$w \cdot l = A$$

or

$$w(w + 8) = 33.$$

Simplifying, we have

$$w^2 + 8w = 33,$$

and collecting the nonzero terms,

$$w^2 + 8w - 33 = 0.$$

Now, we factor and solve for w:

$$(w - 3)(w + 11) = 0$$

$$w - 3 = 0 \text{ and } w + 11 = 0$$

$$w = 3 \text{ and } w = -11.$$

Since a side of a rectangle cannot be negative, the solution -11 is not possible. We call such a solution **extraneous**. The solution $w = 3$ gives a width of 3 inches. Since the length is 8 more, the length is 11 inches. To check, we observe that the area is $3 \cdot 11 = 33$.

EXAMPLE 3.19.   A triangle has an area of $6\frac{7}{8}$ square meters. The height is 3 meters more than the base. Find the base and height.

Solution.   If the base is b, then the height is $b + 3$. The diagram is

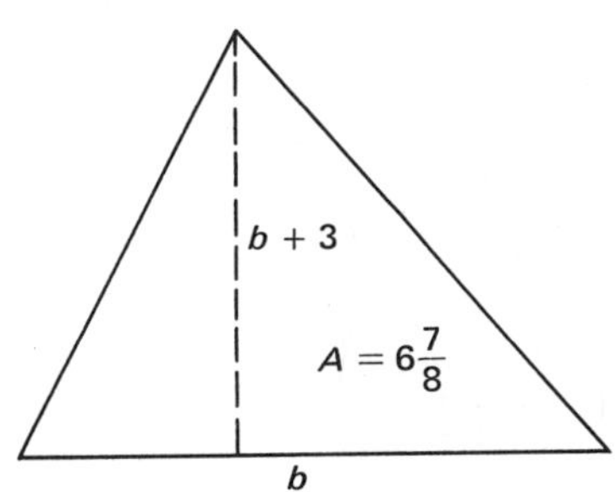

Since the area is one-half of the base times the height, the equation is

$$\frac{1}{2}(b \cdot h) = A$$

or

$$\frac{1}{2}b(b + 3) = \frac{55}{8}.$$

Recall from Section 2.2 that we may multiply both sides of the equation by 8:

$$8\left(\frac{1}{2}b\right)(b + 3) = 8\left(\frac{55}{8}\right)$$

$$4b(b + 3) = 55$$

$$4b^2 + 12b - 55 = 0.$$

Finally, we factor and solve for b:

$$(2b - 5)(2b + 11) = 0$$

$$2b - 5 = 0 \text{ and } 2b + 11 = 0$$

$$b = \frac{5}{2} \text{ and } b = -\frac{11}{2}.$$

The solution $-\frac{11}{2}$ is extraneous. Thus the base is $\frac{5}{2}$, or $2\frac{1}{2}$ meters. The height is 3 more, or $5\frac{1}{2}$ meters. To check, the area is $\frac{1}{2}(2\frac{1}{2})(5\frac{1}{2}) = \frac{1}{2}(\frac{5}{2})(\frac{11}{2}) = \frac{55}{8} = 6\frac{7}{8}$.

EXAMPLE 3.20.   A trapezoid has a bottom base $2\frac{1}{2}$ feet more than its top base. The height is equal to the top base. The area is $\frac{7}{8}$ of a square foot. Find the top and bottom bases.

Solution.   If the top base is b, the diagram is

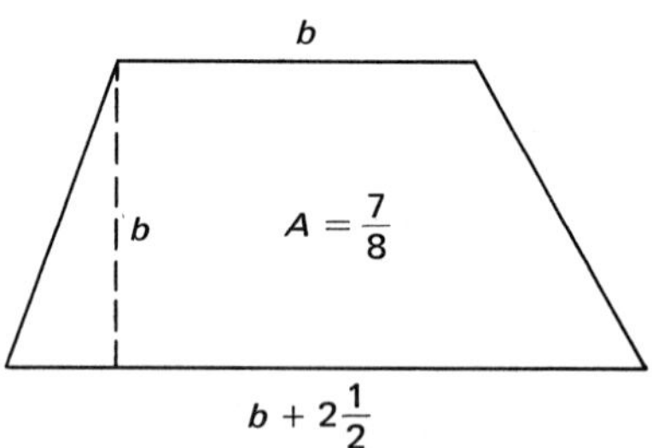

Since the area is one-half of the sum of the bases times the height, we have

$$\frac{1}{2}(b + B)h = A$$

$$\frac{1}{2}\left(b + b + \frac{5}{2}\right)b = \frac{7}{8}$$

$$\frac{1}{2}b\left(2b + \frac{5}{2}\right) = \frac{7}{8}$$

$$8\left(\frac{1}{2}b\right)\left(2b + \frac{5}{2}\right) = 8\left(\frac{7}{8}\right)$$

$$4b\left(2b + \frac{5}{2}\right) = 7$$

$$8b^2 + 10b - 7 = 0$$

$$(2b - 1)(4b + 7) = 0$$

$$2b - 1 = 0 \text{ and } 4b + 7 = 0$$

$$b = \frac{1}{2} \text{ and } b = -\frac{7}{4}.$$

The solution $-\frac{7}{4}$ is extraneous. Thus the top base, and also the height, are each $\frac{1}{2}$ foot. The bottom base is $\frac{1}{2} + 2\frac{1}{2}$, or 3 feet. You should check these answers using the area formula.

We may use the same method to find the dimensions of a box, given the volume and one dimension, and the relationship between the two other dimensions. We will consider a box to be a three-dimensional figure where all the sides are rectangles.

EXAMPLE 3.21.   The base of a box is a rectangle 5 centimeters longer than it is wide. The box is 3 centimeters high. The volume is 42 cubic centimeters. Find the width and length of the box.

Solution.   If the width of the base is w, the length is $w + 5$, and a diagram of the box is

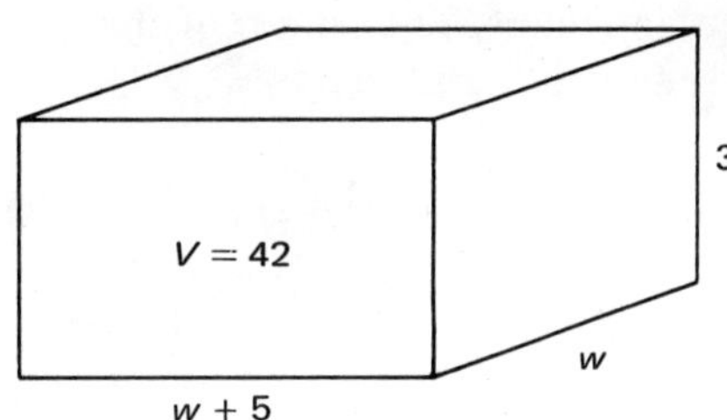

The formula for the volume of a box is the width times the length times the height. Thus,

$$w \cdot l \cdot h = V$$

$$w(w + 5)(3) = 42.$$

Recall from Section 2.3 that we can simplify this equation by dividing both sides by 3:

$$\frac{w(w + 5)(3)}{3} = \frac{42}{3}$$

$$w(w + 5) = 14$$

$$w^2 + 5w - 14 = 0$$

$$(w - 2)(w + 7) = 0$$

$$w - 2 = 0 \text{ and } w + 7 = 0$$

$$w = 2 \text{ and } w = -7.$$

The solution -7 is extraneous. The width of the base is 2 centimeters and the length is 7 centimeters. You should check these answers.

Exercise 3.5

1. The length of a rectangle is 3 centimeters more than its width. The area is 70 square centimeters. Find the width and length.

2. A rectangle has an area of $8\frac{3}{4}$ square inches. The length is 1 inch more than the width. Find the width and length.

3. The area of a parallelogram is $2\frac{2}{9}$ square meters. The base is $\frac{1}{3}$ of a meter more than the height. Find the base and height.

4. The area of a parallelogram is 1 square foot. The base is $1\frac{1}{2}$ feet more than the height. Find the base and height.

5. A triangle has a base 4 inches more than its height. The area is $10\frac{1}{2}$ square inches. Find the base and height.

6. A triangle has a base 1 meter less than its height. The area is $\frac{3}{8}$ of a square meter. Find the base and height.

7. A trapezoid has an area of 15 square yards. The bottom base is 4 yards more than the top base. The height is the same as the top base. Find the top and bottom bases.

8. A trapezoid has an area of 10 square feet. The bottom base is 3 feet more than the top base. The height is the same as the bottom base. Find the top and bottom bases.

9. The volume of a box is 9 cubic meters. The length of the base is 3 meters more than the width. The height of the box is $\frac{1}{2}$ meter. Find the width and length.

10. The volume of a box is $10\frac{1}{2}$ cubic feet. The length of the base is 2 feet more than the width. The box is 2 feet high. Find the width and length.

Self-test

1. Translate into a mathematical expression: two-thirds of five less than a number.

1. _______________

2. Three more than two-fifths of a number is one. Find the number.

2. _______________

3. A solution is 16% salt. Pure water is added to make 8 quarts of a solution which is 5% salt. How many quarts of water are added?

3. _______________

4. Two cities are 120 miles apart. Two cars leave at the same time, one from each city, heading towards one another. One car travels at an average rate of 4 miles per hour more than the other. If they meet each other after $1\frac{1}{2}$ hours, what is the average rate of each car?

4. _______________

5. A triangle has a height of 2 inches more than its base. The area is $2\frac{5}{8}$ square inches. Find the base and height.

5. _______________

Unit 4
Quadratic Equations

INTRODUCTION

In Unit 2 you learned how to solve linear equations. You also learned how to solve equations with a second-degree term, if the nonzero terms could be collected into an expression you could factor. In this unit you will learn how to solve equations with a second-degree term regardless of whether or not the expression of nonzero terms can be factored. You can then solve a much wider range of applied problems. An extension of one of the applications in Unit 3 is in this unit.

OBJECTIVES

When you have finished this unit you should be able to:

1. Write a quadratic equation in standard form and identify a, b, and c.
2. Solve equations of the form $ax^2 + c = 0$ when the equation has real number solutions, or say if the equation does not have real number solutions.
3. Use the quadratic formula to solve quadratic equations when the equation has real number solutions, or say if the equation does not have real number solutions.
4. Use the quadratic formula to solve geometric problems.

Section 4.1
Standard Form of Quadratic Equations

In Sections 2.4 and 2.5 you learned a method of solving equations by factoring. The first two examples of this unit are review examples of this type. If you have any difficulty understanding these examples, then you should review the material in Unit 2.

EXAMPLE 4.1. Solve $5x + 2 = 3x^2$.

Solution. We must collect all the nonzero terms on one side of the equation. The resulting expression is easier to factor if the second-degree term is positive. Therefore, we will collect the terms on the right-hand side of the equation:

$$5x + 2 = 3x^2$$

$$0 = 3x^2 - 5x - 2$$

$$0 = (x - 2)(3x + 1)$$

$$x - 2 = 0 \text{ and } 3x + 1 = 0$$

$$x = 2 \text{ and } 3x = -1$$

$$x = 2 \text{ and } x = -\frac{1}{3}.$$

The solutions are 2 and $-\frac{1}{3}$.

EXAMPLE 4.2. Solve $x^2 = 2x$.

Solution. Recall from Unit 2 that we must never divide out a variable. We must collect the terms and factor. This time we will collect the terms on the left-hand side of the equation:

$$x^2 = 2x$$

$$x^2 - 2x = 0$$

$$x(x - 2) = 0$$

$$x = 0 \text{ and } x - 2 = 0$$

$$x = 0 \text{ and } x = 2.$$

The solutions are 0 and 2.

In both of the preceding examples, when the nonzero terms have been collected on one side, we have an expression of the form $ax^2 + bx + c$. In Example 4.1, we have the expression $3x^2 - 5x - 2$. In this expression, $3x^2$ is the **second-degree** or x^2**-term**, $-5x$ is the **first-degree** or **x-term**, and -2 is the **constant term**. In Example 4.2, no constant term appears in the expression $x^2 - 2x$. In this case we say the constant term is zero.

A **quadratic equation** is an equation which can be written in the form

$$ax^2 + bx + c = 0$$

where $a \neq 0$. The form $ax^2 + bx + c = 0$ is called the **standard form** of the quadratic equation.

EXAMPLE 4.3. Write $2x^2 + 5x = 1$ in standard form, and identify a, b, and c.

Solution. Writing the equation in standard form, we have

$$2x^2 + 5x - 1 = 0,$$

thus $a = 2$, $b = 5$, and $c = -1$. Observe that we must take minus signs into account in identifying a, b, and c.

EXAMPLE 4.4. Write $3x^2 = x$ in standard form and identify a, b, and c.

Solution. In standard form, we have

$$3x^2 - x = 0$$

or

$$3x^2 - 1 \cdot x + 0 = 0.$$

We see that $a = 3$ and $b = -1$, taking the sign into account. Also, $c = 0$ since the constant term is zero.

EXAMPLE 4.5. Write $x^2 = 4$ in standard form and identify a, b, and c.

Solution.   In standard form, we have

$$x^2 - 4 = 0$$

or

$$1 \cdot x^2 + 0 \cdot x - 4 = 0.$$

Thus $a = 1$, $b = 0$ since no x-term appears in the original equation, and $c = -4$. It is possible for b, or c, or both, to be zero in a quadratic equation. However, a is never zero since otherwise the equation would have no x^2-term and thus would not be a quadratic equation.

EXAMPLE 4.6. Write $4 = x^2 - 3x$ in standard form and identify a, b, and c.

Solution.   In standard form,

$$-x^2 + 3x + 4 = 0,$$

so $a = -1$, $b = 3$, and $c = 4$. We have followed standard convention by collecting on the left, even though this convention gives a negative x^2-term.

Sometimes we must simplify the expressions on each side of an equation, and combine like terms, to derive the standard form.

EXAMPLE 4.7. Write $x(x - 3) = x + 2$ in standard form and identify a, b, and c.

Solution.   Starting with

$$x(x - 3) = x + 2,$$

we simplify the left-hand side, giving

$$x^2 - 3x = x + 2.$$

Then collecting like terms on the left-hand side, we have

$$x^2 - 4x - 2 = 0.$$

Therefore, $a = 1$, $b = -4$, and $c = -2$.

Exercise 4.1

Write in standard form and identify a, b, and c:

1. $x^2 + 1 = -2x$ 2. $3 = 3x^2 + x$ 3. $3x = 2x^2$ 4. $4x^2 = x$

5. $x^2 = 1$ 6. $5 = 2x^2$ 7. $2 - x^2 = 3 - x$ 8. $x^2 + 2x - 3 = 2x^2 - 3x + 2$

9. $x(2x - 3) = 2(x + 1)$ 10. $-(2x^2 + 4) = x(x - 2)$

<table>
<tr><td>Section
4.2</td><td># Equations of the Form $ax^2 + c = 0$</td></tr>
</table>

We usually solve a quadratic equation by writing the equation in standard form and factoring the quadratic expression if it can be factored. There is one type of quadratic equation which is not usually easiest to solve by factoring, and which is not solved by first writing the equation in standard form. This type is a quadratic equation where $b = 0$; that is, the x-term does not appear.

EXAMPLE 4.8. Solve $x^2 - 4 = 0$.

Solution.   First we solve for the x^2-term, writing

$$x^2 = 4.$$

Now we will take the square root of each side of the equation. Since the square of either a positive or a negative number is positive, we must allow for both the positive and negative square roots of 4. We write

$$x^2 = 4$$

$$x = \pm\sqrt{4}$$

$$x = \pm 2.$$

The solutions are 2 and -2.

Of course, we could have solved the equation in Example 4.8 by factoring:

$$x^2 - 4 = 0$$

$$(x + 2)(x - 2) = 0$$

$$x + 2 = 0 \text{ and } x - 2 = 0$$

$$x = -2 \text{ and } x = 2.$$

However, the square root method is usually faster and easier. Also, the square root method works for equations with no x-term which cannot be solved by factoring.

EXAMPLE 4.9. Solve $x^2 - 7 = 0$.

Solution.   First we solve for the x^2-term:

$$x^2 = 7.$$

Now we take the square root of each side of the equation:

$$x^2 = 7$$

$$x = \pm\sqrt{7}.$$

The solutions are $\sqrt{7}$ and $-\sqrt{7}$. Observe that, when the solutions contain square roots, the equation could not have been solved by factoring into binomial factors where the constants are integers.

A Note on Simplifying Square Roots

If a square root involves only perfect squares, then we write the square root as a whole number or a fraction. For example,

$$\sqrt{9} = 3$$

$$\sqrt{\frac{4}{25}} = \frac{2}{5}.$$

In this unit we will often encounter square roots which do not involve only perfect squares, but which can be simplified. One type of square root which is easily simplified involves a fraction where the denominator is a perfect square. For example,

$$\sqrt{\frac{7}{25}} = \sqrt{\frac{1}{25} \cdot 7} = \sqrt{\frac{1}{25}}\,\sqrt{7} = \frac{1}{5}\sqrt{7}.$$

We may write the simplified square root as $\frac{1}{5}\sqrt{7}$ or $\frac{\sqrt{7}}{5}$. Some other examples of this kind of simplification are:

$$\sqrt{\frac{5}{16}} = \frac{1}{4}\sqrt{5}\text{ , or }\frac{\sqrt{5}}{4}$$

$$\sqrt{\frac{11}{9}} = \frac{1}{3}\sqrt{11}\text{ , or }\frac{\sqrt{11}}{3}.$$

We will consider one other type of square root which is easily simplified. This type is the square root of a number where the number has a factor which is a perfect square. For example,

$$\sqrt{50} = \sqrt{25 \cdot 2} = \sqrt{25}\,\sqrt{2} = 5\sqrt{2}.$$

We can find factors which are perfect squares simply by trying out the perfect squares, 4, 9, 16, 25, and so on. For example, to simplify $\sqrt{108}$, we might try

$$\sqrt{108} = \sqrt{4 \cdot 27} = 2\sqrt{27}.$$

However, $\sqrt{27}$ also can be simplified:

$$\sqrt{108} = 2\sqrt{27} = 2\sqrt{9 \cdot 3} = 2 \cdot 3\sqrt{3} = 6\sqrt{3}.$$

Some other examples of this kind of simplification are:

$$\sqrt{48} = \sqrt{16 \cdot 3} = 4\sqrt{3}$$

$$\sqrt{128} = \sqrt{64 \cdot 2} = 8\sqrt{2}.$$

We will generally simplify square roots whenever possible.

EXAMPLE 4.10.   Solve $4x^2 - 3 = 0$.

Solution.   Solving for the x^2-term,

$$4x^2 - 3 = 0$$

$$4x^2 = 3$$

$$x^2 = \frac{3}{4}.$$

Now we take the square root of each side, remembering that the number on the right-hand side can be either positive or negative, since its square in either case is positive:

$$x^2 = \frac{3}{4}$$

$$x = \pm\sqrt{\frac{3}{4}}$$

$$= \pm\frac{\sqrt{3}}{2}.$$

The solutions are $\dfrac{\sqrt{3}}{2}$ and $-\dfrac{\sqrt{3}}{2}$.

EXAMPLE 4.11.   Solve $2x^2 - 40 = 0$.

Solution.

$$2x^2 - 40 = 0$$

$$2x^2 = 40$$

$$x^2 = 20$$

$$x = \pm\sqrt{20}$$

$$= \pm\sqrt{4 \cdot 5}$$

$$= \pm 2\sqrt{5}.$$

The solutions are $2\sqrt{5}$ and $-2\sqrt{5}$.

It is possible for an equation of the form $ax^2 - c = 0$ to have no real number solutions. This is the case if, when we take the square root of each side, the number under the square root is negative. For example, there is no real number x such that $\sqrt{-4} = x$, since to have such an x would require that $x^2 = -4$. We have $\sqrt{4} = 2$ since $2^2 = 4$, and also $-\sqrt{4} = -2$ since $(-2)^2 = 4$, but no real number x such that $x^2 = -4$.

EXAMPLE 4.12.   Solve $3x^2 + 9 = 0$.

Solution.

$$3x^2 + 9 = 0$$

$$3x^2 = -9$$

$$x^2 = -3$$

$$x = \pm\sqrt{-3}.$$

Since we have the square root of a negative number, there are no real number solutions. In Unit 15, a type of number is defined which provides solutions for this case.

Exercise 4.2

Solve:

1. $x^2 - 16 = 0$ 2. $x^2 - 49 = 0$ 3. $4x^2 - 5 = 0$ 4. $64x^2 - 25 = 0$

5. $9x^2 + 4 = 0$ 6. $4x^2 + 10 = 0$ 7. $x^2 - 75 = 0$ 8. $9x^2 - 32 = 0$

9. $2x^2 - 50 = 0$ 10. $27x^2 - 216 = 0$

Section
4.3

The Quadratic Formula

Any equation of the form $ax^2 + bx + c = 0$ can be solved using the **quadratic formula**. The proof that the quadratic formula gives the solutions to the equation is given at the end of this unit.

The Quadratic Formula   For the quadratic equation in standard form,

$$ax^2 + bx + c = 0,$$

where $a \neq 0$, the solutions are given by

$$x = \frac{-b \pm \sqrt{b^2 - 4ac}}{2a}.$$

EXAMPLE 4.13.   Solve $x^2 - 4x + 3 = 0$.

Solution.   Since the equation is in standard form, and the expression of nonzero terms is easily factored, the best way to solve the equation is by factoring:

$$x^2 - 4x + 3 = 0$$

$$(x - 1)(x - 3) = 0$$

$$x - 1 = 0 \text{ and } x - 3 = 0$$

$$x = 1 \text{ and } x = 3.$$

However, the quadratic formula will give the same solutions. Since $x^2 - 4x + 3 = 0$ is in standard form, $a = 1$, $b = -4$, and $c = 3$. Using the quadratic formula,

$$x = \frac{-b \pm \sqrt{b^2 - 4ac}}{2a},$$

we substitute 1 for a, -4 for b, and 3 for c:

$$x = \frac{-(-4) \pm \sqrt{(-4)^2 - 4(1)(3)}}{2(1)}$$

$$= \frac{4 \pm \sqrt{16 - 12}}{2}$$

$$= \frac{4 \pm \sqrt{4}}{2}$$

$$= \frac{4 \pm 2}{2}.$$

The expression $\dfrac{4 \pm 2}{2}$ means $x = \dfrac{4 + 2}{2}$ and $x = \dfrac{4 - 2}{2}$:

$$x = \frac{4 + 2}{2} = \frac{6}{2} = 3,$$

and

$$x = \frac{4 - 2}{2} = \frac{2}{2} = 1.$$

The solutions are 3 and 1 which, of course, are the same as those we found by factoring. Clearly it is easier to use the method of factoring for this example.

We use the quadratic formula when we have an equation in standard form where the expression of nonzero terms cannot be factored, or is not easily factored.

EXAMPLE 4.14.   Solve $4x^2 - 8x - 21 = 0$.

Solution.   This equation can be solved by factoring; however, you might not readily see the factors. If not, we may use the quadratic formula with $a = 4$, $b = -8$, and $c = -21$:

$$
\begin{aligned}
x &= \frac{-b \pm \sqrt{b^2 - 4ac}}{2a} \\[2mm]
&= \frac{-(-8) \pm \sqrt{(-8)^2 - 4(4)(-21)}}{2(4)} \\[2mm]
&= \frac{8 \pm \sqrt{64 + 336}}{8} \\[2mm]
&= \frac{8 \pm \sqrt{400}}{8} \\[2mm]
&= \frac{8 \pm 20}{8}.
\end{aligned}
$$

$$x = \frac{8 + 20}{8} = \frac{28}{8} = \frac{7}{2}$$

and

$$x = \frac{8 - 20}{8} = \frac{-12}{8} = -\frac{3}{2}.$$

The solutions are $\frac{7}{2}$ and $-\frac{3}{2}$. When $b^2 - 4ac$ is a perfect square, we have an equation which could have been solved by factoring, and we can find solutions which do not involve square roots.

EXAMPLE 4.15.   Solve $x^2 + 2x - 2 = 0$.

Solution.   It will not take much experimentation for you to conclude that the expression of nonzero terms in this equation cannot be factored into binomial factors where the constants are integers. Using the quadratic formula, $a = 1$, $b = 2$, and $c = -2$:

$$x = \frac{-b \pm \sqrt{b^2 - 4ac}}{2a}$$

$$= \frac{-2 \pm \sqrt{2^2 - 4(1)(-2)}}{2(1)}$$

$$= \frac{-2 \pm \sqrt{4 + 8}}{2}$$

$$= \frac{-2 \pm \sqrt{12}}{2}.$$

Recall from Section 4.2 that $\sqrt{12}$ can be simplified:

$$\sqrt{12} = \sqrt{4 \cdot 3} = 2\sqrt{3}.$$

Therefore,

$$x = \frac{-2 \pm \sqrt{12}}{2}$$

$$= \frac{-2 \pm 2\sqrt{3}}{2}.$$

Now, we recall from Section 1.3 that we may reduce this fraction. We must remember to divide out only factors, and not addends, so we factor the numerator:

$$x = \frac{-2 \pm 2\sqrt{3}}{2}$$

$$= \frac{2(-1 \pm \sqrt{3})}{2}$$

$$= -1 \pm \sqrt{3}.$$

The solutions are $-1 + \sqrt{3}$ and $-1 - \sqrt{3}$. When the solutions contain a square root, and cannot be simplified further, it is common to summarize them in one expression by writing

$$-1 \pm \sqrt{3}.$$

Since $b^2 - 4ac$ is not a perfect square, and a square root appears in the solutions, the equation could not have been solved by factoring.

When $b^2 - 4ac$ is negative, a negative number will appear under the square root. In this case, the equation has no real number solutions.

EXAMPLE 4.16.   Solve $2x^2 - 3x + 2 = 0$.

Solution.   The equation is in standard form, and $a = 2$, $b = -3$, and $c = 2$:

$$x = \frac{-b \pm \sqrt{b^2 - 4ac}}{2a}$$

$$= \frac{-(-3) \pm \sqrt{(-3)^2 - 4(2)(2)}}{2(2)}$$

$$= \frac{3 \pm \sqrt{9 - 16}}{4}$$

$$= \frac{3 \pm \sqrt{-7}}{4}.$$

Since $b^2 - 4ac$ is negative, we have the square root of a negative number, and the equation has no real number solutions. The number under the square root, given by $b^2 - 4ac$, discriminates among equations which could be solved by factoring, equations with solutions involving a square root, and equations which have no real number solution. Therefore, $b^2 - 4ac$ is often called the **discriminant** of the quadratic equation.

A quadratic equation which is not given in standard form must be put into standard form before it can be solved by factoring. Similarly, a quadratic equation must be in standard form before we can use the quadratic formula. However, as with the method of factoring, we may collect the nonzero terms on the right-hand side rather than the left-hand side if it is more convenient.

EXAMPLE 4.17.   Solve $2 - x^2 = 4x$.

Solution.   Collecting on the right, we have

$$0 = x^2 + 4x - 2,$$

and so $a = 1$, $b = 4$, and $c = -2$. Therefore,

$$x = \frac{-b \pm \sqrt{b^2 - 4ac}}{2a}$$

$$= \frac{-4 \pm \sqrt{4^2 - 4(1)(-2)}}{2(1)}$$

$$= \frac{-4 \pm \sqrt{16 + 8}}{2}$$

$$= \frac{-4 \pm \sqrt{24}}{2}$$

$$= \frac{-4 \pm \sqrt{4 \cdot 6}}{2}$$

$$= \frac{-4 \pm 2\sqrt{6}}{2}$$

$$= -2 \pm \sqrt{6}.$$

The solutions may be expressed as $-2 \pm \sqrt{6}$, since they are real number solutions but cannot be simplified further.

Exercise 4.3

Use the quadratic formula to solve:

1. $2x^2 = 5x - 2$
2. $4x^2 + 4x = 3$
3. $9x^2 + 6x = 8$
4. $4x^2 + 4 = 17x$

5. $x^2 - 4x - 3 = 0$
6. $x^2 + 2x - 5 = 0$
7. $x^2 + 4x = 4$
8. $6x - x^2 = 2$

9. $2x^2 = 2x + 1$
10. $3x^2 + 4x = 9$
11. $7 = 2x^2 + 3x$
12. $3x^2 = 5x + 3$

13. $2x^2 - 5x + 5 = 0$
14. $x^2 + x + 1 = 0$
15. $2x(x - 3) = x^2 + 3$

16. $x(x - 4) = 2(x - 5)$

Section 4.4 An Application

In Section 3.5, we solved problems involving geometric figures. The equations used to solve these problems were quadratic equations. The problems in Section 3.5 were chosen so that the quadratic equations could be solved by factoring. However, it is quite possible to encounter problems of the same type which cannot be solved by factoring.

EXAMPLE 4.18.   A rectangle has an area of 8 square feet. Its length is 4 feet more than its width. Find the width and length.

Solution.   Recall that the area of a rectangle is equal to its width times its length. The length is 4 feet more than the width; therefore, if the width is x, then the length is $x + 4$.

$$\begin{array}{|c|}\hline x \quad\quad A = 8 \\ \hline \end{array}$$
$$x + 4$$

$$
\begin{aligned}
w \cdot l &= A \\
x(x + 4) &= 8 \\
x^2 + 4x - 8 &= 0 \\
x &= \frac{-4 \pm \sqrt{4^2 - 4(1)(-8)}}{2(1)} \\
&= \frac{-4 \pm \sqrt{16 + 32}}{2} \\
&= \frac{-4 \pm \sqrt{48}}{2} \\
&= \frac{-4 \pm 4\sqrt{3}}{2} \\
&= -2 \pm 2\sqrt{3} \, .
\end{aligned}
$$

The solution $-2 - 2\sqrt{3}$ is extraneous because it is negative. We will approximate the solution $-2 + 2\sqrt{3}$ to two decimal places. Using either a calculator or a table of square roots (Table I): $-2 + 2\sqrt{3} \approx 1.46$. Thus the width of the rectangle is approximately 1.46 feet. Adding 4 to this solution, the length is approximately 5.46 feet. The solution can be checked by finding $(1.46)(5.46) \approx 7.97$, which is approximately 8, the given area.

In general, we will approximate answers to applied problems to three significant digits. Three significant digits means a three digit answer where the first digit is not zero, regardless of where the decimal point is placed. If the number is one thousand or more, zeros may follow the three significant digits. Thus 1.46, 14.6, and 14,600 are all given to three significant digits.

EXAMPLE 4.19.   The area of a triangle is 1 square inch. Its base is $\frac{1}{2}$ inch more than its height. Find its base and height.

Solution.   Recall that the area of a triangle is equal to one-half of the base times the height. The base is $\frac{1}{2}$ inch more than the height; therefore, if the height is x, then the base is $x + \frac{1}{2}$.

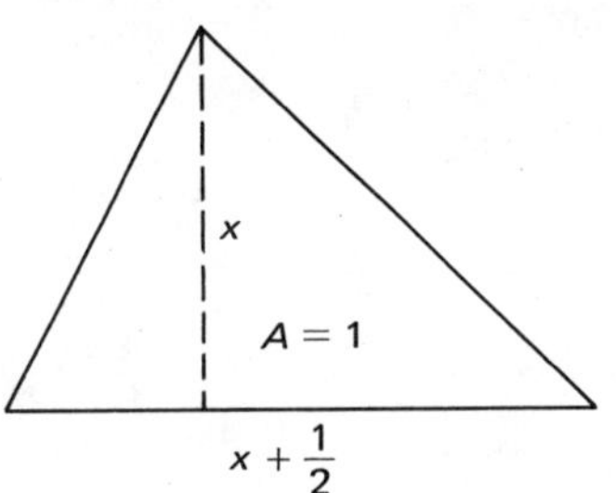

$$\frac{1}{2}(b \cdot h) = A$$

$$\frac{1}{2}\left(x + \frac{1}{2}\right)x = 1$$

$$\frac{1}{2}x^2 + \frac{1}{4}x - 1 = 0$$

$$2x^2 + x - 4 = 0$$

$$x = \frac{-1 \pm \sqrt{1^2 - 4(2)(-4)}}{2(2)}$$

$$= \frac{-1 \pm \sqrt{1 + 32}}{4}$$

$$= \frac{-1 \pm \sqrt{33}}{4}.$$

The solution $\dfrac{-1 - \sqrt{33}}{4}$ is extraneous. We approximate $\dfrac{-1 + \sqrt{33}}{4}$ to three significant digits: $\dfrac{-1 + \sqrt{33}}{4} \approx 1.19$. The height is 1.19 inches and the base is 1.69 inches. To check, $\frac{1}{2}(1.19)(1.69) \approx 1.006$, which is approximately 1.

EXAMPLE 4.20.   The area of a trapezoid is 5 square centimeters. The bottom base is 4 centimeters more than the top base, and the height is equal to the top base. Find the top and bottom bases.

Solution.   Recall that the area of a trapezoid is equal to one-half of the sum of the bases times the height. If the top base is x, the bottom base is $x + 4$, and the height is also x.

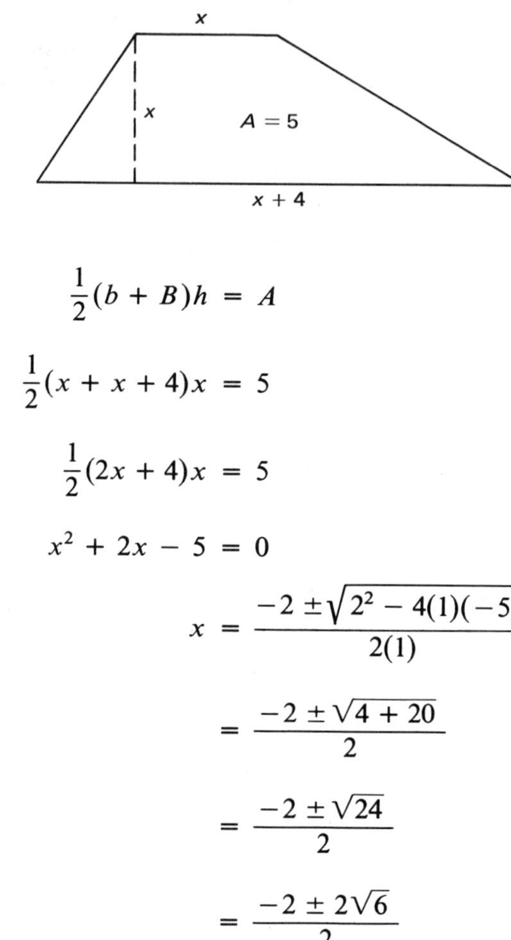

$$\frac{1}{2}(b + B)h = A$$

$$\frac{1}{2}(x + x + 4)x = 5$$

$$\frac{1}{2}(2x + 4)x = 5$$

$$x^2 + 2x - 5 = 0$$

$$x = \frac{-2 \pm \sqrt{2^2 - 4(1)(-5)}}{2(1)}$$

$$= \frac{-2 \pm \sqrt{4 + 20}}{2}$$

$$= \frac{-2 \pm \sqrt{24}}{2}$$

$$= \frac{-2 \pm 2\sqrt{6}}{2}$$

$$= -1 \pm \sqrt{6}.$$

The solution $-1 - \sqrt{6}$ is extraneous. The solution $-1 + \sqrt{6}$ may be approximated: $-1 + \sqrt{6} \approx 1.45$. The top base is 1.45 centimeters and the bottom base is 5.45 centimeters. Also, the height is 1.45 centimeters. To check, $\frac{1}{2}(1.45 + 5.45)(1.45) \approx 5.003$ which is approximately 5.

The type of equation we solved in Section 4.2 may also appear in applied problems.

EXAMPLE 4.21.   The area of a circle is given by the formula $A = \pi r^2$, where r is the radius of the circle, and π (pi) is a constant approximately equal to 3.1416. Find the radius of a circle with an area of 27π square meters.

Solution.   Since $\pi r^2 = A$, we have

$$\pi r^2 = 27\pi.$$

Dividing both sides by π,

$$r^2 = 27$$

$$r = \pm\sqrt{27}$$

$$r = \pm 3\sqrt{3}\,.$$

Since the radius cannot be negative, $-3\sqrt{3}$ is extraneous. The radius is $3\sqrt{3}$, or 5.20 meters to three significant digits.

Exercise 4.4

Approximate solutions to three significant digits:

1. The area of a rectangle is 14 square feet. The length is 2 feet more than the width. Find the width and length.

2. A rectangle has a length 5 inches more than its width. Its area is 5 square inches. Find its width and length.

3. A rectangular board has an area of $4\frac{1}{2}$ square meters. The width is $\frac{1}{2}$ meter less than the length. What are the width and length?

4. A traffic island is in the shape of a triangle with an area of 18 square feet. The height is 1 foot more than the base. What are the base and height?

5. The area of a trapezoidal mirror is 2 square meters. The height is equal to the top base, and the bottom base is 1 meter more. What are the top and bottom bases?

6. A trapezoid has a height equal to its bottom base. Its top base is 2 inches less than its bottom base. Its area is 10 square inches. Find the top and bottom bases.

7. The base of a box is a rectangle 2 feet longer than it is wide. The height is 3 feet. The volume of the box is 6 cubic feet. What are the width and length of the base of the box?

8. You want to build a crate that will hold $8\frac{1}{2}$ cubic feet. Its length is to be 3 feet more than its width, and its height must be $\frac{1}{2}$ foot. What should the width and length be?

9. A circle has an area of 9π square inches. Find its radius.

10. A circular pond has an area of 35 square meters. What is its radius? (Use $\pi \approx 3.1416$, or the π key on your calculator. If you do not have a calculator which can find the square root of a decimal, use Table I to find the square root of the nearest whole number to the decimal.)

Proof of the Quadratic Formula

To prove the quadratic formula, we need a method called **completing the square**. Consider the expressions

$$x^2 + 2x + 1$$

$$x^2 + 4x + 4$$

$$x^2 + 6x + 9$$

$$x^2 - 2x + 1$$

$$x^2 - 4x + 4$$

and so on. These trinomials are called **perfect squares** because

$$x^2 + 2x + 1 = (x + 1)^2$$

$$x^2 + 4x + 4 = (x + 2)^2$$

$$x^2 + 6x + 9 = (x + 3)^2$$

$$x^2 - 2x + 1 = (x - 1)^2$$

$$x^2 - 4x + 4 = (x - 2)^2$$

and so on.

Completing the square means, given an x^2-term with coefficient 1 and an x-term, to find a constant term which will make a trinomial which is a perfect square. For example, if we have

$$x^2 - 6x,$$

the completed square is

$$x^2 - 6x + 9$$

since

$$x^2 - 6x + 9 = (x - 3)^2.$$

To find the constant term, we take one-half of the coefficient of the x-term, and then square the result. Given

$$x^2 - 6x,$$

we take $\frac{1}{2}(-6) = -3$, and $(-3)^2 = 9$. Given

$$x^2 + 10x,$$

we take $\frac{1}{2}(10) = 5$, and $5^2 = 25$. In this case the completed square is

$$x^2 + 10x + 25$$

since

$$x^2 + 10x + 25 = (x + 5)^2.$$

Observe that the number in the right-hand expression is one-half of the coefficient of the x-term.

This method also works for coefficients which are odd numbers or fractions. To complete the square for

$$x^2 + 7x,$$

we take $\frac{1}{2}(7) = \frac{7}{2}$, and $(\frac{7}{2})^2 = \frac{49}{4}$. The completed square is

$$x^2 + 7x + \frac{49}{4}$$

since

$$x^2 + 7x + \frac{49}{4} = \left(x + \frac{7}{2}\right)^2.$$

To complete the square for

$$x^2 - \frac{3}{2}x,$$

we take $\frac{1}{2}(-\frac{3}{2}) = -\frac{3}{4}$, and $(-\frac{3}{4})^2 = \frac{9}{16}$. The completed square is

$$x^2 - \frac{3}{2}x + \frac{9}{16}$$

since

$$x^2 - \frac{3}{2}x + \frac{9}{16} = \left(x - \frac{3}{4}\right)^2.$$

Now, to prove the quadratic formula, we start with the quadratic equation in standard form,

$$ax^2 + bx + c = 0.$$

We make the coefficient of the x^2-term 1 by dividing each term by a:

$$x^2 + \frac{b}{a}x + \frac{c}{a} = 0.$$

The completion of the square for $x^2 + \frac{b}{a}x$ is $\frac{1}{2}\left(\frac{b}{a}\right) = \frac{b}{2a}$, and $\left(\frac{b}{2a}\right)^2 = \frac{b^2}{4a^2}$. Adding $\frac{b^2}{4a^2}$ to each side, we have

$$x^2 + \frac{b}{a}x + \frac{b^2}{4a^2} + \frac{c}{a} = \frac{b^2}{4a^2},$$

and substituting the perfect square,

$$\left(x + \frac{b}{2a}\right)^2 + \frac{c}{a} = \frac{b^2}{4a^2}.$$

Subtracting $\frac{c}{a}$ from both sides,

$$\left(x + \frac{b}{2a}\right)^2 = \frac{b^2}{4a^2} - \frac{c}{a}$$

$$\left(x + \frac{b}{2a}\right)^2 = \frac{b^2}{4a^2} - \frac{4ac}{4a^2}$$

$$\left(x + \frac{b}{2a}\right)^2 = \frac{b^2 - 4ac}{4a^2}.$$

Now, we take the square root of each side:

$$x + \frac{b}{2a} = \pm\sqrt{\frac{b^2 - 4ac}{4a^2}}.$$

Finally, solving for x and simplifying,

$$x = -\frac{b}{2a} \pm \frac{\sqrt{b^2 - 4ac}}{2a}$$

or

$$x = \frac{-b \pm \sqrt{b^2 - 4ac}}{2a},$$

which is the quadratic formula.

Self-test

1. Write $2x^2 = x + 4$ in standard form and identify a, b, and c.

1. ___________________

2. Solve $9x^2 - 5 = 0$.

2. ___________________

Use the quadratic formula to solve:

3. $4x^2 = 5x + 6$

3. ___________________

4. $x^2 + 8x + 6 = 0$

4. ___________________

5. A rectangle has an area of 10 square inches. Its length is 2 inches more than its width. Find its width and length to three significant digits.

5. ___________________

Unit 5

Cumulative Review

INTRODUCTION

In this unit you should make certain you remember all the material in all of the preceding units.

OBJECTIVE

When you have finished this unit you should be able to demonstrate that you can fulfill every objective of each preceding unit.

To prepare for this unit you should review the Self-Tests for Units 1 through 4. Do each problem of each Self-Test over again. If you cannot do a problem, or even if you have the slightest difficulty, you should:

1. Find out from the answer section the Objective for the unit to which the problem relates.
2. Review all the material in the section which has the same number as the objective, and redo all the Exercises for the section.
3. Try the Self-Test for the unit again.

Repeat these steps until you can do each problem in each Self-Test for each of Units 1 through 4 easily and accurately.

Self-test

1. Factor $3x^2 + 10x - 8$.

2. Reduce $\dfrac{6 - 15x}{3}$.

Solve for x:

3. $4(2x - 3) = 6 - x$

4. $3x = 2x^2$

5. $12x^2 = 3$

1. _______________

2. _______________

3. _______________

4. _______________

5. _______________

6. $4x + 5 = 2x^2$

6. _______________________

7. One-third of five less than a number is three-halves. Find the number.

7. _______________________

8. Eighteen grams of an herb mixture is made up of marjoram which costs $8\frac{1}{2}¢$ a gram and thyme which costs $5\frac{1}{2}¢$ a gram. The mixture sells for $6\frac{1}{2}¢$ a gram. How many grams of each herb does it contain?

8. _______________________

9. A car leaves a garage and travels at an average rate of 30 miles per hour. One hour later, another car leaves the same garage and travels along the same route at an average rate of 42 miles per hour. How long will it take for the second car to overtake the first?

9. _______________________

10. A box is 3 feet longer than it is wide. Its height is $\frac{2}{3}$ of a foot. If the volume is 7 cubic feet, find the width and length to three significant digits.

10. _______________________

Unit 6

Linear Functions

INTRODUCTION

In the preceding units, you have learned how to solve some basic types of equations in one variable. In general, solutions to equations in one variable are distinct numbers. In this unit, you will learn some properties of linear equations in two variables. Solutions to equations in two variables are ordered pairs of numbers, and a linear equation in two variables has infinitely many such solutions. The solutions may be indicated by a graph in the Cartesian coordinate system. You will learn about the Cartesian coordinate system, how to find a linear equation in two variables, and how to draw the graph of a linear equation in two variables.

OBJECTIVES

When you have finished this unit you should be able to:

1. Find the distance between two given points.
2. Find the slope of a line containing two given points.
3. Find the equation of a line given the slope and a point, or given two points.
4. Draw the graph of a line given the equation.

Section 6.1 — The Distance between Two Points

To begin this unit, we introduce two concepts with which you may already be familiar. The first is the **Cartesian coordinate system**, named for its inventor, René Descartes (1596-1650). The Cartesian coordinate system is constructed from two **number lines**. To construct a number line we need to establish an **origin**, a **direction**, and a **unit** on a line.

We start with an ordinary line, and choose a point to be the origin. The number 0 is associated with the origin.

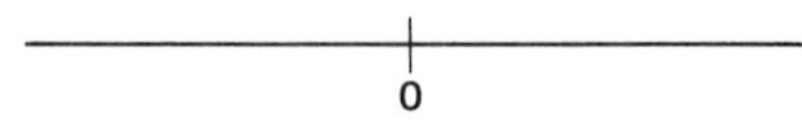

Then we decide the direction, the side of the origin to which the numbers increase. The direction is indicated by an arrow. If the line is horizontal, the direction is usually chosen to the right.

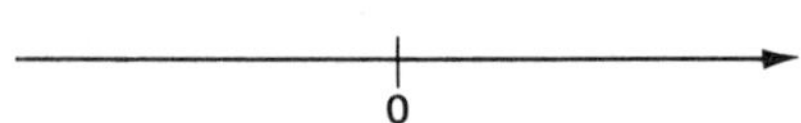

Finally, we decide the unit, the distance from the origin of the point associated with 1. The point associated with 2 is then one unit beyond 1 in the positive direction, 3 is one unit beyond 2, and so on. In the negative direction, one unit is -1, the next -2, and so on.

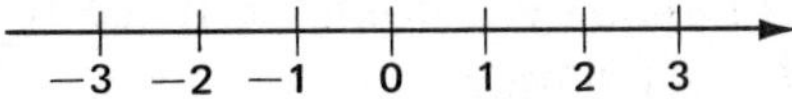

The number line can of course be extended indefinitely in both the positive and negative directions.

To construct the Cartesian coordinate system, we place two number lines at right angles, so that the first is horizontal and the second is vertical. They are placed so that the origin of each is at their point of intersection. The horizontal line is directed to the right and the vertical line is directed upwards. We will usually choose the same unit on each line. The two number lines are called **axes**. We will distinguish between the two axes by calling the horizontal axis the **x-axis** and the vertical axis the **y-axis**. (In more advanced work other letters are often used.)

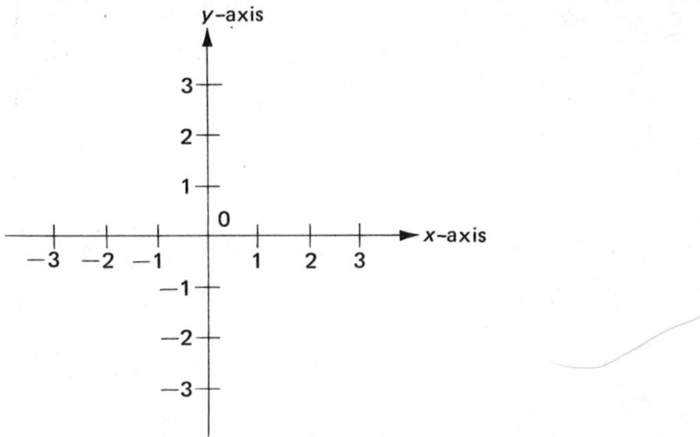

We can now locate any point in the plane by a number on the x-axis and a number on the y-axis. These numbers are called the **coordinates** of the point. We represent the coordinates by an **ordered pair** (x, y), where the x-coordinate is always the first number and the y-coordinate is always the second number. Thus the ordered pair $(2, 3)$ and the ordered pair $(3, 2)$ do not represent the same point. This representation is called an ordered pair because the order of the numbers is important.

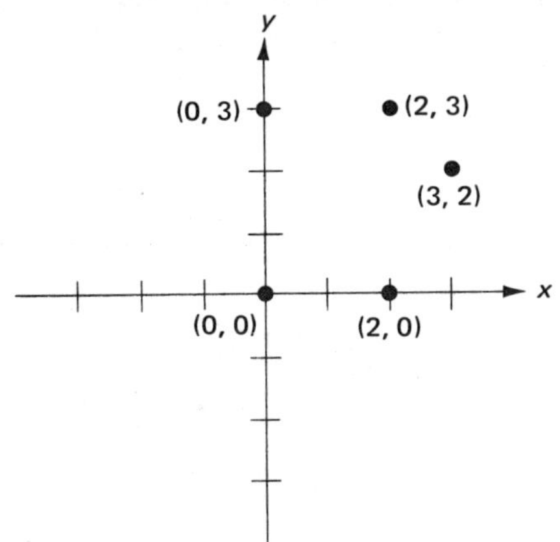

Observe that the coordinates of the origin are (0, 0). All points on the x-axis have y-coordinate 0, and can be represented as $(x, 0)$. All points on the y-axis have x-coordinate 0, and can be represented as $(0, y)$.

We can now find the distance between two points if the two points have either the same x-coordinate or the same y-coordinate. If the two points have the same x-coordinate, then the distance between the points is

$$d = y_2 - y_1$$

where y_1 and y_2 are the y-coordinates of the points, and $y_1 < y_2$. This symbol means y_1 is less than y_2 or, in the Cartesian coordinate system, y_1 is below y_2.

EXAMPLE 6.1.   Find the distance between (1, 3) and (1, 8).

Solution.

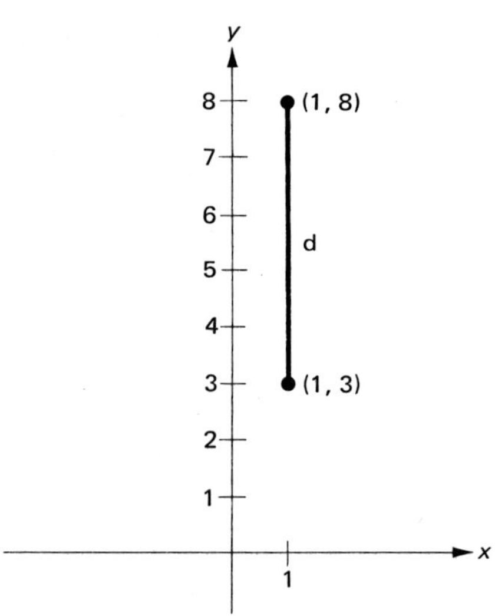

$$d = y_2 - y_1 = 8 - 3 = 5.$$

If the two points have the same y-coordinate, then the distance between them is

$$d = x_2 - x_1$$

where x_1 and x_2 are the x-coordinates of the points, and $x_1 < x_2$, meaning x_1 is less than x_2, or x_1 is to the left of x_2.

EXAMPLE 6.2.   Find the distance between (4, 2) and $(-3, 2)$.

Solution.

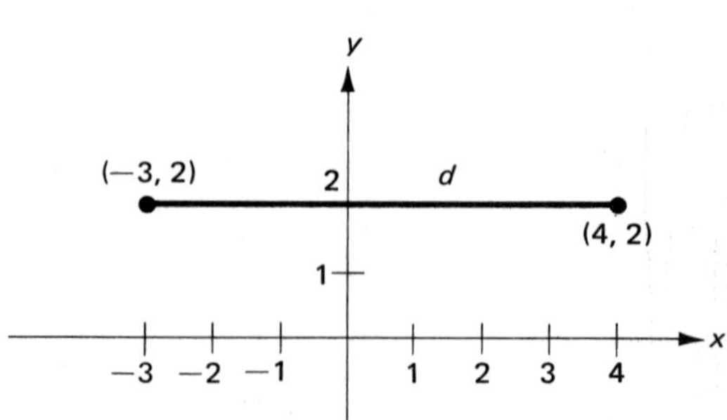

$$d = x_2 - x_1 = 4 - (-3) = 7.$$

To find distances between points which have neither coordinate the same we need the **Pythagorean theorem**, named for the Greek philosopher and mathematician Pythagoras (c. 540 B.C.). The Pythagorean theorem was known to more ancient cultures, in particular, the Babylonians (c. 2000 B.C.; well-known stories of the ancient Egyptians using the Pythagorean theorem are so far unsubstantiated). The theorem was probably first proved by Pythagoras, or one of his followers, all of whom credited their discoveries to Pythagoras.

You may recall from geometry that the Pythagorean theorem relates the squares of the sides of a right triangle. A **right triangle** is a triangle with two sides perpendicular, so that they form a **right angle**. The side opposite the right angle, and not forming part of it, is called the **hypotenuse**. If we call the hypotenuse c, and the other two sides a and b, the Pythagorean theorem may be stated:

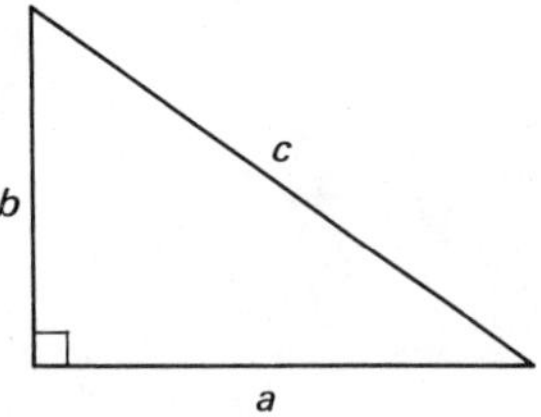

$$c^2 = a^2 + b^2.$$

The Pythagorean theorem may be applied to finding the distance between two points in the Cartesian coordinate system. We will call the first point (x_1, y_1) and the second point (x_2, y_2), where $x_1 \neq x_2$ and $y_1 \neq y_2$ so that the points have neither coordinate the same. For simplicity, we will start with $x_1 < x_2$ and $y_1 < y_2$.

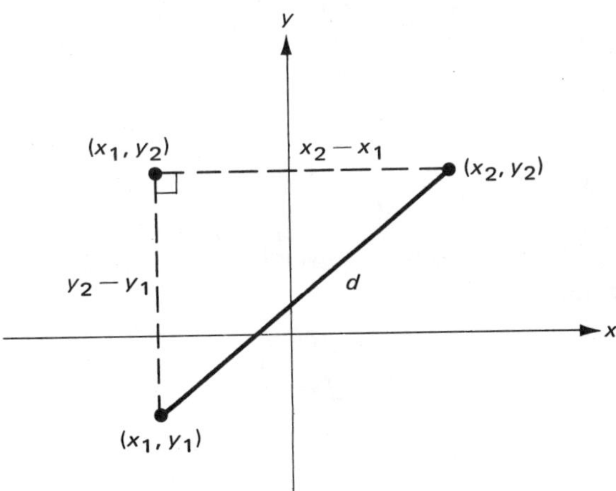

We locate a third point which has coordinates (x_1, y_2). The three points then form a right triangle. The points on the vertical side have the same x-coordinate, x_1, and the points on the horizontal side have the same y-coordinate, y_2. The distance between the two vertical points is $y_2 - y_1$ and the distance between the two horizontal points is $x_2 - x_1$. Thus, using the Pythagorean theorem, the distance d between the two points (x_1, y_1) and (x_2, y_2) is

$$d^2 = (x_2 - x_1)^2 + (y_2 - y_1)^2,$$

and

$$d = \sqrt{(x_2 - x_1)^2 + (y_2 - y_1)^2} \; .$$

This formula is called the **distance formula**. Since the horizontal and vertical distances in the distance formula are squared, and the square of either a negative number or a positive number is positive, it does not matter which x-coordinate is smaller or which y-coordinate is smaller. The distance formula applies to any two points in the plane.

EXAMPLE 6.3.   Find the distance between $(-1, 2)$ and $(5, -3)$.

Solution.

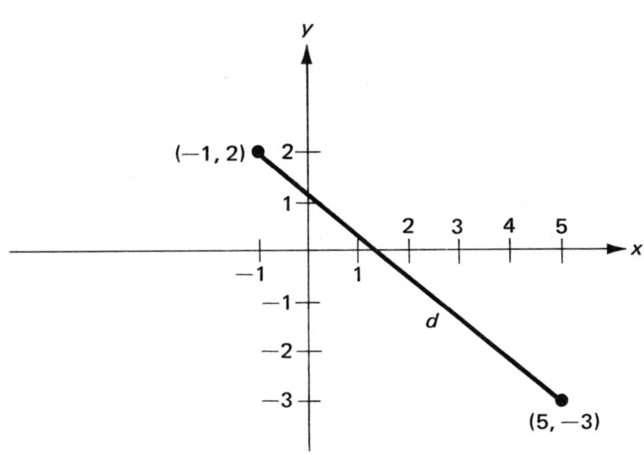

$$d = \sqrt{[5 - (-1)]^2 + (-3 - 2)^2}$$

$$= \sqrt{6^2 + (-5)^2}$$

$$= \sqrt{36 + 25}$$

$$= \sqrt{61} \; .$$

Observe that it does not matter which point we take to be the first point and which the second. You could just as well write

$$d = \sqrt{(-1 - 5)^2 + [2 - (-3)]^2}$$

$$= \sqrt{(-6)^2 + 5^2}$$

$$= \sqrt{36 + 25}$$

$$= \sqrt{61} \; .$$

If desired, the square root may be approximated using a calculator or a table of square roots. We will use approximations in applied situations. However, you should always simplify a square root by extracting any possible factors which are perfect squares.

EXAMPLE 6.4.   Find the distance between $(6, -1)$ and $(0, 7)$.

Solution.

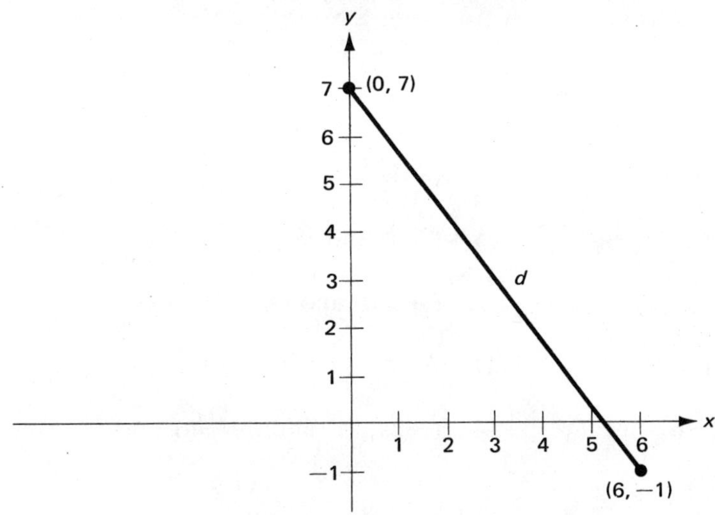

$$d = \sqrt{(6 - 0)^2 + (-1 - 7)^2}$$

$$= \sqrt{6^2 + (-8)^2}$$

$$= \sqrt{36 + 64}$$

$$= \sqrt{100}$$

$$= 10.$$

EXAMPLE 6.5.   Find the distance between $(5, 4)$ and $(4, -3)$.

Solution.

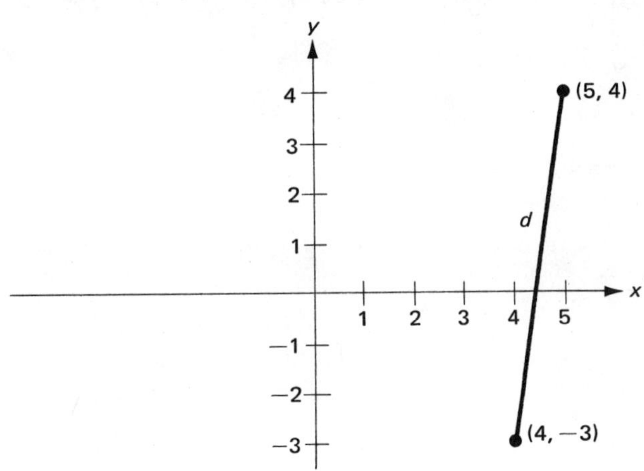

$$d = \sqrt{(5 - 4)^2 + [4 - (-3)]^2}$$

$$= \sqrt{1^2 + 7^2}$$

$$= \sqrt{1 + 49}$$

$$= \sqrt{50}$$

$$= \sqrt{(25)(2)}$$

$$= 5\sqrt{2} \, .$$

Exercise 6.1

Find the distance between the pairs of points:

1. (2, 1) and (2, 6)
2. (−3, 2) and (−3, −5)
3. (7, 4) and (−1, 4)

4. (−2, −2) and (−6, −2)
5. (2, 3) and (5, 7)
6. (4, 1) and (2, 5)

7. (−1, 2) and (0, −3)
8. (−3, 0) and (4, −1)
9. (2, −5) and (−3, 7)

10. (−3, 3) and (1, −3)
11. (−1, −3) and (−4, −2)
12. (−6, −2) and (−1, −4)

13. (−2, 1) and (−6, −3)
14. (−2, −1) and (3, −6)
15. (4, −2) and (−4, −4)

16. (−3, −3) and (−6, 1)

<table>
<tr><td>Section
6.2</td><td><h1 align="center">The Slope of a Line</h1></td></tr>
</table>

You may recall from geometry that two points determine a line; that is, given two points, we can draw exactly one line joining them. The locations of the two points in the plane determine how steeply the line rises or falls. We can measure this incline by using the vertical and horizontal distances between the points.

Suppose (x_1, y_1) is one point on a line and (x_2, y_2) is a second point on the line. For simplicity, we will again start with $x_1 < x_2$ and $y_1 < y_2$.

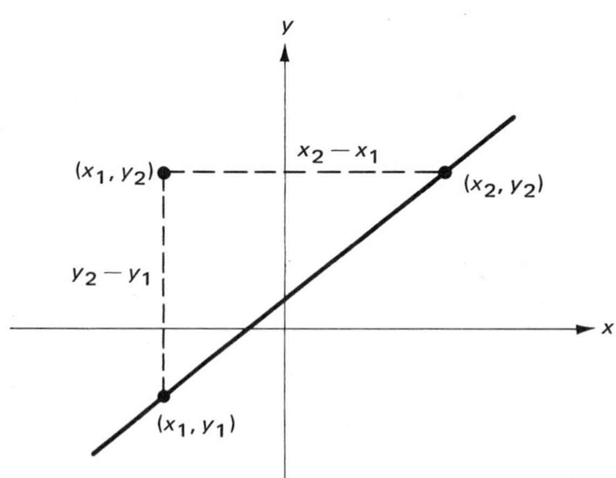

As in the distance formula, we locate the point (x_1, y_2), the vertical distance is $y_2 - y_1$, and the horizontal distance is $x_2 - x_1$. We define the **slope m** of the line to be

$$m = \frac{y_2 - y_1}{x_2 - x_1}.$$

EXAMPLE 6.6.   Find the slope of the line determined by $(-1, 3)$ and $(2, 5)$.

Solution.

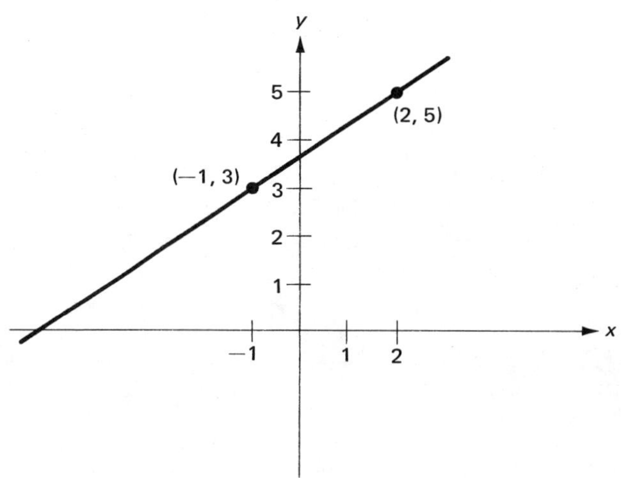

$$m = \frac{5 - 3}{2 - (-1)}$$

$$m = \frac{2}{3}.$$

The slope of the line is $\frac{2}{3}$. This number means the line rises 2 units for every 3 units from left to right. The vertical distance is often called the "rise" and the horizontal distance the "run." The slope of a line is then "rise over run."

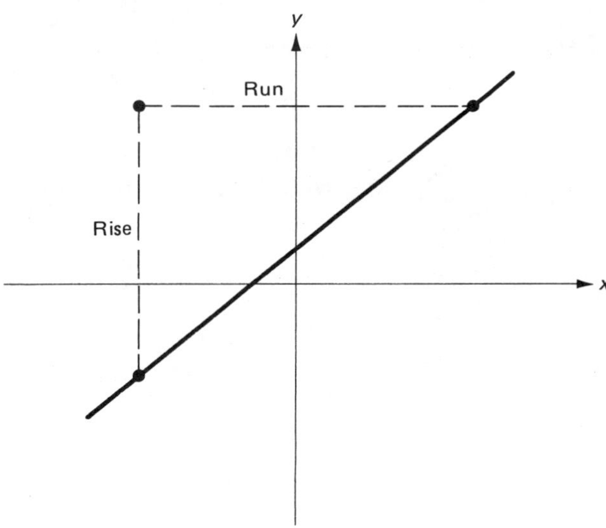

A line has a negative "rise" if it falls as it goes from left to right. When this is the case, we may use a negative vertical distance.

EXAMPLE 6.7.   Find the slope of the line determined by $(2, 7)$ and $(4, 3)$.

Solution.

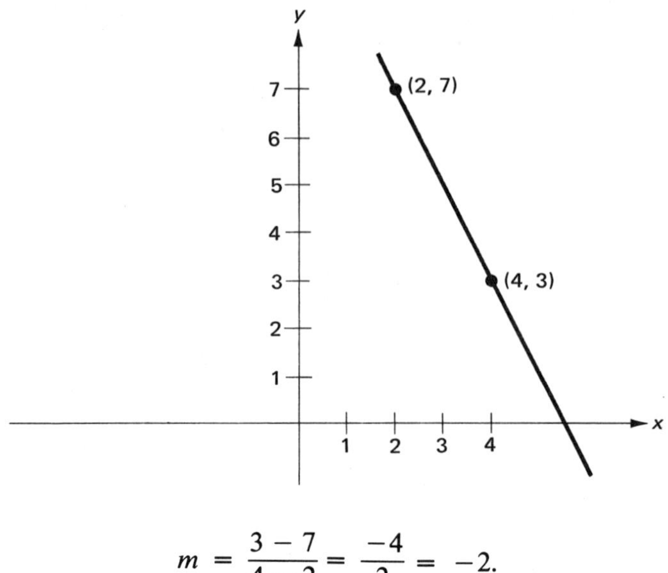

$$m = \frac{3 - 7}{4 - 2} = \frac{-4}{2} = -2.$$

This line drops 4 units for every 2 units from left to right or, equivalently, drops 2 units for every 1 unit from left to right. Of course, you can take the points in the opposite order and get the same slope:

$$m = \frac{7 - 3}{2 - 4} = \frac{4}{-2} = -2.$$

However, once the order is chosen, you cannot change between the numerator and the denominator. If you do, you will get the wrong sign. The slope of a line which rises from left to right is always positive and the slope of a line which falls from left to right is always negative.

EXAMPLE 6.8.   Find the slope of the line determined by $(-1, -4)$ and $(5, 6)$.

Solution.

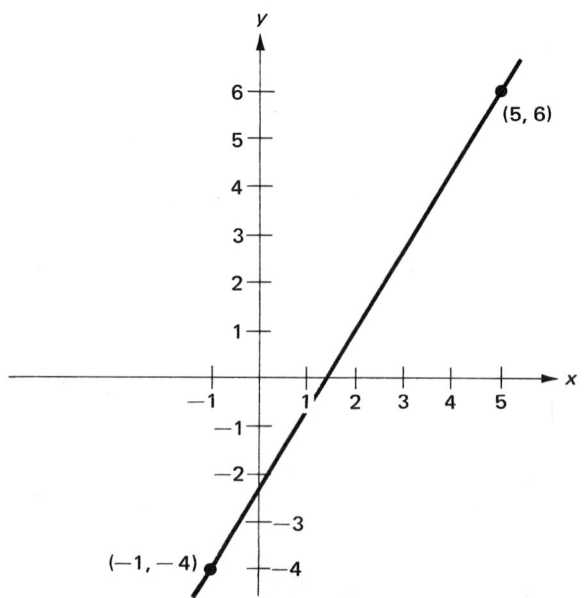

$$m = \frac{6 - (-4)}{5 - (-1)} = \frac{10}{6} = \frac{5}{3}.$$

EXAMPLE 6.9.   Find the slope of the line determined by $(-2, 3)$ and $(4, -1)$.

Solution.

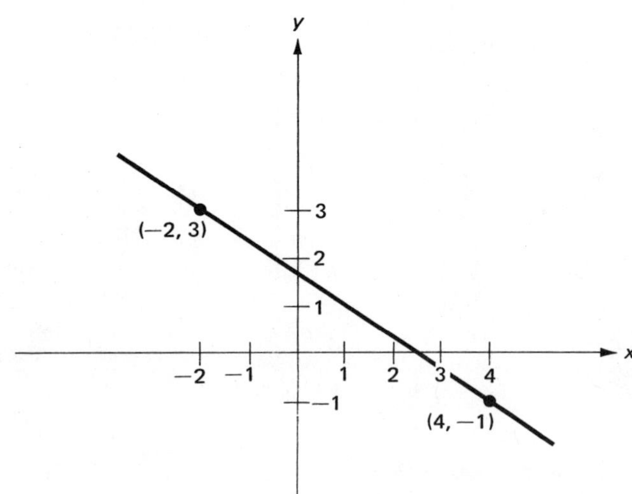

$$m = \frac{-1 - 3}{4 - (-2)} = \frac{-4}{6} = -\frac{2}{3}.$$

EXAMPLE 6.10.   Find the slope of the line determined by $(-3, 4)$ and $(5, 4)$.

Solution.

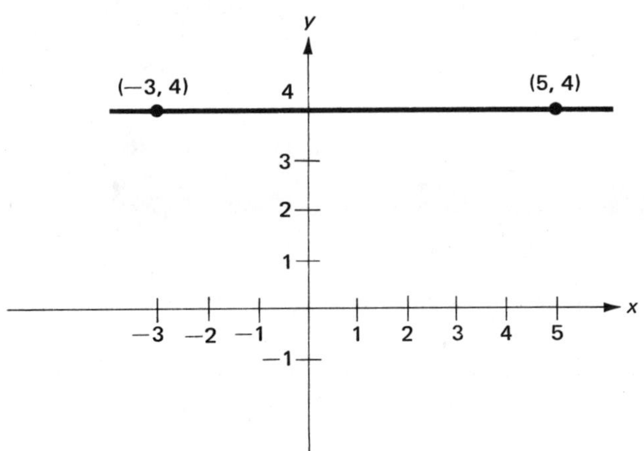

$$m = \frac{4 - 4}{5 - (-3)} = \frac{0}{8} = 0.$$

This line is horizontal, and neither rises nor falls. The slope of a horizontal line always is zero.

EXAMPLE 6.11.   Find the slope of the line determined by $(-2, 6)$ and $(-2, -1)$.

Solution.

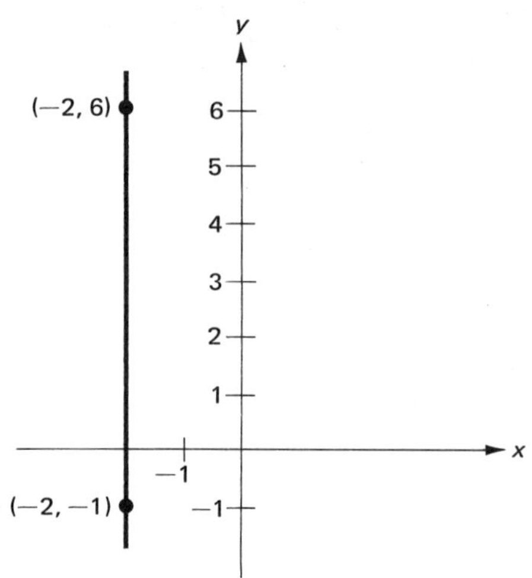

$$m = \frac{6 - (-1)}{-2 - (-2)} = \frac{7}{0}.$$

Since we cannot divide by zero, we say the slope of this line is **undefined**. The line is vertical, and rises, or falls, so steeply that it has no run. The slope of a vertical line always is undefined.

Exercise 6.2

Find the slope of the line determined by the pairs of points:

1. (3, 1) and (5, 6)

2. (6, 8) and (1, 5)

3. (2, 8) and (4, 4)

4. (6, 3) and (4, 6)

5. (−1, 5) and (4, −2)

6. (−5, 2) and (3, −8)

7. (−6, 0) and (2, −8)

8. (5, 8) and (0, −2)

9. (7, −3) and (−5, 3)

10. (2, −1) and (−3, −2)

11. (0, −6) and (2, 0)

12. (4, 0) and (0, 6)

13. (5, 3) and (5, −4)

14. (4, −2) and (6, −2)

15. (−3, −3) and (2, −3)

16. (−1, 0) and (−1, 5)

Section 6.3	Equations of Lines

If we know the slope of a line and one point on the line, we can find an equation involving x and y which represents every point on the line. This equation is called the **equation of the line**.

Suppose a line has slope m and contains the point (x_0, y_0). Let (x, y) represent any other point on the line. Then we know from the slope formula that

$$m = \frac{y - y_0}{x - x_0}.$$

Multiplying both sides by $x - x_0$, we have

$$m(x - x_0) = y - y_0$$

or

$$y - y_0 = m(x - x_0).$$

This equation is called the **point-slope form** of the equation of the line. We usually simplify the point-slope form to the **general form**

$$ax + by + c = 0.$$

EXAMPLE 6.12.   Find the equation of the line with slope $-\frac{1}{2}$ and containing the point $(-3, 2)$, and write the equation in general form.

Solution.   Since we know the line goes through the point $(-3, 2)$, and goes down 1 unit for every 2 units across, we can draw its graph:

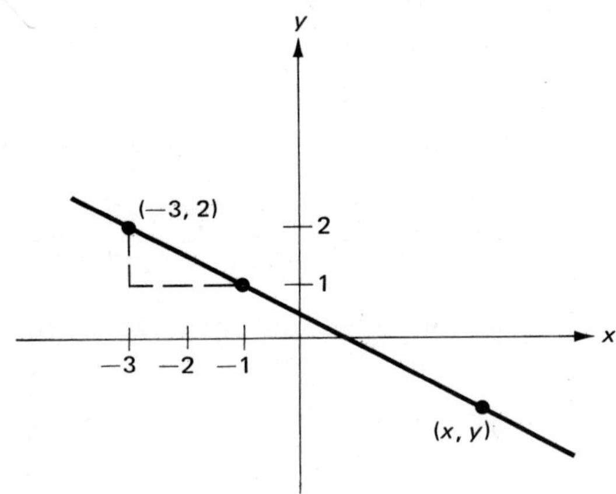

Let (x, y) be any other point on the line. Then,

$$-\frac{1}{2} = \frac{y - 2}{x - (-3)}$$

$$\frac{-1}{2} = \frac{y - 2}{x + 3}$$

$$-1(x + 3) = 2(y - 2)$$

$$-x - 3 = 2y - 4$$

$$-x - 2y + 1 = 0$$

or

$$x + 2y - 1 = 0$$

is the general form of the equation of the line.

Since two points determine a line and we can find the slope of a line given two points on it, we can find the equation of a line given two points on the line.

EXAMPLE 6.13.   Find the equation of the line containing the points $(-3, -1)$ and $(2, 6)$, and write the equation in general form.

Solution.

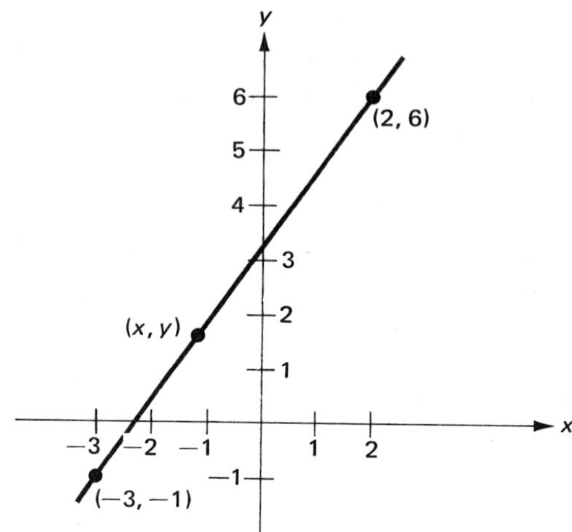

First, we find the slope of the line using the two given points:

$$m = \frac{6 - (-1)}{2 - (-3)} = \frac{7}{5}.$$

Then, we can find the equation of the line using the slope and either given point. Suppose we use the point $(2, 6)$:

$$\frac{7}{5} = \frac{y - 6}{x - 2}$$

$$7(x - 2) = 5(y - 6)$$

$$7x - 14 = 5y - 30$$

$$7x - 5y + 16 = 0$$

is the general form of the equation of the line. Of course, we could just as well have used the slope and the other point $(-3, -1)$:

$$\frac{7}{5} = \frac{y - (-1)}{x - (-3)}$$

$$\frac{7}{5} = \frac{y + 1}{x + 3}$$

$$7(x + 3) = 5(y + 1)$$

$$7x + 21 = 5y + 5$$

$$7x - 5y + 16 = 0.$$

EXAMPLE 6.14.   Find the equation of the line containing the points $(4, -3)$ and $(-3, -3)$.

Solution.

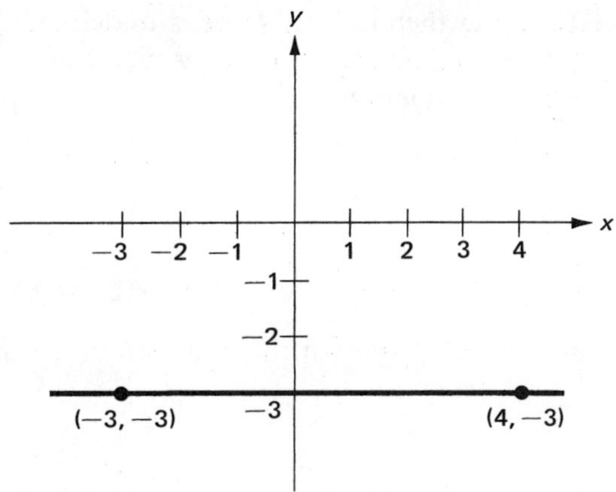

Since the line is horizontal, its slope is zero. Thus,

$$0 = \frac{y - (-3)}{x - 4}$$

$$0 = \frac{y + 3}{x - 4}$$

$$0(x - 4) = y + 3$$

$$0 = y + 3$$

or

$$y + 3 = 0.$$

Observe that no x-term appears in the equation of a horizontal line. This equation is often written in the form $y = -3$. It expresses the fact that the y-coordinate of every point on the line is -3.

EXAMPLE 6.15.   Find the equation of the line containing the points $(2, -4)$ and $(2, 5)$.

Solution.

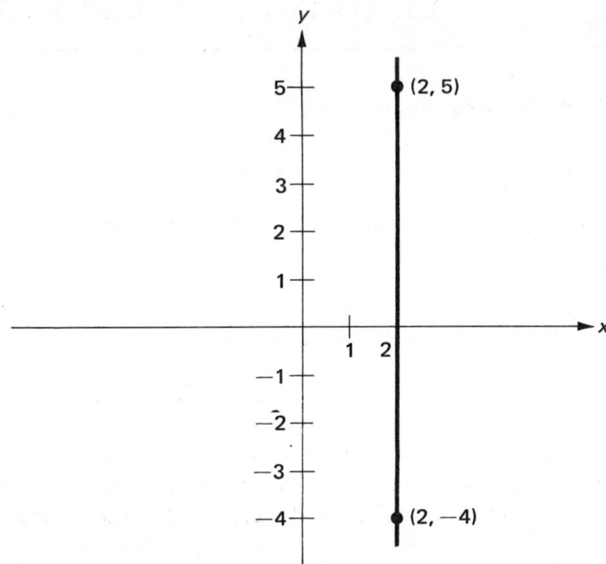

The line is vertical and therefore its slope is undefined. Thus we cannot write an equation using the method of the preceding examples. However, the x-coordinate of every point on the line is 2. Thus we can write the equation

$$x = 2$$

or

$$x - 2 = 0.$$

Observe that no y-term appears in the equation of a vertical line.

Exercise 6.3

Find the equation of the line with the given slope and containing the given point, and write the equation in general form:

1. 4 and $(1, 3)$

2. -2 and $(4, 2)$

3. -1 and $(-3, -2)$

4. 2 and $(-1, -4)$

5. $\dfrac{3}{4}$ and $(1, -2)$

6. $\dfrac{5}{2}$ and $(-4, 3)$

7. $-\dfrac{2}{3}$ and $(-3, -5)$

8. $-\dfrac{3}{2}$ and $(3, -1)$

Find the equation of the line containing the given points, and write the equation in general form:

9. $(2, 1)$ and $(3, 5)$

10. $(4, 2)$ and $(2, 6)$

11. $(3, -2)$ and $(5, 1)$

12. $(-1, 5)$ and $(3, -1)$

13. $(-1, -3)$ and $(-5, -1)$

14. $(4, -3)$ and $(-4, -5)$

15. $(4, 0)$ and $(0, -6)$

16. $(0, 2)$ and $(3, 0)$

17. $(3, 2)$ and $(-5, 2)$

18. $(-4, -1)$ and $(3, -1)$

19. $(-2, 3)$ and $(-2, -4)$

20. $(6, -3)$ and $(6, 0)$

<table>
<tr><td>Section
6.4</td><td align="center"><h1>Graphs of Linear Equations</h1></td></tr>
</table>

A solution of the equation

$$ax + by + c = 0$$

is an ordered pair (x, y) for which the equation is true. For example, given the equation

$$4x + 2y - 12 = 0,$$

one solution is $(2, 2)$ since it is true that

$$4(2) + 2(2) - 12 = 0.$$

Clearly there are many other solutions to the equation. A graph indicating the set of solutions of the equation is called the **graph of the equation** or the **graph of the line**.

Often it is convenient to draw the graph of a line by finding its intercepts. The **intercepts** of a graph are the points where the graph crosses the axes. When a graph crosses the y-axis, the

x-coordinate of the point is zero. The point is called the **y-intercept** and has the form $(0, y)$. Similarly, the point where the graph crosses the x-axis has y-coordinate zero, is called the **x-intercept**, and has the form $(x, 0)$.

EXAMPLE 6.16.   Find the y-intercept, x-intercept, and draw the graph of $4x + 2y - 12 = 0$.

Solution.   When $x = 0$,

$$2y - 12 = 0$$

$$2y = 12$$

$$y = 6.$$

The y-intercept is $(0, 6)$. When $y = 0$,

$$4x - 12 = 0$$

$$4x = 12$$

$$x = 3.$$

The x-intercept is $(3, 0)$. The graph is

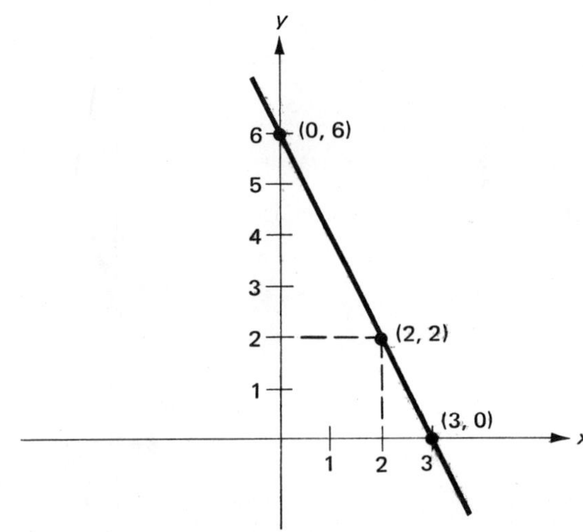

Observe that $(2, 2)$, the solution we illustrated previously, is a point of the graph. The graph indicates the entire solution set, and continues indefinitely in both directions.

EXAMPLE 6.17.   Find the y-intercept, x-intercept, and draw the graph of $7x - 2y + 5 = 0$.

Solution.   When $x = 0$,

$$-2y + 5 = 0$$

$$-2y = -5$$

$$y = \frac{-5}{-2}$$

$$y = \frac{5}{2}.$$

Thus the y-intercept is $(0, \frac{5}{2})$. When $y = 0$,

$$7x + 5 = 0$$

$$7x = -5$$

$$x = -\frac{5}{7}.$$

Thus the x-intercept is $(-\frac{5}{7}, 0)$. In this case, we might like to find one other point as a "check point." Suppose $x = 1$. Then,

$$7 - 2y + 5 = 0$$

$$-2y + 12 = 0$$

$$-2y = -12$$

$$y = \frac{-12}{-2}$$

$$y = 6.$$

Thus the point $(1, 6)$ is on the graph. The graph is

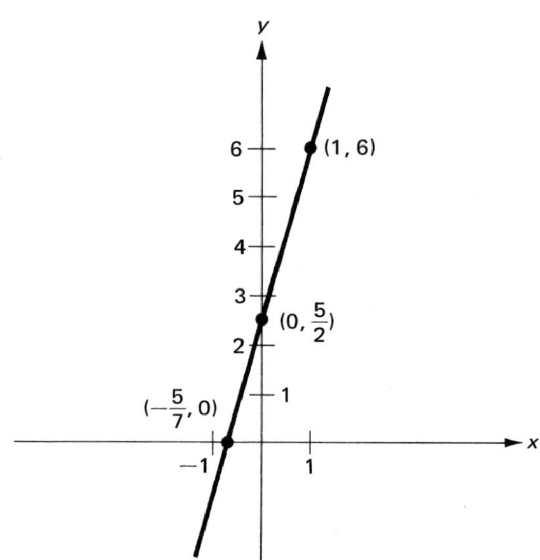

If, in the equation $ax + by + c = 0$, the constant term c does not appear, then the intercepts both are the point $(0, 0)$. In this case we must find a point other than an intercept to draw the graph.

EXAMPLE 6.18.   Find the y-intercept, x-intercept, and draw the graph of $2x + 8y = 0$.

Solution.   If $x = 0$,

$$8y = 0$$

$$y = 0.$$

Thus the y-intercept is $(0, 0)$. If $y = 0$,

$$2x = 0$$

$$x = 0.$$

Thus the x-intercept is $(0, 0)$. To find another point, suppose $x = 4$. Then,

$$8 + 8y = 0$$

$$8y = -8$$

$$y = -1.$$

Thus the point $(4, -1)$ is on the graph. The graph is

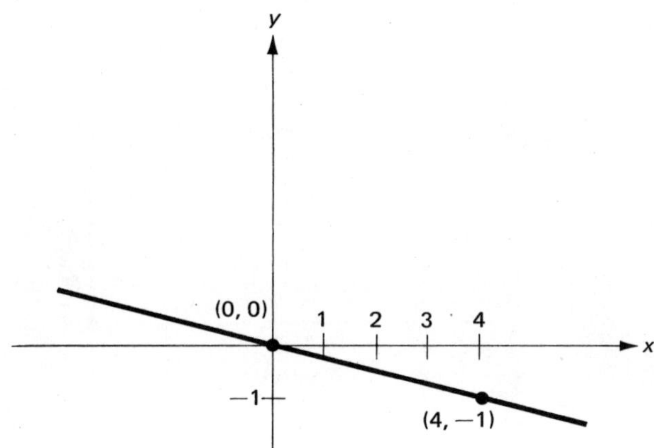

We chose to use $x = 4$ rather than $x = 1$ in the preceding example because it gives an easier solution which does not involve fractions, and because $x = 1$ would give a solution very close to the double intercept $(0, 0)$.

In cases where either the x-term or the y-term does not appear, there is only one intercept and the other intercept does not exist. If the x-term does not appear, the line is horizontal, parallel to the x-axis, and there is no x-intercept. If the y-term does not appear, the line is vertical, parallel to the y-axis, and there is no y-intercept.

EXAMPLE 6.19.   Find the y-intercept, x-intercept, and draw the graph of $y = -5$.

Solution.   If $x = 0$, $y = -5$. Thus the y-intercept is $(0, -5)$. Since y cannot be zero, there is no x-intercept. The graph is .

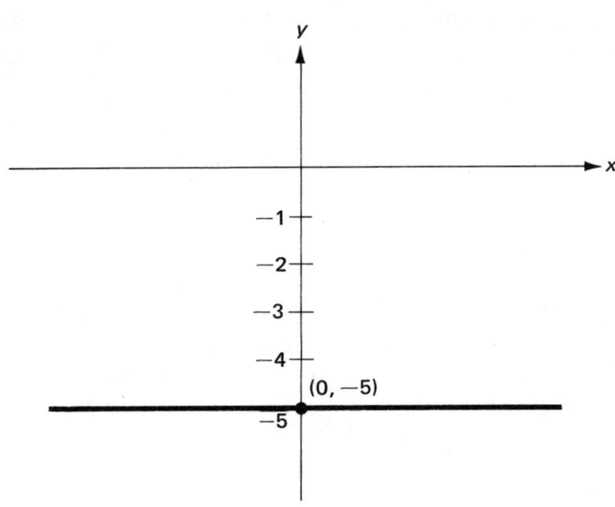

EXAMPLE 6.20.   Find the y-intercept, x-intercept, and draw the graph of $2x = 3$.

Solution.   Since x cannot be zero, there is no y-intercept. If $y = 0$,

$$2x = 3$$

$$x = \frac{3}{2}.$$

Thus the x-intercept is $(\frac{3}{2}, 0)$. The graph is

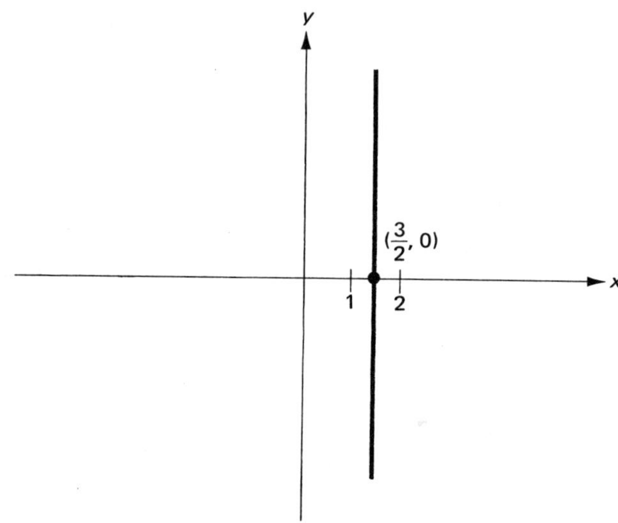

From the first of the two preceding examples we see that the x-axis itself is a horizontal line, and its equation is $y = 0$. From the second example, we see that the y-axis is a vertical line, and its equation is $x = 0$.

Any linear equation of the form

$$ax + by + c = 0,$$

where the y-term is present, can be solved for y. The resulting equation is of the form

$$y = mx + b.$$

Any linear equation which can be written in this form is a **linear function**. Any linear equation in which the y-term is present is a linear function. In an equation of the form

$$ax + c = 0,$$

no y-term appears. Such an equation cannot be solved for y and is not considered to be a linear function.

We will use the symbol $f(x)$, which we read "f of x," to find values of functions. For a linear function which has been solved for y, we write

$$f(x) = mx + b.$$

The $f(x)$ symbol (remember to say "f of x") provides a compact form for finding coordinates of points of the graph of a function. For example, if $x = 0$, we write

$$f(0) = m(0) + b = b,$$

which gives the point $(0, b)$. If $x = 1$, we write

$$f(1) = m(1) + b = m + b,$$

which gives the point $(1, m + b)$.

The linear function is also called the **slope-intercept** form of a line because m is the slope and $(0, b)$ is the y-intercept. We have shown above that the point $(0, b)$ is a point of the graph of the function. Because $x = 0$, this point is the y-intercept. To show that the slope is m, we use the points $(0, b)$ and $(1, m + b)$. The slope is

$$\frac{(m + b) - b}{1 - 0} = \frac{m}{1} = m.$$

EXAMPLE 6.21.   Find the y-intercept, slope, and draw the graph of $4x + 2y - 12 = 0$.

Solution.   We solve the equation for y:

$$4x + 2y - 12 = 0$$

$$2y = -4x + 12$$

$$y = \frac{-4x}{2} + \frac{12}{2}$$

$$y = -2x + 6.$$

The equation is now in slope-intercept form, with $m = -2$ and $b = 6$. Therefore, the y-intercept is $(0, 6)$ and the slope is -2. The line goes down 2 units for every unit across. The graph is

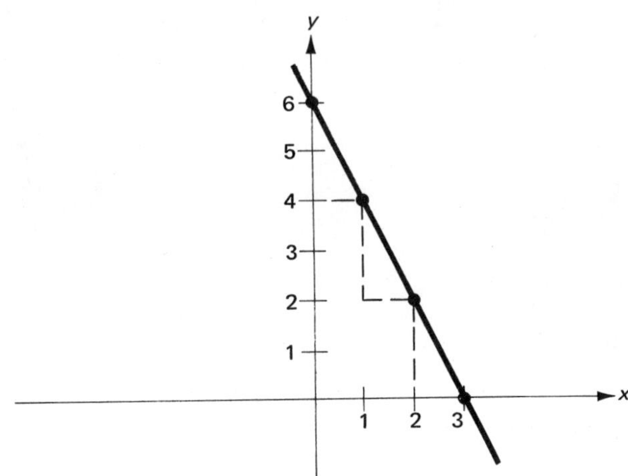

Using the $f(x)$ symbol, we can find the coordinates of other points on the line. For example,

$$f(1) = -2(1) + 6 = 4, \text{ gives } (1, 4).$$

$$f(2) = -2(2) + 6 = 2, \text{ gives } (2, 2).$$

$$f(3) = -2(3) + 6 = 0, \text{ gives } (3, 0).$$

Exercise 6.4

Find the y-intercept, x-intercept, and draw the graph:

1. $x - 2y + 4 = 0$

2. $3x + y - 6 = 0$

3. $-2x + 5y + 10 = 0$

4. $-7x + 2y - 14 = 0$

5. $5x + 6y + 12 = 0$

6. $-3x + 5y + 9 = 0$

7. $3x + 2y = 0$

8. $2x - 5y = 0$

9. $y - 3 = 0$

10. $2y + 7 = 0$

11. $3x - 4 = 0$

12. $x + 2 = 0$

Find the y-intercept, slope, and draw the graph:

13. $y = 4x - 6$

14. $y = -2x + 2$

15. $5x + 2y - 10 = 0$

16. $4x - 3y - 12 = 0$

Self-test

1. Find the slope of the line determined by (5, 2) and $(-3, -4)$.

1. ________________

2. Find the distance between $(1, -2)$ and $(-1, 4)$.

2. ________________

3. Find the y-intercept, x-intercept, and draw the graph of $3x - 2y - 12 = 0$.

y - intercept   ________________

x - intercept   ________________

graph:

4. Find the equation of the line with slope -3 and containing the point $(-5, 1)$, and write the equation in general form.

4. ________________

5. Find the equation of the line containing $(-2, 0)$ and $(3, -2)$, and write the equation in general form.

5. ________________

INTRODUCTION

The quadratic function relates two variables, where the first variable is contained in quadratic expression. In this unit, you will learn some properties of quadratic functions, and how to draw a parabola, which is the graph of a quadratic function. You will learn an application of one important property, the vertex of the parabola. Also, you will learn the general definition of a function, and how to determine whether or not a graph is the graph of a function.

OBJECTIVES

When you have finished this unit you should be able to:

1. Find the vertex of a parabola, and say whether the vertex is a maximum or minimum.
2. Draw the graph of a quadratic function.
3. Given an appropriate formula, solve problems describing the motion of an object due to gravity.
4. Determine whether or not a given graph is the graph of a function.

<table>
<tr><td>Section
7.1</td><td>The Vertex of a Parabola</td></tr>
</table>

The equation

$$y = ax^2 + bx + c,$$

where $a \neq 0$, is the general form of the **quadratic function**. The simplest such equation is

$$y = x^2.$$

Observe that if $x = 0$, then $y = 0$, and if $y = 0$, then $x = 0$. Thus the y-intercept and x-intercept both are (0, 0). We will draw the graph of $y = x^2$ by finding some other points.

Recall the $f(x)$ symbol from Section 6.4. We may write

$$f(x) = x^2,$$

and so

$$f(1) = 1^2 = 1, \text{ gives } (1, 1).$$
$$f(2) = 2^2 = 4, \text{ gives } (2, 4).$$
$$f(-1) = (-1)^2 = 1, \text{ gives } (-1, 1).$$
$$f(-2) = (-2)^2 = 4, \text{ gives } (-2, 4).$$

The graph is

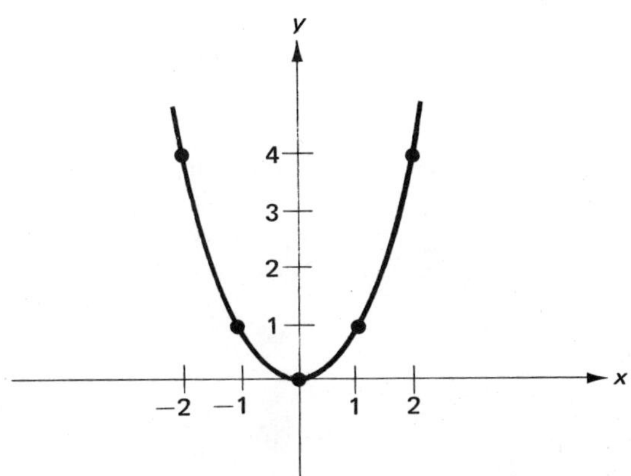

The graph of a quadratic function is called a **parabola**. A very important property of the parabola above is that it decreases to a minimum point, where it turns and then increases. The minimum point is called the **vertex** of the parabola. The vertex of the parabola which is the graph of $y = x^2$ is clearly the point (0, 0).

The vertex of a parabola also can be a maximum point. The graph of the quadratic function $y = -x^2$ is

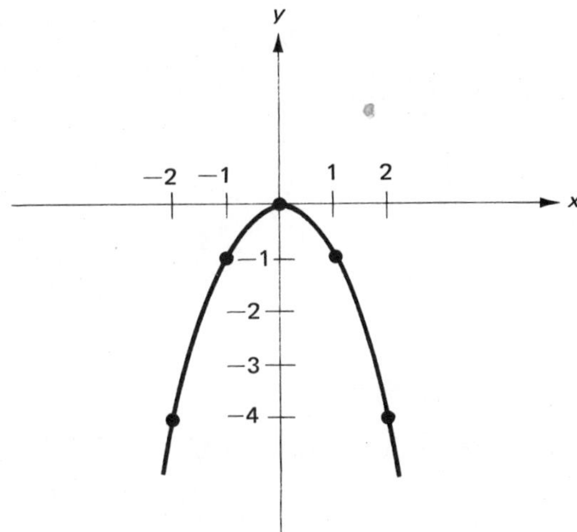

Here the vertex is again at the point (0, 0), but it is a maximum; that is, the graph increases to the vertex, where it turns and then decreases.

If a quadratic function is in the general form

$$y = ax^2 + bx + c,$$

then the x-coordinate of the vertex is found by the formula

$$x = -\frac{b}{2a}.$$

Moreover, if a is positive, then the vertex is a minimum. If a is negative, then the vertex is a maximum. The proof of these facts is given at the end of this unit.

EXAMPLE 7.1.   Find the vertex of $y = x^2 - 4$, and say whether it is a maximum or a minimum.

Solution.   The general form is

$$y = 1 \cdot x^2 + 0 \cdot x - 4,$$

so $a = 1$ and $b = 0$. The x-coordinate of the vertex is

$$x = -\frac{b}{2a}$$

$$= -\frac{0}{2(1)}$$

$$= 0.$$

Then,

$$f(0) = 0^2 - 4 = -4.$$

Therefore the vertex is $(0, -4)$. Since $a = 1$ is positive, the vertex is a minimum.

EXAMPLE 7.2.   Find the vertex of $y = 4x - x^2$, and say whether it is a maximum or a minimum.

Solution.   The general form is

$$y = -1 \cdot x^2 + 4x + 0,$$

so $a = -1$ and $b = 4$. The x-coordinate of the vertex is

$$x = -\frac{b}{2a}$$

$$= -\frac{4}{2(-1)}$$

$$= -\frac{4}{-2}$$

$$= 2.$$

Then,

$$f(2) = 4(2) - 2^2$$

$$= 8 - 4$$

$$= 4,$$

so the vertex is $(2, 4)$. Since $a = -1$ is negative, the vertex is a maximum.

Although there is a formula to find the y-coordinate of the vertex as well as the x-coordinate, it is usually easier to substitute in the equation of the quadratic function than to remember the formula. In more complicated cases, however, you must be very careful with the arithmetic.

EXAMPLE 7.3.   Find the vertex of $y = 2 - 3x - x^2$, and say whether it is a maximum or a minimum.

Solution.   The general form is

$$y = -1 \cdot x^2 - 3 \cdot x + 2,$$

so $a = -1$ and $b = -3$. The x-coordinate of the vertex is

$$x = -\frac{b}{2a}$$

$$= -\frac{-3}{2(-1)}$$

$$= -\frac{-3}{-2}$$

$$= -\frac{3}{2}.$$

Then,

$$f\left(-\frac{3}{2}\right) = 2 - 3\left(-\frac{3}{2}\right) - \left(-\frac{3}{2}\right)^2$$

$$= 2 + \frac{9}{2} - \frac{9}{4}$$

$$= \frac{8 + 18 - 9}{4}$$

$$= \frac{17}{4}.$$

Therefore the vertex is $\left(-\frac{3}{2}, \frac{17}{4}\right)$ and is a maximum since a is negative.

Exercise 7.1

Find the vertex, and say whether it is a maximum or a minimum:

1. $y = x^2 - 3$

2. $y = 4 - 2x^2$

3. $y = x - 2x^2$

4. $y = x^2 + 3x$

5. $y = x^2 - 4x + 5$

6. $y = x^2 + 2x + 1$

7. $y = 3x^2 - 2x - 2$

8. $y = 2x^2 + 3x - 1$

9. $y = 4 - 3x - x^2$

10. $y = 1 + 2x - 3x^2$

<table>
<tr><td>Section
7.2</td><td><h1 style="text-align:center">Graphs of Quadratic Functions</h1></td></tr>
</table>

Consider the properties of these three parabolas:

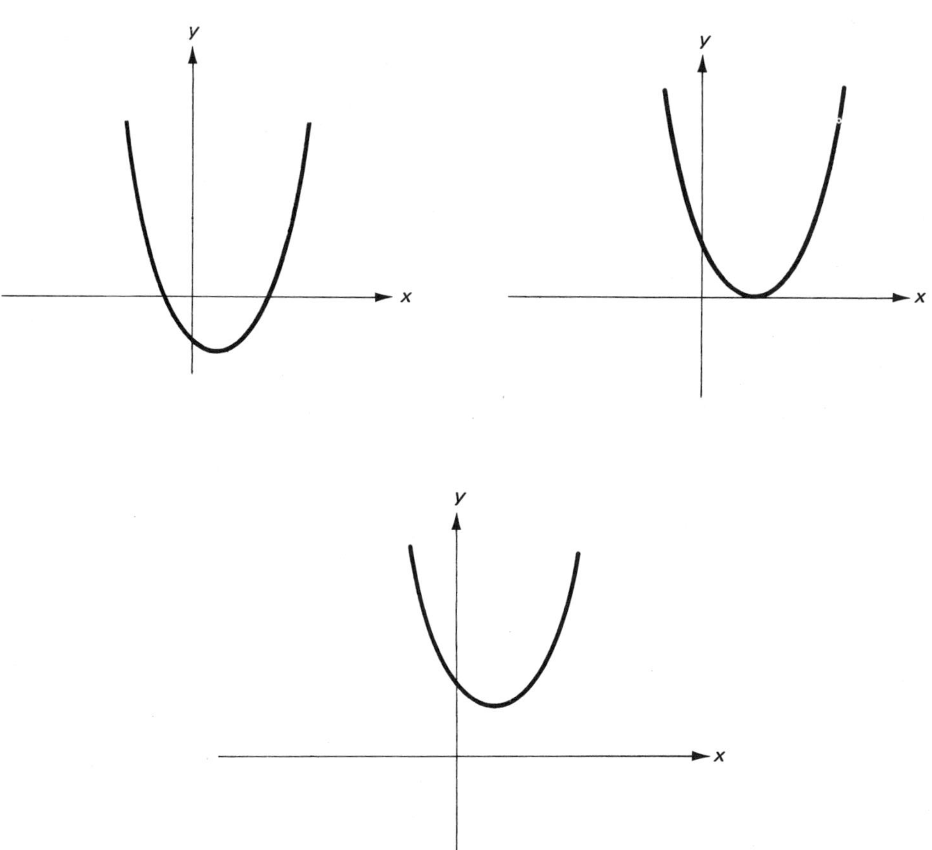

Each of the parabolas above has a vertex which is a minimum. Each has one y-intercept. However, the first has two x-intercepts, the second one, and the third none. Similarly, if the vertex is a maximum, the parabola has one y-intercept, but may have two x-intercepts, one, or none.

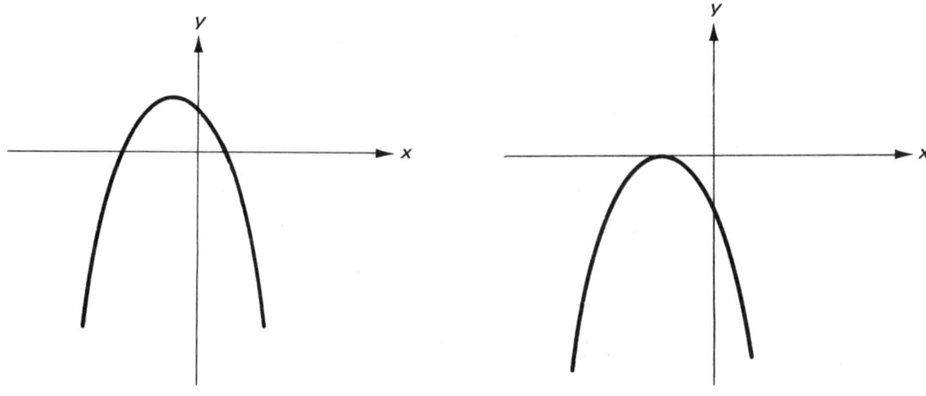

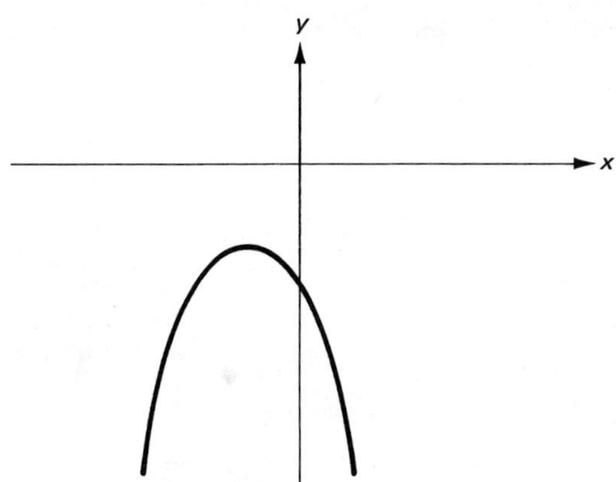

The graph of a quadratic function has one y-intercept, and may have two, one, or no x-intercepts.

If the graph of a quadratic function has two x-intercepts, and the y-intercept, x-intercepts, and vertex are all distinct points, we can use these four points to draw the graph.

EXAMPLE 7.4.   Find the y-intercept, x-intercepts, vertex, and draw the graph of $y = x^2 - 3x + 2$.

Solution.   If $x = 0$,

$$f(0) = 0^2 - 3(0) + 2 = 2,$$

so the y-intercept is $(0, 2)$.

If $y = 0$, we have

$$0 = x^2 - 3x + 2$$

$$0 = (x - 1)(x - 2)$$

$$x - 1 = 0 \text{ and } x - 2 = 0$$

$$x = 1 \text{ and } x = 2.$$

The x-intercepts are $(1, 0)$ and $(2, 0)$.

To find the vertex, we use

$$x = -\frac{b}{2a}$$

$$= -\frac{-3}{2(1)}$$

$$= \frac{3}{2}.$$

$$f\left(\frac{3}{2}\right) = \left(\frac{3}{2}\right)^2 - 3\left(\frac{3}{2}\right) + 2$$

$$= \frac{9}{4} - \frac{9}{2} + 2$$

$$= \frac{9 - 18 + 8}{4}$$

$$= -\frac{1}{4}.$$

The vertex is $(\frac{3}{2}, -\frac{1}{4})$, and is a minimum since a is positive. The graph is

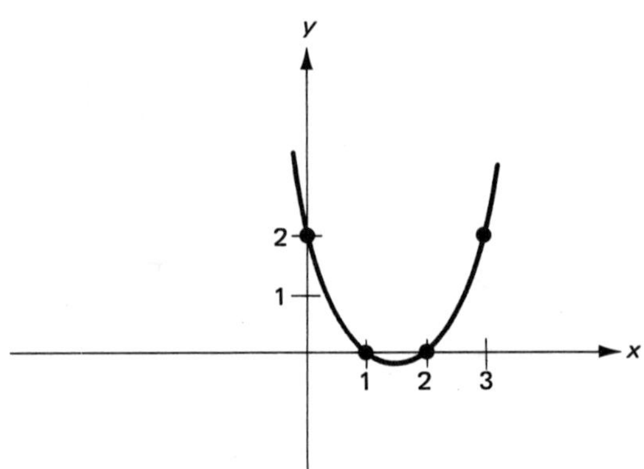

You may have observed that the points on either side of the vertex of a parabola are symmetrical. Thus, if we calculate the point at $x = 3$, we get

$$f(3) = 3^2 - 3(3) + 2$$
$$= 9 - 9 + 2$$
$$= 2.$$

The point $(3, 2)$ serves as a check point.

The x-intercepts of the graph of a quadratic function may involve approximations, but in this case they still exist and there are two. We can approximate them using the quadratic formula.

EXAMPLE 7.5.   Find the y-intercept, x-intercepts, vertex, and draw the graph of $y = x^2 + 2x - 4$.

Solution.   The y-intercept is easily found to be $(0, -4)$. To find the x-intercepts, if $y = 0$ we have

$$0 = x^2 + 2x - 4.$$

Using the quadratic formula,

$$x = \frac{-2 \pm \sqrt{4 - 4(1)(-4)}}{2(1)}$$

$$= \frac{-2 \pm \sqrt{4 + 16}}{2}$$

$$= \frac{-2 \pm \sqrt{20}}{2}$$

$$= \frac{-2 \pm 2\sqrt{5}}{2}$$

$$= -1 \pm \sqrt{5}.$$

We may use a calculator or table of square roots to approximate these solutions:

$$-1 + \sqrt{5} \approx -1 + 2.2 = 1.2$$

$$-1 - \sqrt{5} \approx -1 - 2.2 = -3.2,$$

so the x-intercepts are approximately $(1.2, 0)$ and $(-3.2, 0)$. We have approximated only to one decimal place because a point cannot be plotted to greater accuracy than a tenth of a unit.

To find the vertex,

$$x = -\frac{2}{2(1)} = -1,$$

and

$$f(-1) = (-1)^2 + 2(-1) - 4 = -5,$$

so the vertex is $(-1, -5)$. The graph is

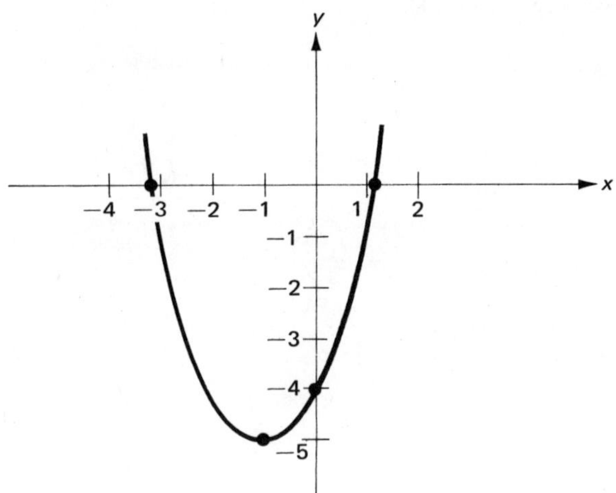

If, in the quadratic formula, the number under the square root is negative, then it cannot be approximated by a decimal and the graph of the quadratic function has no x-intercepts.

EXAMPLE 7.6.   Find the y-intercept, x-intercepts, vertex, and draw the graph of $y = 2x^2 - 2x + 3$.

Solution.   The y-intercept is $(0, 3)$.

If $y = 0$,

$$0 = 2x^2 - 2x + 3,$$

and

$$x = \frac{2 \pm \sqrt{4 - 4(2)(3)}}{2(2)}$$

$$= \frac{2 \pm \sqrt{-20}}{4}.$$

Therefore there are no x-intercepts. To find the vertex,

$$x = -\frac{-2}{2(2)} = \frac{1}{2},$$

and

$$f\left(\frac{1}{2}\right) = 2\left(\frac{1}{2}\right)^2 - 2\left(\frac{1}{2}\right) + 3$$

$$= 2\left(\frac{1}{4}\right) - 1 + 3$$

$$= \frac{1}{2} + 2$$

$$= \frac{5}{2}.$$

The vertex is $(\frac{1}{2}, \frac{5}{2})$. However, we have only two points, the y-intercept and the vertex. To find some other points, we use

$$f(x) = 2x^2 - 2x + 3,$$

and so

$$f(1) = 2(1)^2 - 2(1) + 3 = 3, \text{ gives } (1, 3).$$

$$f(2) = 2(2)^2 - 2(2) + 3 = 7, \text{ gives } (2, 7).$$

$$f(-1) = 2(-1)^2 - 2(-1) + 3 = 7, \text{ gives } (-1, 7).$$

The graph is

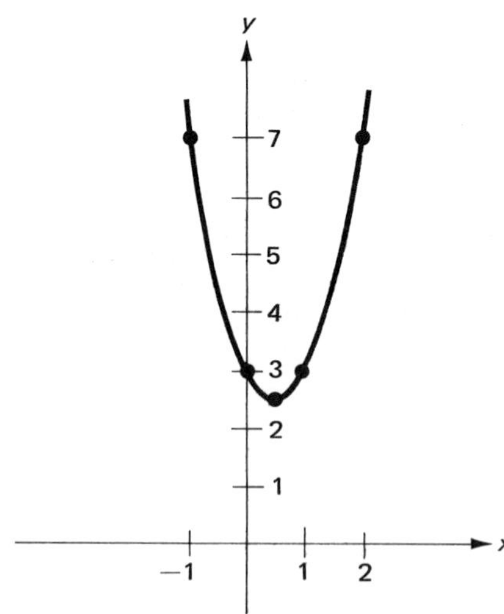

Finally, the graph of a quadratic function may have one x-intercept. In this case, the x-intercept comes from a double root of a quadratic equation. Moreover, the vertex will also be the same point as the one x-intercept.

EXAMPLE 7.7.   Find the y-intercept, x-intercepts, vertex, and draw the graph of $y = x^2 + 4x + 4$.

Solution.   The y-intercept is $(0, 4)$.
       If $y = 0$,

$$0 = x^2 + 4x + 4$$

$$0 = (x + 2)(x + 2)$$

$$x + 2 = 0$$

$$x = -2.$$

There is one x-intercept, $(-2, 0)$. Since the x-coordinate of the vertex is

$$x = -\frac{4}{2(1)} = -2,$$

the vertex is also $(-2, 0)$. We have two points on the graph, the y-intercept $(0, 4)$ and the point $(-2, 0)$ which represents both x-intercepts and the vertex. To find another point, we choose a value of x on the other side of the vertex from the y-intercept:

$$f(-4) = (-4)^2 + 4(-4) + 4 = 4, \text{ gives } (-4, 4).$$

The graph is

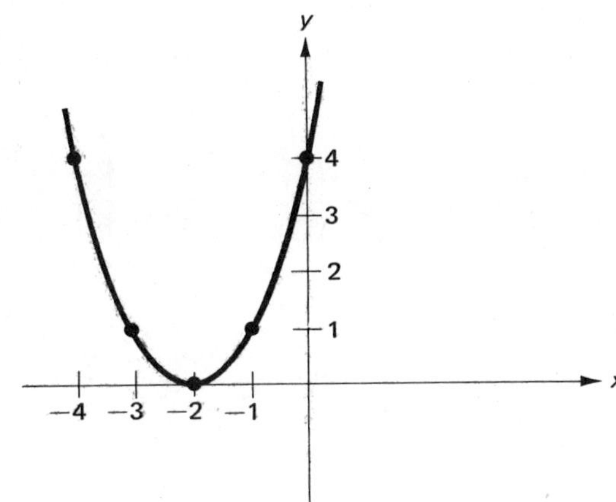

Observe that we have also included the symmetrical points $(-1, 1)$ and $(-3, 1)$.

If a is negative, and so the vertex is a maximum, we have the same three cases described above. There can be two, one, or no x-intercepts. We use exactly the same methods.

EXAMPLE 7.8.   Find the y-intercept, x-intercepts, vertex, and draw the graph of $y = -2x^2 - x + 1$.

Solution.   The y-intercept is $(0, 1)$.
       If $y = 0$,

$$0 = -2x^2 - x + 1$$

or

$$2x^2 + x - 1 = 0$$

$$(2x - 1)(x + 1) = 0$$

$$2x - 1 = 0 \text{ and } x + 1 = 0$$

$$x = \frac{1}{2} \text{ and } x = -1.$$

The x-intercepts are $(\frac{1}{2}, 0)$ and $(-1, 0)$.
Using the vertex formula,

$$ x = -\frac{-1}{2(-2)} = -\frac{1}{4}, $$

and

$$ f\left(-\frac{1}{4}\right) = -2\left(-\frac{1}{4}\right)^2 - \left(-\frac{1}{4}\right) + 1 $$

$$ = -2\left(\frac{1}{16}\right) + \frac{1}{4} + 1 $$

$$ = -\frac{1}{8} + \frac{1}{4} + 1 $$

$$ = \frac{-1 + 2 + 8}{8} $$

$$ = \frac{9}{8}. $$

The vertex is $(-\frac{1}{4}, \frac{9}{8})$, and is a maximum. The graph is

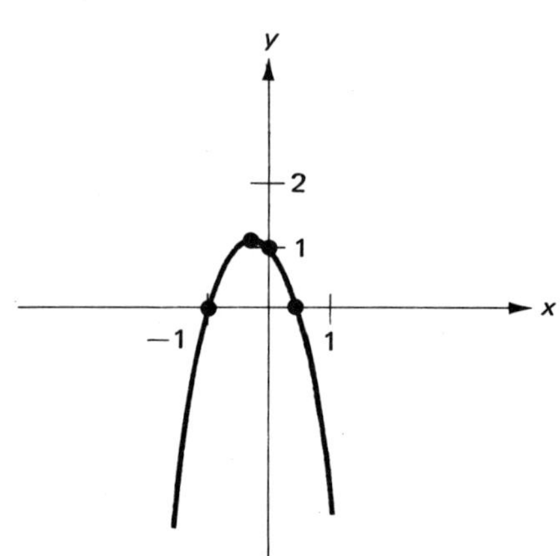

Exercise 7.2

Find the y-intercept, x-intercepts, vertex, and draw the graph:

1. $y = x^2 - 4$

2. $y = 2x^2 + 3$

3. $y = 1 - 2x^2$

4. $y = -x^2 - 2$

5. $y = x^2 + 2x + 1$

6. $y = x^2 - 6x + 9$

7. $y = x^2 - 6x + 8$

8. $y = x^2 - 2x - 3$

9. $y = 15 + 2x - x^2$

10. $y = 5 - 4x - x^2$

11. $y = x^2 - 2x - 2$

12. $y = -2x^2 + 3x + 1$

13. $y = x^2 - 2x + 2$

14. $y = 2x^2 + 3x + 3$

15. $y = -x^2 + x - 1$

16. $y = -3x^2 - 3x - 2$

<table><tr><td>Section
7.3</td><td><h1>An Application</h1></td></tr></table>

The fact that a parabola has either a maximum or minimum point has a great many applications. You will see some of them if you take a beginning calculus course. We will consider just one application here.

Suppose an object such as a ball or stone is tossed straight upward into the air. Suppose further that the only force which influences its motion is the force of gravity; that is, we will ignore such forces as air friction, wind, and so forth. Then the equation of the motion of the object is

$$s = -16t^2 + v_0 t + s_0$$

where s is the height of the object at any time t, v_0 is a constant which represents the velocity with which the object was thrown, and s_0 is a constant which represents the height from which the object was thrown. The way the equation is stated assumes that s and s_0 are in feet, t is in seconds, and v_0 is in feet per second.

We observe that the equation is a quadratic function with t and s in place of x and y. Regardless of the values of v_0 and s_0, $a = -16$, and so the graph of the function is a parabola whose vertex is a maximum. The vertex represents the highest point reached by the object, where it turns and falls back to earth. The t-coordinate of the vertex represents the time, in seconds, when the object reaches its maximum height, and the s-coordinate is the maximum height, in feet, the object reaches.

EXAMPLE 7.9. A ball is thrown into the air according to the equation $s = -16t^2 + 48t$. How high will it rise and at what time will it reach that height?

Solution. The t-coordinate of the vertex is

$$t = -\frac{48}{2(-16)}$$

$$= -\frac{48}{-32}$$

$$= \frac{3}{2}.$$

Substituting $t = \frac{3}{2}$ in the equation,

$$s = -16\left(\frac{3}{2}\right)^2 + 48\left(\frac{3}{2}\right)$$

$$= -16\left(\frac{9}{4}\right) + 48\left(\frac{3}{2}\right)$$

$$= -36 + 72$$

$$= 36.$$

The ball will rise 36 feet and reach that height in $\frac{3}{2}$ seconds.

We can also find when the object leaves the ground and when it returns. Since s is the height of the object at any time, $s = 0$ when the object is on the ground.

EXAMPLE 7.10.   When does the ball in Example 7.9 leave the ground and when does it return?

Solution.   When $s = 0$,

$$0 = -16t^2 + 48t$$

or

$$16t^2 - 48t = 0$$

$$16t(t - 3) = 0$$

$$t = 0 \text{ and } t - 3 = 0$$

$$t = 0 \text{ and } t = 3.$$

The ball leaves the ground at 0 seconds and returns at 3 seconds. Observe that the time the ball takes to reach its highest point is exactly one-half the total time. In other words, the time it takes the ball to go up and the time it takes the ball to come down are exactly the same. This situation occurs when the height s_0 from which the object is thrown is 0.

EXAMPLE 7.11.   A ball is thrown into the air according to the equation $s = -16t^2 + 24t + 32$. Find its maximum height, when it reaches that height, and also when it leaves and returns to the ground.

Solution.   The t-coordinate of the vertex is

$$t = -\frac{24}{2(-16)}$$

$$= -\frac{24}{-32}$$

$$= \frac{3}{4}.$$

The s-coordinate is

$$s = -16\left(\frac{3}{4}\right)^2 + 24\left(\frac{3}{4}\right) + 32$$

$$= -16\left(\frac{9}{16}\right) + 24\left(\frac{3}{4}\right) + 32$$

$$= -9 + 18 + 32$$

$$= 41.$$

Thus the ball reaches a height of 41 feet in $\frac{3}{4}$ second.
When $s = 0$,

$$0 = -16t^2 + 24t + 32$$

$$16t^2 - 24t - 32 = 0$$

$$8(2t^2 - 3t - 4) = 0.$$

Thus $2t^2 - 3t - 4 = 0$, and

$$t = \frac{3 \pm \sqrt{9 - 4(2)(-4)}}{2(2)}$$

$$= \frac{3 \pm \sqrt{9 + 32}}{4}$$

$$= \frac{3 \pm \sqrt{41}}{4}$$

$$t \approx -.85 \text{ and } 2.35.$$

We cannot consider a negative answer, as that would represent a negative time, so $t = -.85$ is extraneous. However, the original equation tells us that the ball never left the ground but was thrown from a height s_0 of 32 feet. The ball returns to the ground at approximately 2.35 seconds. Since the ball was thrown from a point above the ground, it takes less time to reach its maximum height than to fall back to the ground.

We may also consider objects thrown straight downward towards the earth. In this case, we take the velocity v_0 with which the object is thrown to be negative.

EXAMPLE 7.12.   A stone is thrown downward according to the equation $s = -16t^2 - 64t + 48$. At what time will it reach the ground?

Solution.   When $s = 0$,

$$0 = -16t^2 - 64t + 48$$

$$16t^2 + 64t - 48 = 0$$

$$16(t^2 + 4t - 3) = 0$$

$$t^2 + 4t - 3 = 0$$

$$t = \frac{-4 \pm \sqrt{16 - 4(1)(-3)}}{2(1)}$$

$$= \frac{-4 \pm \sqrt{16 + 12}}{2}$$

$$= \frac{-4 \pm \sqrt{28}}{2}$$

$$= \frac{-4 \pm 2\sqrt{7}}{2}$$

$$= -2 \pm \sqrt{7}.$$

Since $t = -2 - \sqrt{7}$ is clearly negative, we use $t = -2 + \sqrt{7} \approx .646$. Thus the stone reaches the ground in approximately .646 seconds. It does not make sense to ask when the stone reaches its highest point. The stone was thrown downward, so the highest point is the point $s_0 = 48$ from which it was thrown. However, were you to go through the process of finding the t-coordinate of the vertex, you would find it is negative and thus extraneous.

If an object is simply allowed to drop from a height, then the initial velocity v_0 is zero.

EXAMPLE 7.13.   A slate falls off a roof 64 feet from the ground according to the equation $s = -16t^2 + 64$. How long will it take to reach the ground?

Solution.   When $s = 0$,

$$0 = -16t^2 + 64$$

$$16t^2 = 64$$

$$t^2 = 4$$

$$t = \pm 2.$$

The solution $t = -2$ is extraneous. Using the solution $t = 2$, it will take 2 seconds for the slate to reach the ground.

Exercise 7.3

If an object is thrown into the air according to the given formula, how high will it rise and when will it reach that height:

1. $s = -16t^2 + 64t$
2. $s = -16t^2 + 32t + 16$
3. $s = -16t^2 + 80t + 50$

4. $s = -16t^2 + 28t + 4$

If an object is thrown into the air according to the given formula, when will it leave the ground and when will it return:

5. $s = -16t^2 + 64t$
6. $s = -16t^2 + 36t$
7. $s = -16t^2 + 24t + 40$

8. $s = -16t^2 + 56t + 48$

When will an object reach the ground if it is thrown downward according to the formula:

9. $s = -16t^2 - 32t + 16$
10. $s = -16t^2 - 40t + 100$

11. A ball is thrown into the air from a height of 4 feet with a velocity of 24 feet per second. How high will it rise, how long will it take to reach that height, and how long will it take to return to the ground?

12. You are standing on the observation deck of the Empire State Building, approximately 1400 feet above the street, when you drop your guidebook. How long will it take for the book to hit the street?

<table><tr><td>Section
7.4</td><td><h1>The Definition of a Function</h1></td></tr></table>

Recall from Section 7.1 the equation $y = x^2$ and its graph:

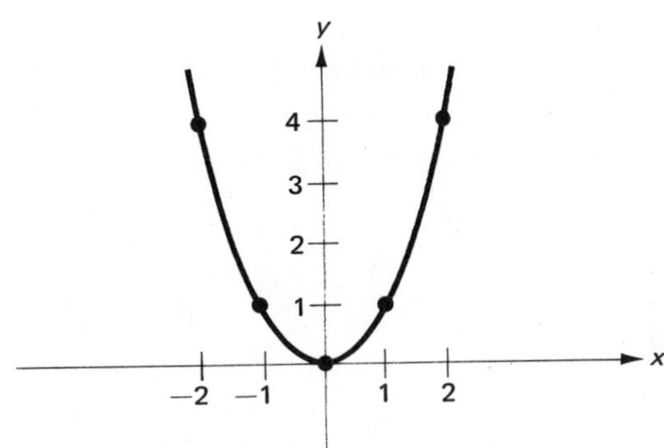

Now, we draw the graph of $y^2 = x$. If $x = 0$, $y = 0$ gives $(0, 0)$. If $x = 1$,

$$y^2 = 1$$

$$y = \pm 1$$

gives $(1, 1)$ and $(1, -1)$. If $x = 4$,

$$y^2 = 4$$

$$y = \pm 2$$

gives $(4, 2)$ and $(4, -2)$. The graph is

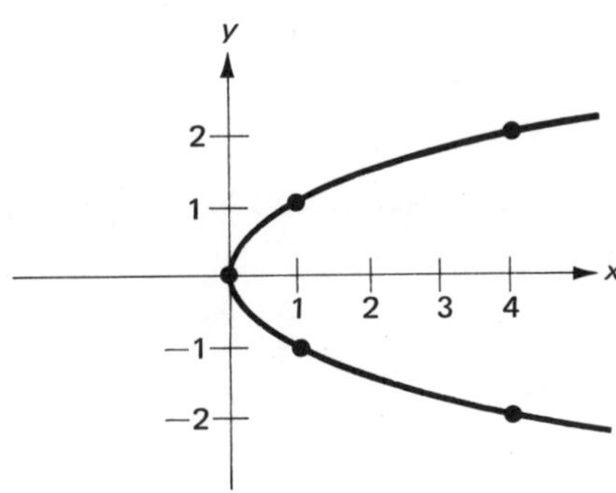

We did not use the $f(x)$ symbol to calculate the points for this graph. The $f(x)$ symbol is used only when there is just one value of y for each x. However, for $y^2 = x$, there are generally two values of y for each x. For example, if $x = 1$, $y = 1$ and $y = -1$.

> **Definition:**   A relation in two variables x and y is a **function** if for each x-value there is just one y-value.

We see that $y = x^2$ is a function. However, $y^2 = x$ is not a function because for the first coordinate x we can get two values of the second coordinate y.

Often the easiest way to tell whether or not a relation is a function is by its graph. If we can construct a vertical line anywhere at all through the graph, and the vertical line crosses the graph only once, then there is just one y-value and the graph is the graph of a function. If there is any place where the vertical line crosses the graph more than once, then there is a value of x that has more than one y-value and the graph is not the graph of a function. For example, we can draw a vertical line through the graph of $y = x^2$ at any value of x and the line will cross the graph just once, so $y = x^2$ is a function. But a vertical line to the right of the y-axis crosses the graph of $y^2 = x$ twice, so $y^2 = x$ is not a function.

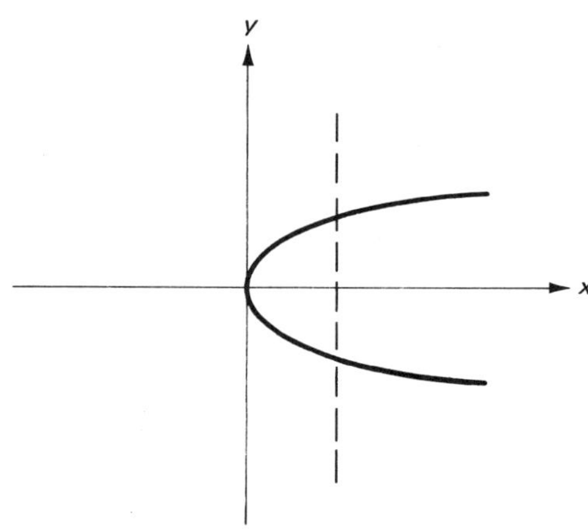

We may apply these concepts to the general linear equation we studied in Unit 6,

$$ax + by + c = 0.$$

The linear function, in the form

$$y = mx + b,$$

includes all cases of the general linear equation except $ax + c = 0$, where the y-term does not appear. The graph of the linear function $y = mx + b$ is an oblique line:

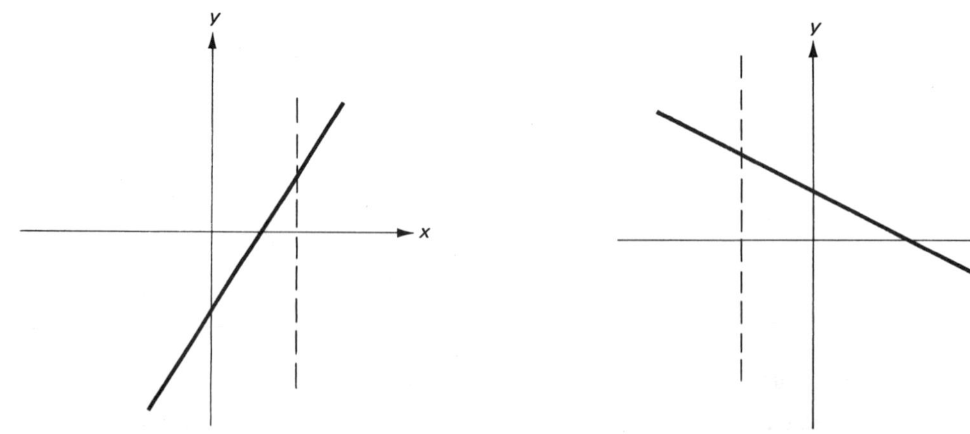

unless $m = 0$, when we have a horizontal line:

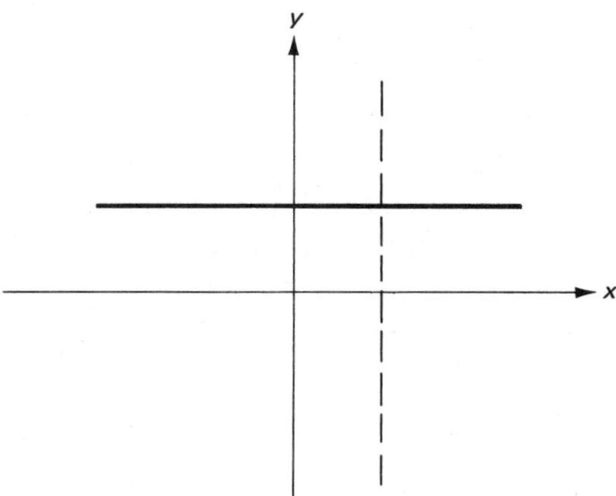

In any of these cases, the vertical line test tells us we have the graph of a function. Thus $y = mx + b$ is correctly called the linear function.

In the case of $ax + c = 0$, the graph is a vertical line:

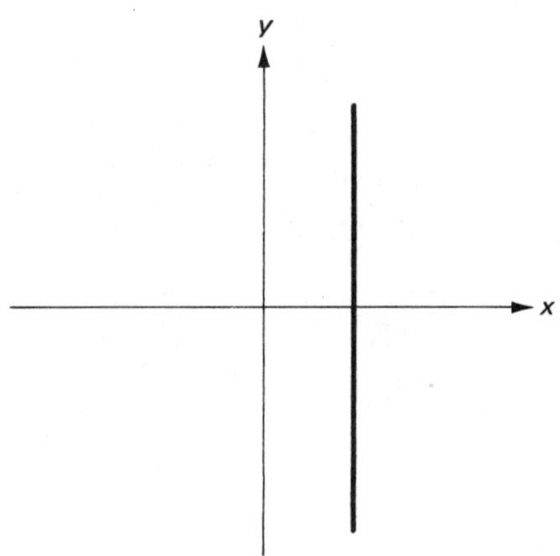

Here there are an infinite number of y-values for the one value of x. The only way we could draw a vertical line through this graph is to draw the graph itself. The vertical line touches the graph at every point, which is certainly more than just one point. Thus $ax + c = 0$ is not a function.

> ***Definition:***   The **domain** of a function is the set of all x-values for which a y-value is defined.

The domain of any linear function (remember that a vertical line is not a linear function) is all real numbers. The domain of any quadratic function is also all real numbers. For each of these types of functions, there is a y-value for every real number x.

> ***Definition:***   The **range** of a function is the set of all y-values of the function.

The range of any linear function, except a horizontal line, is all real numbers. However, the range of a horizontal line is the one y-value of all the points of the line. For example, the range of line $y = 2$ is just the y-value $y = 2$:

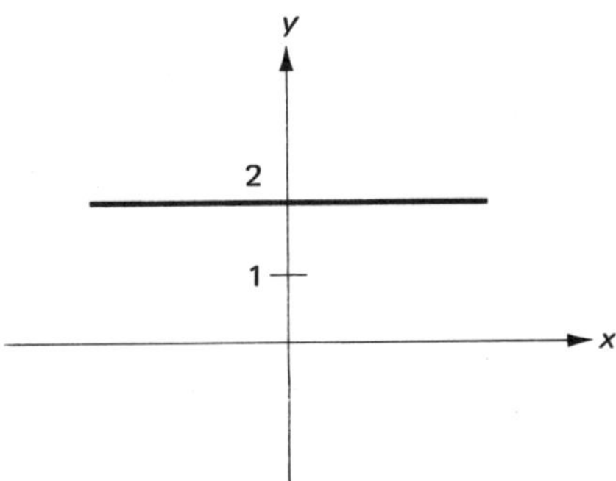

The range of a quadratic function where the vertex is a minimum is all y-values greater than or equal to the y-value of the vertex. For example, the range of $y = x^2$ is all positive real numbers and zero, which we write $y \geqslant 0$:

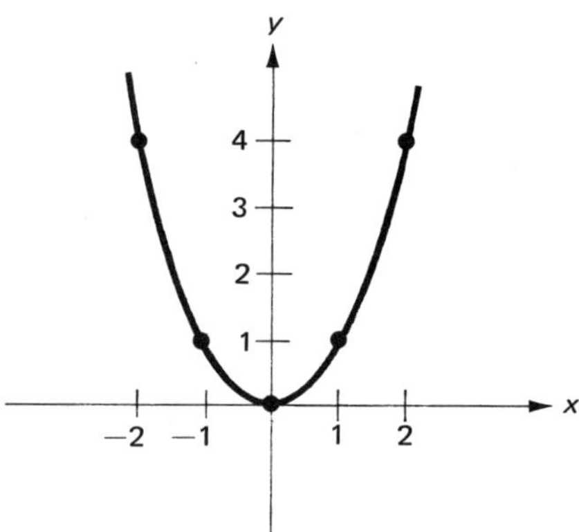

Similarly, the range of a quadratic function where the vertex is a maximum is all y-values less than or equal to the y-value of the vertex. For example, the range of $y = -x^2$ is all negative real numbers and zero, which we write $y \leqslant 0$:

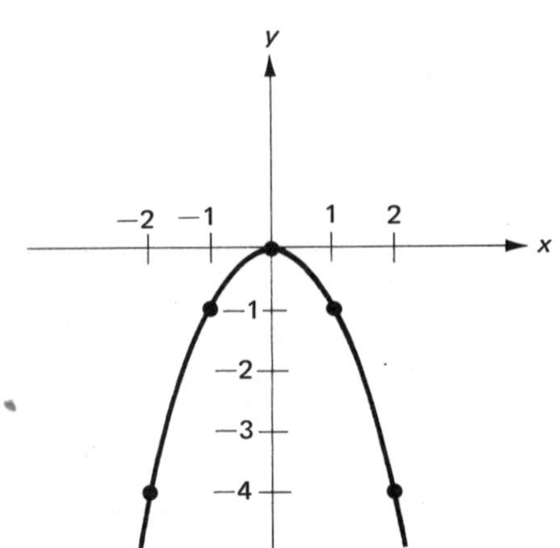

Exercise 7.4

Say whether or not each graph is the graph of a function:

1. 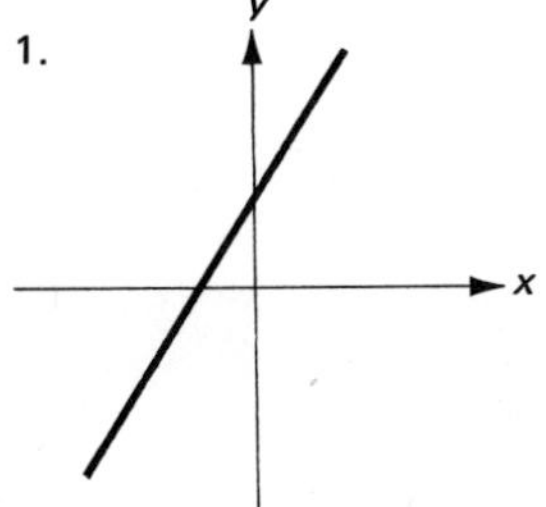   2. 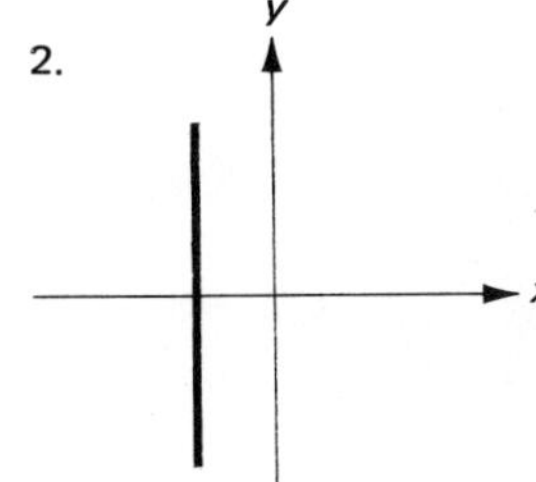   3. 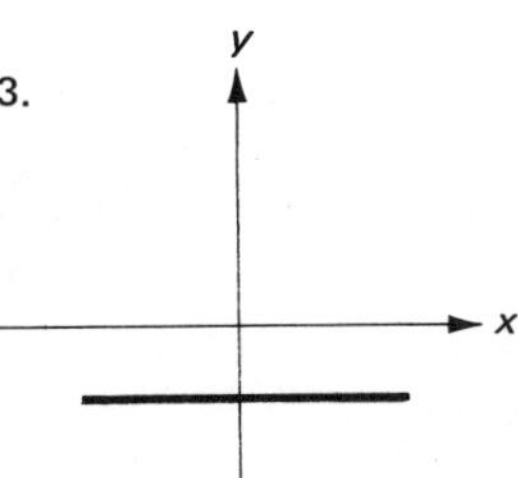   4. 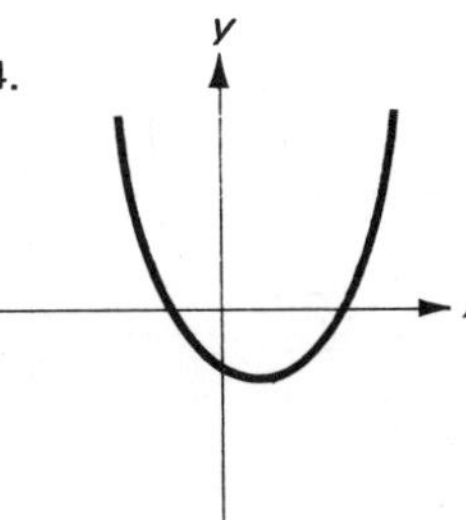

5. 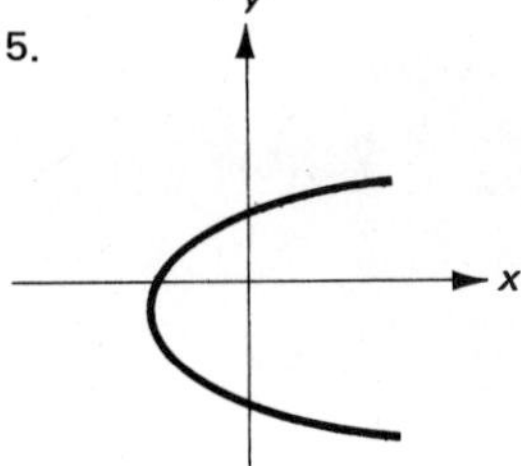   6. 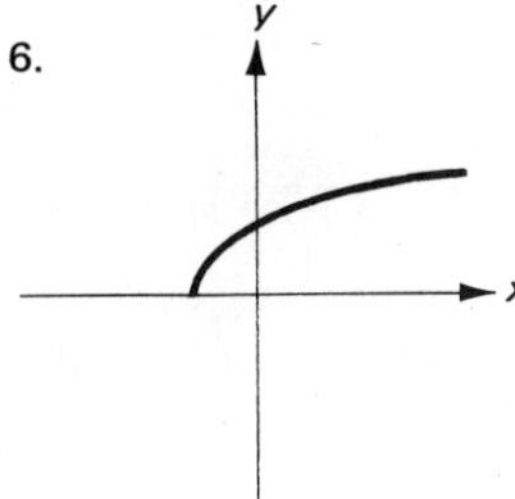   7. 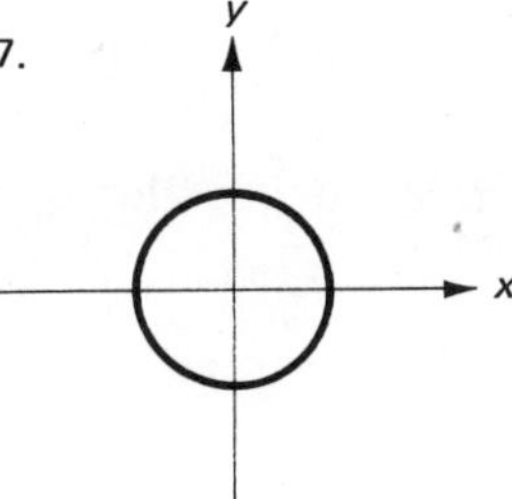   8. 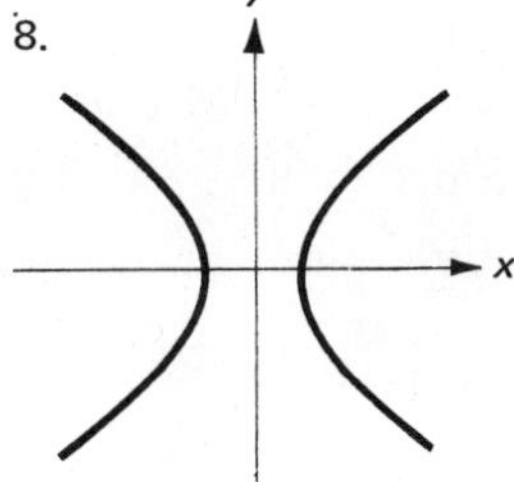

9. 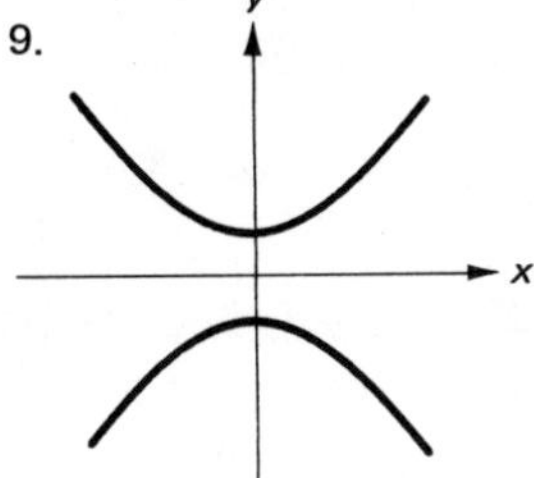   10. 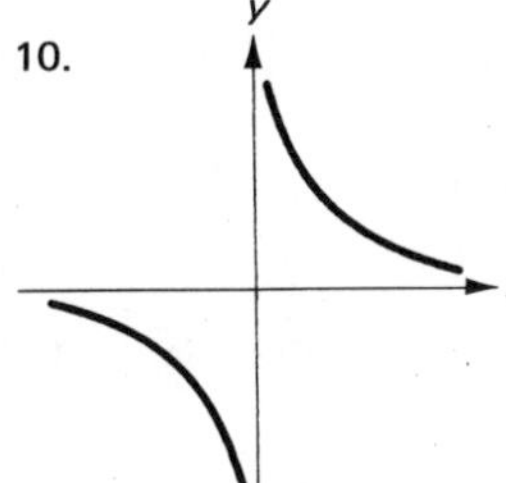   11. 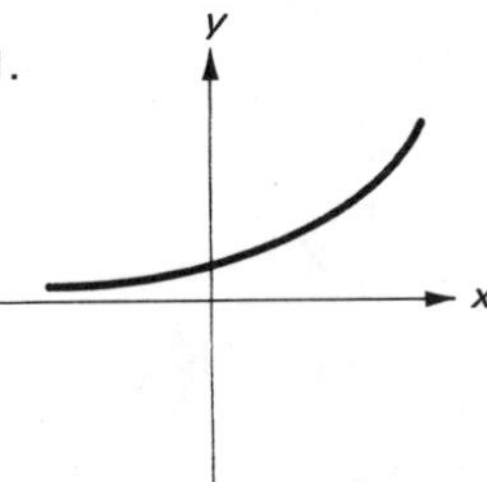   12. 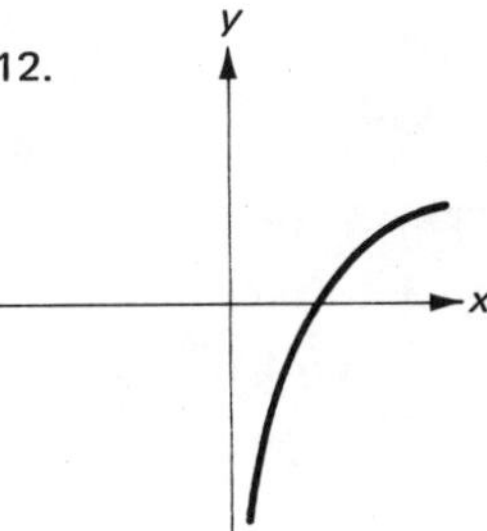

13. 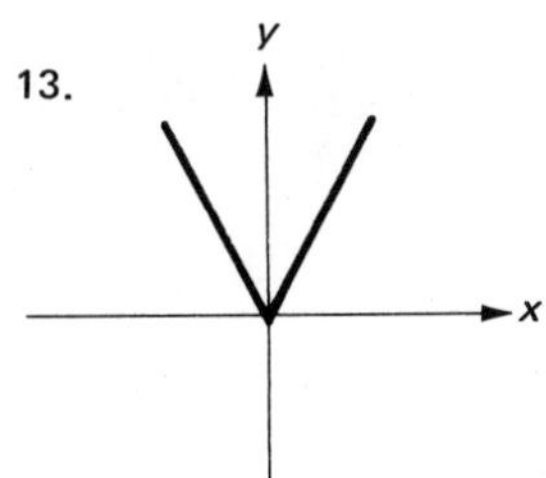   14. 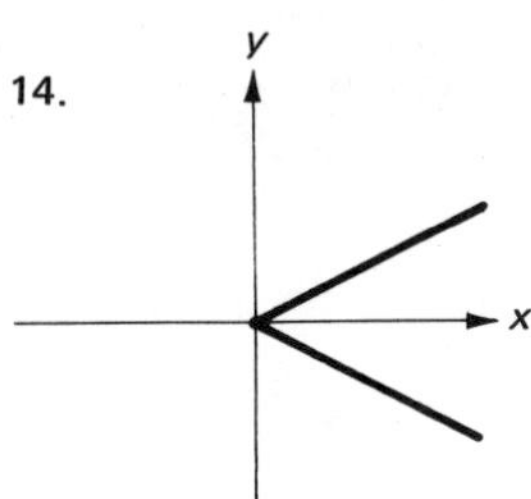   15. 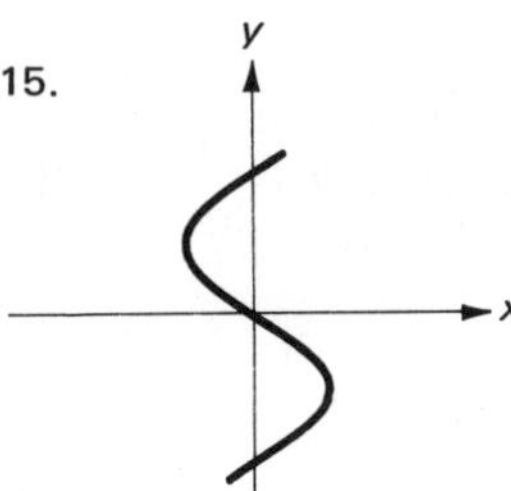   16. 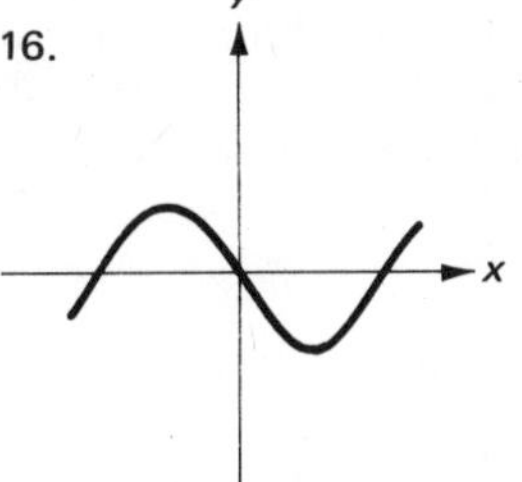

Proof of the Vertex Formula

We use the method of completing the square from Unit 4. Starting with the quadratic function

$$y = ax^2 + bx + c$$

and dividing through by a, we get

$$\frac{y}{a} = x^2 + \frac{b}{a}x + \frac{c}{a}.$$

The completion of the square for $x^2 + \dfrac{b}{a}x$ is $\dfrac{b^2}{4a^2}$ so adding $\dfrac{b^2}{4a^2}$ to both sides we have

$$\frac{y}{a} + \frac{b^2}{4a^2} = x^2 + \frac{b}{a}x + \frac{b^2}{4a^2} + \frac{c}{a}$$

$$\frac{y}{a} + \frac{b^2}{4a^2} = \left(x + \frac{b}{2a}\right)^2 + \frac{c}{a}.$$

Solving again for y,

$$\frac{y}{a} = \left(x + \frac{b}{2a}\right)^2 + \frac{c}{a} - \frac{b^2}{4a^2}$$

$$y = a\left(x + \frac{b}{2a}\right)^2 + c - \frac{b^2}{4a}.$$

Since $\left(x + \dfrac{b}{2a}\right)^2$ is a perfect square, this quantity must always be positive or zero. It will be a minimum when it is zero, that is, when

$$x + \frac{b}{2a} = 0$$

or

$$x = -\frac{b}{2a}.$$

This is the x-coordinate of the vertex of the parabola.

To establish the other facts, if a is positive, the quantity $a\left(x + \dfrac{b}{2a}\right)^2$ is positive. When $x \neq \dfrac{-b}{2a}$, the y-values are larger than the value at $\dfrac{-b}{2a}$, and the vertex is a minimum.

If a is negative, then the quantity $a\left(x + \dfrac{b}{2a}\right)^2$ is negative. When $x \neq \dfrac{-b}{2a}$, the y-values are smaller than the value at $\dfrac{-b}{2a}$, and the vertex is a maximum.

Finally, if $x = \dfrac{-b}{2a}$, then $x + \dfrac{b}{2a} = 0$, and

$$y = c - \frac{b^2}{4a}.$$

This formula gives the y-coordinate of the vertex.

Self-test

1. Which of these graphs are graphs of functions:

a. 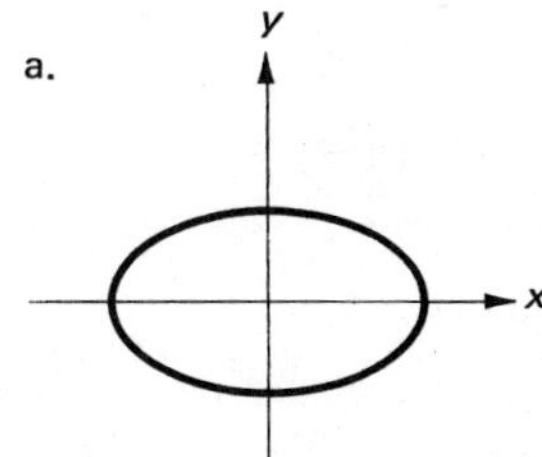   b. 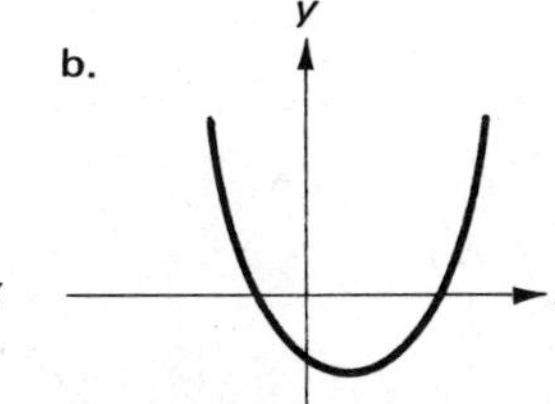 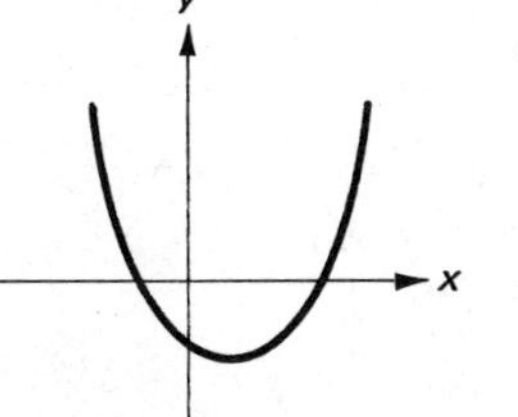   c. 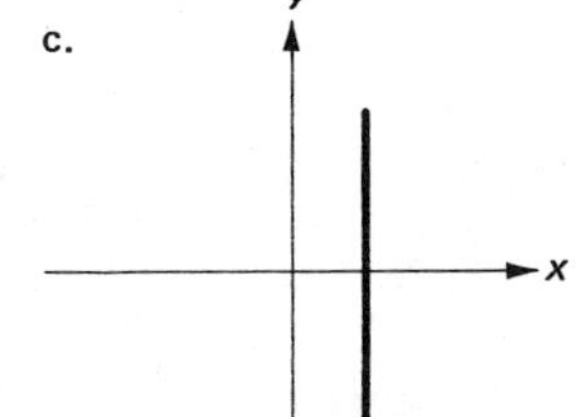   d. 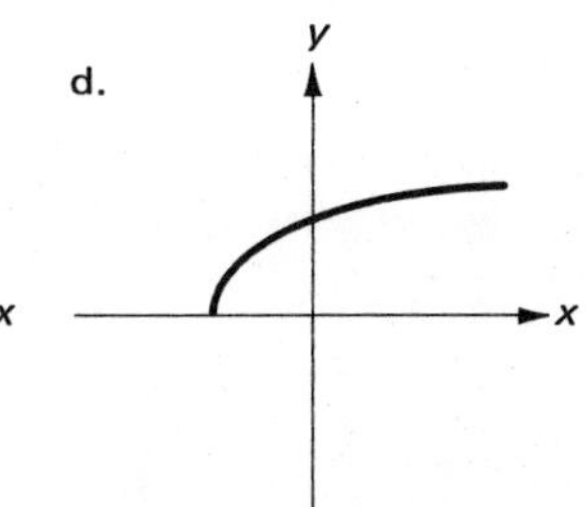

1. _________________

2. Find the vertex of $y = -2x^2 - 3x + 4$, and say whether it is a maximum or minimum.

2. _________________

3. Find the y-intercept, x-intercepts, vertex, and draw the graph of $y = x^2 + 2x - 3$.

y-intercept_________________

x-intercepts_________________

vertex_________________

4. Find the y-intercept, x-intercepts, vertex, and draw the graph of $y = -x^2 + 2x - 2$.

y-intercept_______________________

x-intercepts_______________________

vertex _______________________

graph:

5. A ball is tossed into the air according to the formula $s = -16t^2 + 64t + 48$. How long will it take to return to the ground?

5._______________________

INTRODUCTION

In Unit 1 you learned how to remove a common factor and how to factor a trinomial. In other units you have used factoring to solve equations, solve applied problems, and draw graphs. Clearly, factoring is an important tool of algebra. In this unit you will learn how to do some specific types of factoring which go beyond the methods you have learned so far.

OBJECTIVES

When you have finished this unit you should be able to:

1. Identify a perfect trinomial square, and write it as the square of a binomial.
2. Factor sums or differences of odd powers.
3. Factor differences of even powers, and state that sums of some even powers cannot be factored.
4. Given an appropriate polynomial in four terms, factor the polynomial by grouping.

Section
8.1

Perfect Trinomial Squares

A Note on Exponents

In the preceding units we have assumed you were familiar with simple products such as

$$(2x)(3x) = 6x^2.$$

Before we begin the study of special types of factoring, we will review three basic rules for exponents.

> **Rule 1:** If m and n are positive integers, then $x^m \cdot x^n = x^{m+n}$.

We use this rule when we write

$$x \cdot x = x^1 \cdot x^1 = x^{1+1} = x^2.$$

Some other examples of Rule 1 are

$$x(x^2) = x^1(x^2) = x^{1+2} = x^3,$$

and

$$(x^3)(x^3) = x^{3+3} = x^6.$$

Rule 2: If m and n are positive integers with m larger than n, and $x \neq 0$, then $\dfrac{x^m}{x^n} = x^{m-n}.$

For example,

$$\frac{x^5}{x} = \frac{x^5}{x^1} = x^{5-1} = x^4,$$

and

$$\frac{x^6}{x^3} = x^{6-3} = x^3.$$

Rule 3:   If m and n are positive integers, then $(x^m)^n = x^{mn}.$

For example,

$$(x^3)^2 = x^{(3)(2)} = x^6.$$

Observe that $(x^3)^2$ is the same as $(x^3)(x^3)$ and both are equal to x^6. These rules will be discussed in more detail and expanded in Unit 12.

We now turn to some special forms of products and factoring. In Unit 1, we factored trinomials such as

$$x^2 + 5x + 6 = (x + 2)(x + 3).$$

Occasionally, it can happen that the two factors of a trinomial are identical. For example,

$$x^2 + 6x + 9 = (x + 3)(x + 3)$$

$$= (x + 3)^2.$$

In this case we say the trinomial $x^2 + 6x + 9$ is a **perfect square**, and it can be written as the **square of a binomial** $(x + 3)^2$.

In general, consider the square of the binomial $(x + k)^2$:

$$(x + k)^2 = (x + k)(x + k)$$

$$= x^2 + kx + kx + k^2$$

$$= x^2 + 2kx + k^2.$$

If the coefficient of the x^2-term of a trinomial is 1, we look at the coefficient of the x-term and at the constant term. We are looking for a number k such that the coefficient of the x-term is $2k$, and the constant term is k^2. If we can find such a number k, then the trinomial is a perfect square, and it can be written as the square of a binomial $(x + k)^2$.

EXAMPLE 8.1.   Show that the trinomial $x^2 + 10x + 25$ is a perfect square, and write the trinomial as the square of a binomial.

Solution.   The coefficient of the x^2-term is 1, so we look at the x-term. The coefficient of the x-term is 10, so

$$2k = 10$$
$$k = 5.$$

Therefore,

$$k^2 = 25.$$

Thus k^2 is the constant term. Since there is a number $k = 5$ such that $2k$ is the coefficient of the x-term and k^2 is the constant term, the trinomial is a perfect square. Moreover, we can write the trinomial as the square of the binomial $(x + k)^2$, and so

$$x^2 + 10x + 25 = (x + 5)^2.$$

To check, we multiply $(x + 5)^2 = (x + 5)(x + 5) = x^2 + 10x + 25$. You should always check factoring by multiplying the factors to see that you get back the original expression.

EXAMPLE 8.2.   Show that the trinomial $x^2 - 3x + \frac{9}{4}$ is a perfect square, and write the trinomial as the square of a binomial.

Solution.   Since the coefficient of the x^2-term is 1, we write

$$2k = -3$$
$$k = -\frac{3}{2}$$

and

$$k^2 = \frac{9}{4}.$$

Since $2k$ is the coefficient of the x-term and k^2 is the constant term, the trinomial is a perfect square. The square of the binomial is $(x + k)^2$, thus

$$x^2 - 3x + \frac{9}{4} = \left(x - \frac{3}{2}\right)^2.$$

You should check this example by multiplying the factored form to see that it is the same as the original expression.

EXAMPLE 8.3.   Show that the trinomial $x^2 + 8x + 8$ is not a perfect square.

Solution.   Since the coefficient of the x^2-term is 1, we write

$$2k = 8$$
$$k = 4$$

and

$$k^2 = 16.$$

But the constant term of the trinomial is not 16. Thus the trinomial $x^2 + 8x + 8$ is not a perfect square. Observe, however, that the trinomial $x^2 + 8x + 16$ is a perfect square.

The method we have used in the preceding examples is called **completing the square**. The method of completing the square is used in the proof of the quadratic formula, Unit 4, and the vertex formula, Unit 7.

In the case where the coefficient of the x^2-term is 1, we may also start by looking at the constant term. Assuming the constant term is positive, we may write it as k^2. Then we find k, and the coefficient of the x-term should be $2k$. In this case, the trinomial is a perfect square and can be written as the square of the binomial $(x + k)^2$.

EXAMPLE 8.4.   Show that the trinomial $x^2 - 16x + 64$ is a perfect square, and write the trinomial as the square of a binomial.

Solution.   Since the coefficient of the x^2-term is 1, we may look at the constant term 64, and write

$$k^2 = 64$$

$$k = 8 \text{ or } k = -8$$

and

$$2k = 16 \text{ or } 2k = -16.$$

The coefficient of the x-term is -16, therefore the trinomial is a perfect square, and $k = -8$. Thus

$$x^2 - 16x + 64 = (x - 8)^2.$$

More generally, suppose the coefficient of the x^2-term is not 1 or, indeed, we have a trinomial containing variables other than x. Consider the square of the binomial $(P + Q)^2$, where P and Q are monomials, and may contain variables or constants or both. Then

$$(P + Q)^2 = (P + Q)(P + Q)$$

$$= P^2 + PQ + PQ + Q^2$$

$$= P^2 + 2PQ + Q^2.$$

We assume both the P^2- and Q^2- terms are positive, and find P and Q. We need to consider only the positive values of P and Q. Then the middle term of the trinomial should be $2PQ$ or $-2PQ$. In either of these cases, the trinomial is a perfect square and can be written as the square of the binomial $(P + Q)^2$ or $(P - Q)^2$.

EXAMPLE 8.5.   Show that the trinomial $4x^2 + 20x + 25$ is a perfect square, and write the trinomial as the square of a binomial.

Solution.   The P^2-term is $4x^2$ and the Q^2-term is 25. Therefore,

$$P = 2x \text{ and } Q = 5$$

$$2PQ = 2(2x)(5) = 20x.$$

Since the middle term of the trinomial is $20x$, the trinomial is a perfect square and can be written as the square of the binomial $(P + Q)^2$:

$$4x^2 + 20x + 25 = (2x + 5)^2.$$

These examples should also be checked by multiplication.

EXAMPLE 8.6. Show that the trinomial $9x^2 - 15xy + \frac{25}{4}y^2$ is a perfect square, and write the trinomial as the square of a binomial.

Solution. The P^2-term is $9x^2$ and the Q^2-term is $\frac{25}{4}y^2$, so

$$P = 3x \text{ and } Q = \frac{5}{2}y$$

$$2PQ = 2(3x)\left(\frac{5}{2}y\right) = 15xy.$$

Since the middle term of the trinomial is $-15xy$, the trinomial is a perfect square and can be written as the square of the binomial $(P - Q)^2$:

$$9x^2 - 15xy + \frac{25}{4}y^2 = \left(3x - \frac{5}{2}y\right)^2.$$

EXAMPLE 8.7. Show that the trinomial $-4x^2 + 12x - 9$ is a perfect square, and write the trinomial as the square of a binomial.

Solution. In this example, the P^2-term and the Q^2-term both are negative. We can make them both positive by factoring -1 from the trinomial:

$$-4x^2 + 12x - 9 = -1(4x^2 - 12x + 9)$$

or

$$-4x^2 + 12x - 9 = -(4x^2 - 12x + 9).$$

Now the P^2-term is $4x^2$ and the Q^2-term is 9, so

$$P = 2x \text{ and } Q = 3$$

$$2PQ = 2(2x)(3) = 12x$$

and

$$-2PQ = -12x.$$

Therefore,

$$-(4x^2 - 12x + 9) = -(2x - 3)^2.$$

Observe that the P^2-term and Q^2-term must have the same sign in order for a trinomial to be a perfect square. For example, the trinomial $4x^2 + 12x - 9$ cannot be a perfect square.

Exercise 8.1

Identify the trinomials which are perfect squares, and write those which are as the square of a binomial:

1. $x^2 + 4x + 4$
2. $x^2 - 12x + 36$
3. $x^2 - 2x + 2$
4. $x^2 + 6x + 12$

5. $x^2 - 5x + \frac{25}{4}$
6. $x^2 + 9x + \frac{81}{2}$
7. $x^2 - \frac{3}{2}x + \frac{9}{16}$
8. $x^2 + 2\sqrt{3}\,x + 3$

9. $4x^2 + 6x + 9$
10. $25x^2 - 30x + 9$
11. $16x^2 - 12x + \frac{9}{4}$
12. $2x^2 + \sqrt{2}\,x + \frac{1}{4}$

13. $x^2 + 16xy + 64y^2$ 14. $4x^2 - 6x + \frac{9}{4}y^2$ 15. $4x^2 - 3xy + \frac{9}{16}y^2$

16. $\frac{1}{4}x^2y^2 - 3xyz + 9z^2$ 17. $-4x^2 + 20x - 25$ 18. $-x^2 + 7x - \frac{49}{4}$

19. $-9x^2 - 24xy + 16y^2$ 20. $-25x^2 - 40xy - 16y^2$

| Section 8.2 | Sums or Differences of Odd Powers |

In this section we consider expressions of the form $P^n + Q^n$ or $P^n - Q^n$, where P and Q are monomials, and n is an odd number greater than 1. In particular, if $n = 3$, we have **the sum of two cubes** $P^3 + Q^3$ or **the difference of two cubes** $P^3 - Q^3$. Each of these expressions can be factored:

$$P^3 + Q^3 = (P + Q)(P^2 - PQ + Q^2)$$

$$P^3 - Q^3 = (P - Q)(P^2 + PQ + Q^2).$$

Observe that the middle term of the second factor is PQ or $-PQ$, not $2PQ$ or $-2PQ$. Also, the positive middle term PQ goes with the difference $P^3 - Q^3$, while the negative middle term $-PQ$ goes with the sum $P^3 + Q^3$. The formulas can be proved by multiplying the factors on the right-hand side. For the first formula,

$$(P + Q)(P^2 - PQ + Q^2) = (P + Q)(P^2) + (P + Q)(-PQ) + (P + Q)(Q^2)$$

$$= P(P^2) + Q(P^2) + P(-PQ) + Q(-PQ) + P(Q^2) + Q(Q^2)$$

$$= P^3 + P^2Q - P^2Q - PQ^2 + PQ^2 + Q^3$$

$$= P^3 + Q^3.$$

It should be clear from this proof why the negative middle term $-PQ$ must go with the sum $P^3 + Q^3$. The proof of the second formula is left for you to do.

EXAMPLE 8.8.   Factor $x^3 + 27$.

Solution.   In this expression, the P^3-term is x^3 and the Q^3-term is 27. Therefore,

$$P = x \text{ and } Q = 3.$$

Using the formula

$$P^3 + Q^3 = (P + Q)(P^2 - PQ + Q^2),$$

the factored expression is

$$x^3 + 27 = (x + 3)(x^2 - 3x + 9).$$

Observe that the trinomial $x^2 - 3x + 9$ cannot be factored, either by the methods of section 1 of this unit, or by the methods of Unit 1.

EXAMPLE 8.9.   Factor $8x^3 - 125y^3$.

Solution.   The P^3-term is $8x^3$ and the Q^3-term is $125y^3$, so

$$P = 2x \text{ and } Q = 5y.$$

Using the formula

$$P^3 - Q^3 = (P - Q)(P^2 + PQ + Q^2),$$

the factored expression is

$$8x^3 - 125y^3 = (2x - 5y)(4x^2 + 10xy + 25y^2).$$

You should check these examples by multiplying the two factors to see that you get the original expression.

These expressions also can be factored by using long division of polynomials. To use division, we must know one of the factors. We then divide the expression by the known factor to obtain the other factor.

EXAMPLE 8.10.   Use division to factor $x^3 - 125$.

Solution.   Since $P = x$ and $Q = 5$, we know that one factor is $x - 5$. We must divide $x^3 - 125$ by the factor $x - 5$. To start the division, we divide x^3 by x, multiply back, and subtract:

$$
\begin{array}{r}
x^2 \\
x - 5 \overline{)\, x^3 - 125} \\
\underline{x^3 - 5x^2 } \\
5x^2 - 125
\end{array}
$$

Now, we divide $5x^2$ by x, multiply back, and subtract:

$$
\begin{array}{r}
x^2 + 5x \\
x - 5 \overline{)\, x^3 - 125} \\
\underline{x^3 - 5x^2 } \\
5x^2 - 125 \\
\underline{5x^2 - 25x } \\
25x - 125
\end{array}
$$

Finally, we divide $25x$ by x, multiply back, and subtract:

$$
\begin{array}{r}
x^2 + 5x + 25 \\
x - 5 \overline{)\, x^3 - 125} \\
\underline{x^3 - 5x^2 } \\
5x^2 - 125 \\
\underline{5x^2 - 25x } \\
25x - 125 \\
\underline{25x - 125} \\
0
\end{array}
$$

Since the remainder is zero, $x - 5$ is indeed a factor of $x^3 - 125$, and $x^2 + 5x + 25$ is the other factor. Therefore,

$$x^3 - 125 = (x - 5)(x^2 + 5x + 25).$$

Of course, you would get the same answer using the formula for the difference of two cubes, and using the formula is faster. Division is useful if you are not sure of the second factor in the formula.

More important, the division method can be used to factor any expression of the form $P^n + Q^n$ or $P^n - Q^n$, where P and Q are monomials, and n is an odd number greater than 1. We only need to know that, if n is odd, $P^n + Q^n$ always has as one factor $P + Q$. Similarly, if n is odd, $P^n - Q^n$ always has as one factor $P - Q$.

EXAMPLE 8.11.   Factor $x^5 + 32y^5$.

Solution.   Since $P = x$ and $Q = 2y$, we know that one factor is $x + 2y$. Therefore, we divide $x^5 + 32y^5$ by $x + 2y$:

$$
\begin{array}{r}
x^4 - 2x^3y + 4x^2y^2 - 8xy^3 + 16y^4 \\ \hline
x + 2y \overline{\smash{\big)}\ x^5 \qquad\qquad\qquad\qquad\quad + 32y^5} \\
\underline{x^5 + 2x^4y} \qquad\qquad\qquad\qquad\quad \\
-2x^4y \qquad\qquad\qquad + 32y^5 \\
\underline{-2x^4y - 4x^3y^2} \qquad\qquad\quad \\
4x^3y^2 \qquad\qquad + 32y^5 \\
\underline{4x^3y^2 + 8x^2y^3} \qquad\quad \\
-8x^2y^3 \qquad + 32y^5 \\
\underline{-8x^2y^3 - 16xy^4} \quad \\
16xy^4 + 32y^5 \\
\underline{16xy^4 + 32y^5} \\
0
\end{array}
$$

The factored expression is

$$x^5 + 32y^5 = (x + 2y)(x^4 - 2x^3y + 4x^2y^2 - 8xy^3 + 16y^4).$$

Exercise 8.2

Factor:

1. $x^3 + 1$

2. $x^3 - 1$

3. $x^3 - 27$

4. $x^3 + 8$

5. $8x^3 + 1$

6. $8x^3 - 27$

7. $64x^3 + 125y^3$

8. $1000x^3 - 27y^3$

9. $\frac{1}{8}x^3 - y^3$

10. $x^3 + \frac{1}{27}y^3$

Use division to factor:

11. $x^5 + 1$

12. $x^5 - 1$

13. $x^5 - 32y^5$

14. $32x^5 + y^5$

15. $x^7 + 1$

16. $x^7 - 128$

<table>
<tr><td>Section
8.3</td><td># Differences of Even Powers</td></tr>
</table>

We saw in Unit 1 that expressions of the form $P^2 - Q^2$ can be considered as trinomials where the coefficient of the middle term is zero.

EXAMPLE 8.12. Factor $x^2 - 4$.

Solution. We may think of the expression as $x^2 + 0x - 4$. Since $-2x + 2x = 0x$, the factored expression is

$$x^2 - 4 = (x + 2)(x - 2).$$

In general,

$$P^2 - Q^2 = (P + Q)(P - Q)$$

since

$$(P + Q)(P - Q) = P^2 - PQ + PQ - Q^2$$

$$= P^2 - Q^2.$$

An expression of the form $P^2 - Q^2$ is called **the difference of two squares.**

EXAMPLE 8.13. Factor $25x^2 - 16y^2$.

Solution. Since the P^2-term is $25x^2$ and the Q^2-term is $16y^2$,

$$P = 5x \text{ and } Q = 4y.$$

Therefore, the factored expression is

$$25x^2 - 16y^2 = (5x + 4y)(5x - 4y).$$

You should check this example by multiplying the two factors to see that the product is the original expression.

It is important to remember that **the sum of two squares** $P^2 + Q^2$ cannot be factored. A common error is to write $P^2 + Q^2$ as $(P + Q)(P + Q)$, or $(P + Q)^2$. However, we know from Section 8.1 that $(P + Q)^2 = (P + Q)(P + Q) = P^2 + 2PQ + Q^2$, which is not the same as $P^2 + Q^2$.

The results for $n = 2$ can be extended to expressions of the form $P^n - Q^n$, where P and Q are monomials, and n is an even number greater than 2.

EXAMPLE 8.14. Factor $x^4 - 16y^4$.

Solution. We think of $P^2 - Q^2$, where the P^2-term is x^4 and the Q^2-term is $16y^4$. Then,

$$P = x^2 \text{ and } Q = 4y^2.$$

We have the factored expression

$$x^4 - 16y^4 = (x^2 + 4y^2)(x^2 - 4y^2).$$

The first factor is a sum of two squares and so cannot be factored further. The second factor is a difference of two squares:

$$x^2 - 4y^2 = (x + 2y)(x - 2y).$$

Therefore, we complete the factorization of the original expression by writing

$$x^4 - 16y^4 = (x^2 + 4y^2)(x^2 - 4y^2)$$
$$= (x^2 + 4y^2)(x + 2y)(x - 2y).$$

The expression is now completely factored. You can check by multiplication that the final product is the same as the original expression.

EXAMPLE 8.15.   Factor $x^6 - y^6$.

Solution.   If we think of the P^2-term as x^6 and the Q^2-term as y^6, then

$$P = x^3 \text{ and } Q = y^3.$$

Thus the initial factorization is

$$x^6 - y^6 = (x^3 + y^3)(x^3 - y^3).$$

Each of these factors can be factored again. From Section 8.2,

$$x^3 + y^3 = (x + y)(x^2 - xy + y^2)$$

and

$$x^3 - y^3 = (x - y)(x^2 + xy + y^2).$$

Therefore,

$$x^6 - y^6 = (x^3 + y^3)(x^3 - y^3)$$
$$= (x + y)(x^2 - xy + y^2)(x - y)(x^2 + xy + y^2),$$

and the expression is completely factored. The product can be checked by multiplication.

Exercise 8.3

Identify those expressions which can be factored, and factor completely those which can be factored:

1. $x^2 - 1$ 2. $x^2 + 1$ 3. $x^2 + 25$ 4. $x^2 - 36$

5. $4x^2 - 9y^2$ 6. $49x^2 - 16y^2$ 7. $64x^2 + y^2$ 8. $x^2 - 144y^2$

9. $\frac{1}{100}x^2 - 1$ 10. $\frac{1}{4}x^2 - \frac{1}{9}y^2$ 11. $x^4 + y^4$ 12. $x^4 - y^4$

13. $x^4 - 81y^4$ 14. $16x^4 - 1$ 15. $x^6 - 64$ 16. $x^8 - y^8$

<table>
<tr><td>Section
8.4</td><td><h1>Factoring by Grouping</h1></td></tr>
</table>

Another special method of factoring is called **factoring by grouping**. Factoring by grouping is simply a double use of the distributive property. It can be used to factor certain expressions containing four terms.

EXAMPLE 8.16.   Factor $ax - 2ay + bx - 2by$.

Solution.   We group the four terms into groups of two terms:

$$ax - 2ay + bx - 2by = (ax - 2ay) + (bx - 2by).$$

The terms are grouped so that we can factor out a separate common factor from each group:

$$(ax - 2ay) + (bx - 2by) = a(x - 2y) + b(x - 2y).$$

The binomial $x - 2y$ is now a common factor in each group. Factoring this common factor from each group, we have

$$a(x - 2y) + b(x - 2y) = (x - 2y)(a + b).$$

Therefore,

$$ax - 2ay + bx - 2by = (ax - 2ay) + (bx - 2by)$$
$$= a(x - 2y) + b(x - 2y)$$
$$= (x - 2y)(a + b).$$

The expression is now completely factored, and you can check by multiplying the product on the right-hand side.

EXAMPLE 8.17.   Factor $x^3 + x^2 + x + 1$.

Solution.   We group

$$x^3 + x^2 + x + 1 = (x^3 + x^2) + (x + 1).$$

The first group has a common factor x^2. The second group has no common factor, so we write 1 as its coefficient:

$$(x^3 + x^2) + (x + 1) = x^2(x + 1) + 1(x + 1).$$

Then, factoring the common factor $x + 1$ from each group,

$$x^2(x + 1) + 1(x + 1) = (x + 1)(x^2 + 1).$$

Therefore,

$$x^3 + x^2 + x + 1 = (x^3 + x^2) + (x + 1)$$
$$= x^2(x + 1) + 1(x + 1)$$
$$= (x + 1)(x^2 + 1).$$

Remember that $x^2 + 1$ is a sum of two squares and cannot be factored. Therefore, the expression is completely factored.

In some cases, we must factor a negative common factor from one group in order to have the same binomial factor in each group.

EXAMPLE 8.18.   Factor $x^2 - xy - 2x + 2y$.

Solution.   We group

$$x^2 - xy - 2x + 2y = (x^2 - xy) + (-2x + 2y).$$

Observe that we must factor -2 from the second group to obtain the common binomial factor $x - y$ in each group:

$$(x^2 - xy) + (-2x + 2y) = x(x - y) + (-2)(x - y).$$

Therefore,

$$
\begin{aligned}
x^2 - xy - 2x + 2y &= (x^2 - xy) + (-2x + 2y) \\
&= x(x - y) + (-2)(x - y) \\
&= x(x - y) - 2(x - y) \\
&= (x - y)(x - 2).
\end{aligned}
$$

Sometimes it is necessary to rearrange the terms of an expression in order to have the same binomial factor in each group.

EXAMPLE 8.19.   Factor $xy - 4x + 2y - 2x^2$.

Solution.   If we group

$$xy - 4x + 2y - 2x^2 = (xy - 4x) + (2y - 2x^2),$$

we get

$$(xy - 4x) + (2y - 2x^2) = x(y - 4) + 2(y - x^2).$$

The binomial factor in each group is not the same, so we cannot continue. However, if we rearrange the terms to

$$xy - 2x^2 + 2y - 4x,$$

we can group and factor the expression:

$$
\begin{aligned}
xy - 2x^2 + 2y - 4x &= (xy - 2x^2) + (2y - 4x) \\
&= x(y - 2x) + 2(y - 2x) \\
&= (y - 2x)(x + 2).
\end{aligned}
$$

There are usually several rearrangements of the terms of the expression which will give an expression that can be factored. You should always get the same factors, regardless of the arrangement you use. However, the factors may appear in the opposite order. We could have used arrangements in this example which would give $(x + 2)(y - 2x)$.

Exercise 8.4

Factor completely:

1. $ax - ay + bx - by$

2. $ax - bx - ay + by$

3. $ax + 2y - 2x - ay$

4. $2ax + 2by - bx - 4ay$

5. $x^2 - 3x - xy + 3y$

6. $xy - 2x - 2y + 4$

7. $3xy + 4x - 9y - 12$

8. $2x - 4xy - 3y + 6y^2$

9. $xy - 9x + 3y - 3x^2$

10. $4xy + x - 2y - 2x^2$

11. $x^3 - x^2 + x - 1$

12. $2x^3 + 2x^2 + x + 1$

13. $x^3 - x^2y + x - y$

14. $x - xy^2 + y - x^2y$

15. $x^3 - x^2 - x + 1$

16. $x^4 - x^2 - x^2y^2 + y^2$

Self-test

Factor:

1. $2x^2 + 2y - 4x - xy$

2. $x^2 + 16y^2$

3. $64x^2 - 9y^2$

4. $x^3 - 8y^3$

5. $4x^2 - 18xy + \frac{81}{4}y^2$

1. ______________

2. ______________

3. ______________

4. ______________

5. ______________

Rational Expressions

INTRODUCTION

You have used factoring in solving equations, solving applied problems, and drawing graphs. In this unit you will use factoring to simplify rational expressions, where there are numerators and denominators involving variables. First, you will learn how to reduce rational expressions. Then you can apply the techniques for reducing rational expressions to multiplication and division of rational expressions. Finally, you will learn to draw the graph of a basic form of function where one of the variables is in a denominator.

OBJECTIVES

When you have finished this unit you should be able to:

1. Reduce a rational expression to simplest form.
2. Multiply and divide rational expressions, and reduce to lowest terms.
3. Draw the graph of a function of the form $y = \dfrac{a}{x}$.

Section 9.1

Reducing Rational Expressions

A **positive rational number** is a quotient, or ratio, of two whole numbers where the denominator is not zero. A **rational expression** is a quotient, or ratio, of two polynomials where the denominator is not zero. Thus, if A and B are polynomials, with $B \neq O$, then $\dfrac{A}{B}$ is a rational expression.

In Section 1.3 we reduced expressions by dividing out a constant factor. The expressions in that section are rational expressions. The following example reviews the method we used.

EXAMPLE 9.1.   Reduce $\dfrac{3 + 12x}{3}$.

Solution.   We use the distributive property to factor out the common factor 3 in the numerator:

$$\frac{3 + 12x}{3} = \frac{3(1 + 4x)}{3}.$$

Then, since $\frac{3}{3} = 1$, we have

$$\frac{3 + 12x}{3} = \frac{3(1 + 4x)}{3}$$

$$= \frac{3}{3}(1 + 4x)$$

$$= 1(1 + 4x)$$

$$= 1 + 4x.$$

Observe that $\dfrac{3 + 12x}{3}$ is a rational expression where A is the polynomial $3 + 12x$ and B is the polynomial 3. (Recall that a polynomial can consist of a constant alone.)

It is important to remember that you can divide out only *factors* from a rational expression. For example, it is true that

$$\frac{15}{3} = \frac{3(5)}{3} = \frac{3}{3}(5) = 1(5) = 5.$$

However, clearly we cannot write $\dfrac{16}{3} = \dfrac{3 + 13}{3}$, and then write $\dfrac{3 + 13}{3}$ as $\dfrac{3}{3} + 13 = 1 + 13 = 14$. In the second case 3 is a *addend*. You must never divide addends from a rational expression.

EXAMPLE 9.2.   Reduce $\dfrac{x^2 - 1}{x + 1}$.

Solution.   A common error is to try to divide the x^2-term by the x-term. However, we cannot do this because these terms are addends and not factors. Therefore, we factor the numerator:

$$\frac{x^2 - 1}{x + 1} = \frac{(x + 1)(x - 1)}{x + 1}.$$

Now the binomial $x + 1$ is a factor of both the numerator and the denominator, and may be divided out:

$$\frac{x^2 - 1}{x + 1} = \frac{(x + 1)(x - 1)}{x + 1}$$

$$= \left(\frac{x + 1}{x + 1}\right)(x - 1)$$

$$= 1(x - 1)$$

$$= x - 1.$$

A rational expression is reduced to **lowest terms** when all common factors have been divided out of the numerator and denominator. The expression $x - 1$ has numerator $x - 1$ and denominator understood to be the polynomial 1, and so is a rational expression reduced to lowest terms.

EXAMPLE 9.3.   Reduce $\dfrac{2x(x - 2)}{4(x - 2)}$.

Solution.   This rational expression has a constant common factor 2, and a binomial common factor $x - 2$. Therefore, we write

$$\frac{2x(x - 2)}{4(x - 2)} = \frac{2x(x - 2)}{2(2)(x - 2)}$$

$$= \frac{2}{2}\left(\frac{x}{2}\right)\left(\frac{x - 2}{x - 2}\right)$$

$$= 1\left(\frac{x}{2}\right)(1)$$

$$= \frac{x}{2}.$$

Since the polynomial x and the polynomial 2 have no common factors, the rational expression is reduced to lowest terms.

Usually the polynomials in a rational expression are not given in factored form. In general, we must first factor the polynomials, then identify any common factors, and finally divide out the common factors to reduce the rational expression.

EXAMPLE 9.4.   Reduce $\dfrac{x^2 + 3x}{3x + 9}$.

Solution.   First we factor each polynomial:

$$\frac{x^2 + 3x}{3x + 9} = \frac{x(x + 3)}{3(x + 3)}.$$

We now have a common factor $x + 3$, so we reduce the rational expression by dividing out this common factor:

$$\frac{x^2 + 3x}{3x + 9} = \frac{x(x + 3)}{3(x + 3)}$$

$$= \frac{x}{3}\left(\frac{x + 3}{x + 3}\right)$$

$$= \frac{x}{3}(1)$$

$$= \frac{x}{3}.$$

When you are certain you are dividing out factors, and not addends, you may skip from the first step to the last. The missing steps can be indicated by crossing out the common factors to show they have been divided out. Thus, we may write

$$\frac{x^2 + 3x}{3x + 9} = \frac{x(\cancel{x + 3})}{3(\cancel{x + 3})}$$

$$= \frac{x}{3}.$$

EXAMPLE 9.5.   Reduce $\dfrac{x^2 - 9x + 20}{x^2 - 5x - xy + 5y}$.

Solution.   Observe that the polynomial in the denominator must be factored by grouping:

$$x^2 - 5x - xy + 5y = (x^2 - 5x) + (-xy + 5y)$$

$$= x(x - 5) - y(x - 5)$$

$$= (x - 5)(x - y).$$

Therefore,

$$\frac{x^2 - 9x + 20}{x^2 - 5x - xy + 5y} = \frac{(x - 4)(x - 5)}{(x - 5)(x - y)}$$

$$= \frac{x - 4}{x - y}.$$

The x-terms are addends and not factors, so the expression cannot be reduced further.

EXAMPLE 9.6.   Reduce $\dfrac{x^2 - 4}{x^3 - 8}$.

Solution.   In this example the denominator is the difference of two cubes:

$$x^3 - 8 = (x - 2)(x^2 + 2x + 4).$$

Therefore,

$$\frac{x^2 - 4}{x^3 - 8} = \frac{(x + 2)(x - 2)}{(x - 2)(x^2 + 2x + 4)}$$

$$= \frac{x + 2}{x^2 + 2x + 4}.$$

This expression cannot be reduced further since the polynomial in the denominator cannot be factored further. All the remaining terms are addends and cannot be divided out.

There are two important points to keep in mind in factoring the polynomials of a rational expression. First, we must factor out all common factors, constant and variable, from each polynomial. Second, we must factor each polynomial completely.

EXAMPLE 9.7.   Reduce $\dfrac{2x^3 + 4x^2 + 2x}{x^5 - x}$.

Solution.   We factor out the common factors 2 and x from the polynomial in the numerator, and factor the remaining polynomial completely:

$$2x^3 + 4x^2 + 2x = 2x(x^2 + 2x + 1)$$

$$= 2x(x + 1)(x + 1).$$

Then we factor out the common factor x from the polynomial in the denominator, and factor the remaining polynomial, being sure to factor completely:

$$x^5 - x = x(x^4 - 1)$$

$$= x(x^2 + 1)(x^2 - 1)$$

$$= x(x^2 + 1)(x + 1)(x - 1).$$

Then the expression reduces to

$$\frac{2x^3 + 4x^2 + 2x}{x^5 - x} = \frac{2x(x + 1)(x + 1)}{x(x^2 + 1)(x + 1)(x - 1)}$$

$$= \frac{2(x + 1)}{(x^2 + 1)(x - 1)}.$$

We will generally leave reduced expressions in factored form.

In order to reduce certain rational expressions, it is useful to know that a binomial of the form $Q - P$, where P and Q are monomials, is equal to $(-1)(P - Q)$. To verify this equality, we write

$$(-1)(P - Q) = (-1)(P) + (-1)(-Q)$$

$$= -P + Q$$

$$= Q - P.$$

Thus we may replace a binomial of the form $Q - P$ by $(-1)(P - Q)$.

EXAMPLE 9.8.   Reduce $\dfrac{x^2 - 5x + 6}{8 - x^3}$.

Solution.   If we simply factor the two polynomials, we have

$$\frac{x^2 - 5x + 6}{8 - x^3} = \frac{(x - 2)(x - 3)}{(2 - x)(4 + 2x + x^2)}.$$

However, $x - 2$ and $2 - x$ are not equal. Therefore, they are not common factors and cannot be divided out. To reduce the expression, we replace $8 - x^3$ by $(-1)(x^3 - 8)$. Now we have

$$\frac{x^2 - 5x + 6}{8 - x^3} = \frac{x^2 - 5x + 6}{(-1)(x^3 - 8)}$$

$$= \frac{(x - 2)(x - 3)}{(-1)(x - 2)(x^2 + 2x + 4)}$$

$$= \frac{x - 3}{(-1)(x^2 + 2x + 4)}.$$

We can express the factor -1 in either of two ways, writing

$$-\frac{x - 3}{x^2 + 2x + 4}$$

or

$$\frac{3 - x}{x^2 + 2x + 4}.$$

Exercise 9.1

Reduce:

1. $\dfrac{ax + ay}{2x + 2y}$ 2. $\dfrac{x^2 + 2x}{2x + 4}$ 3. $\dfrac{3x + 9}{x^2 + 6x + 9}$

4. $\dfrac{4x - 4}{x^2 - 4x + 3}$ 5. $\dfrac{x^2 + 4x}{x^2 - 16}$ 6. $\dfrac{2x^2 - 4x}{x^2 - 4}$

7. $\dfrac{x^2 - y^2}{x^2 + 2xy + y^2}$ 8. $\dfrac{x^2 + y^2}{x^4 - y^4}$ 9. $\dfrac{2x^2 - 2y^2}{2x^2 - 4xy + 2y^2}$

10. $\dfrac{4x^4 - 4y^4}{2x^2 - 2y^2}$ 11. $\dfrac{x^2 - 4x + 4}{4 - x^2}$ 12. $\dfrac{3 - 3x^2}{3x - 3}$

13. $\dfrac{x^2 - 3x + 2}{x^2 - 5x + 4}$ 14. $\dfrac{x^2 - x - 6}{x^2 + 5x + 6}$ 15. $\dfrac{2x^2 - 3x - 2}{2x^2 - x - 6}$

16. $\dfrac{2x^2 - 4x + 2}{2x^2 - 3x + 1}$ 17. $\dfrac{x^2 - 2x + xy - 2y}{2x - 4}$ 18. $\dfrac{x^2 - 2x + 2xy - 4y}{2x^2 - x + 4xy - 2y}$

19. $\dfrac{x^2 - 9}{x^3 - 27}$ 20. $\dfrac{64 - x^3}{x^2 - 8x + 16}$ 21. $\dfrac{1 - x^3}{x^2 + x + 1}$

22. $\dfrac{2x^2 + 10x + 50}{2x^3 - 250}$ 23. $\dfrac{3x^2 - 3xy - 6x + 6y}{3x^2 - 15x + 18}$ 24. $\dfrac{ax^2 - axy + 2ax - 2ay}{ax^2 - axy - 2ax + 2ay}$

25. $\dfrac{abx - aby + ax - ay}{abx - aby - ax + ay}$ 26. $\dfrac{a^3b^2 - a^3}{a^3b^3 + a^3}$ 27. $\dfrac{x^4 - a^4}{x^3 - ax^2 + a^2x - a^3}$

28. $\dfrac{x^3 - y^3}{x^4 - y^4}$ 29. $\dfrac{x^4 - x^3 + xy^3 - y^3}{x^3 - x^2 - xy^2 + y^2}$ 30. $\dfrac{x^5 + x^4 - x - 1}{x^3 + x^2 - x - 1}$

Section 9.2   Multiplication and Division

To multiply two rational numbers, we simply multiply the numerators and multiply the denominators. For example,

$$\frac{1 \cdot 1}{2 \cdot 3} \qquad \frac{1}{2} \cdot \frac{1}{3} = \frac{1}{6}.$$

A product should be reduced when possible. For example,

$$\frac{2}{9} \cdot \frac{3}{4} = \frac{2 \cdot 3}{9 \cdot 4} = \frac{6}{36} = \frac{1}{6}.$$

When a product can be reduced, we can divide out common factors before performing the multiplications. This approach simplifies both the multiplication and reducing. The preceding example can be done in this way:

$$\frac{2}{9} \cdot \frac{3}{4} = \frac{1 \cdot \cancel{2}}{\cancel{3} \cdot 3} \cdot \frac{1 \cdot \cancel{3}}{\cancel{2} \cdot 2} = \frac{1 \cdot 1}{3 \cdot 2} = \frac{1}{6}.$$

Multiplication of rational expressions is exactly the same as multiplication of rational numbers. We multiply the two polynomials in the numerators and the two polynomials in the denominators.

EXAMPLE 9.9.   Multiply $\dfrac{x}{x + 3} \cdot \dfrac{x + 2}{x - 3}$.

Solution.   We write the products of the two numerators and the two denominators:

$$\frac{x}{x + 3} \cdot \frac{x + 2}{x - 3} = \frac{x(x + 2)}{(x + 3)(x - 3)}.$$

The resulting expression has no common factors and cannot be reduced. We will generally leave the expression in factored form.

Now, suppose we have a multiplication of rational expressions where some or all of the polynomials in the expressions can be factored. We will always factor each polynomial which can be factored, and reduce by dividing out common factors in the numerators and denominators.

EXAMPLE 9.10.   Multiply $\dfrac{2x + 4}{x^2 + 2x + 1} \cdot \dfrac{x^2 + x}{2x}$.

Solution.   First, we factor each polynomial which can be factored:

$$\frac{2x + 4}{x^2 + 2x + 1} \cdot \frac{x^2 + x}{2x} = \frac{2(x + 2)}{(x + 1)(x + 1)} \cdot \frac{x(x + 1)}{2x}.$$

We have common factors of 2, x, and $x + 1$. We divide out these factors to reduce the resulting expression:

$$\frac{2x + 4}{x^2 + 2x + 1} \cdot \frac{x^2 + x}{2x} = \frac{\cancel{2}(x + 2)}{(\cancel{x + 1})(x + 1)} \cdot \frac{\cancel{x}(\cancel{x + 1})}{2\cancel{x}}$$

$$= \frac{x + 2}{x + 1}.$$

EXAMPLE 9.11.   Multiply $\dfrac{x^2 - 2x - 3}{x^2 - 4} \cdot \dfrac{x^2 + 4x + 4}{x^2 + x - 12}$.

Solution.   A common error is to divide out parts which look alike, such as the x^2-terms or the constant terms. However, these parts are addends and not factors. Before we can divide out any common factors, we must factor each polynomial:

$$\frac{x^2 - 2x - 3}{x^2 - 4} \cdot \frac{x^2 + 4x + 4}{x^2 + x - 12} = \frac{(x + 1)(x - 3)}{(x + 2)(x - 2)} \cdot \frac{(x + 2)(x + 2)}{(x + 4)(x - 3)}.$$

Now we can divide out the common factors, and write the product in factored form:

$$\frac{x^2 - 2x - 3}{x^2 - 4} \cdot \frac{x^2 + 4x + 4}{x^2 + x - 12} = \frac{(x + 1)(x - 3)}{(x + 2)(x - 2)} \cdot \frac{(x + 2)(x + 2)}{(x + 4)(x - 3)}$$

$$= \frac{(x + 1)(x + 2)}{(x - 2)(x + 4)}.$$

The resulting expression cannot be reduced further.

EXAMPLE 9.12.   Multiply $\dfrac{x + y}{x^2 - xy} \cdot \dfrac{x^3 - y^3}{x^2 - 2x + xy - 2y}.$

Solution.   The numerator of the second expression is a difference of cubes, and the denominator must be factored by grouping. Also, we must be careful not to overlook the common factor x in the denominator of the first expression. Factoring each polynomial, and dividing out common factors from the numerators and denominators, the product reduces to

$$\frac{x + y}{x^2 - xy} \cdot \frac{x^3 - y^3}{x^2 - 2x + xy - 2y} = \frac{x + y}{x(x - y)} \cdot \frac{(x - y)(x^2 + xy + y^2)}{(x - 2)(x + y)}$$

$$= \frac{x^2 + xy + y^2}{x(x - 2)}.$$

Since the numerator cannot be factored further, there are no more common factors and the expression cannot be reduced further.

The method of replacing a binomial of the form $Q - P$ by $(-1)(P - Q)$, shown in Section 9.1, is useful in multiplying rational expressions.

EXAMPLE 9.13.   Multiply $\dfrac{y}{y^2 - x^2} \cdot \dfrac{x^2 - 2xy + y^2}{x}.$

Solution.   Replacing $y^2 - x^2$ by $(-1)(x^2 - y^2)$, we can factor and divide out common factors:

$$\frac{y}{y^2 - x^2} \cdot \frac{x^2 - 2xy + y^2}{x} = \frac{y}{(-1)(x^2 - y^2)} \cdot \frac{x^2 - 2xy + y^2}{x}$$

$$= \frac{y}{(-1)(x + y)(x - y)} \cdot \frac{(x - y)(x - y)}{x}$$

$$= \frac{y(x - y)}{(-1)(x + y)x}.$$

This expression may be written in the form

$$- \frac{y(x - y)}{x(x + y)}$$

or

$$\frac{y(y - x)}{x(x + y)}.$$

Also, since $x + y$ is the same as $y + x$, we may write

$$\frac{y(y - x)}{x(y + x)}.$$

Any of these three forms is correct.

To divide two rational numbers, we invert the second number and multiply. For example,

$$\frac{1}{2} \div \frac{3}{4} = \frac{1}{2} \cdot \frac{4}{3} = \frac{1}{2} \cdot \frac{2 \cdot 2}{3} = \frac{2}{3}.$$

To see why "invert and multiply" works, we write the division as a fraction where the numerator and denominator are each rational numbers:

$$\frac{1}{2} \div \frac{3}{4} = \frac{\dfrac{1}{2}}{\dfrac{3}{4}}.$$

Now we multiply the numerator and denominator of the fraction by the reciprocal of the denominator. Then we have a denominator of 1. Since a denominator of 1 leaves only the numerator, we have

$$\frac{1}{2} \div \frac{3}{4} = \frac{\dfrac{1}{2}}{\dfrac{3}{4}} = \frac{\dfrac{1}{2} \cdot \dfrac{4}{3}}{\dfrac{3}{4} \cdot \dfrac{4}{3}} = \frac{\dfrac{1}{2} \cdot \dfrac{4}{3}}{1} = \frac{1}{2} \cdot \frac{4}{3}.$$

Thus division by a rational number is the same as multiplication by its reciprocal, which is also called the multiplicative inverse.

The same method is used to divide rational expressions. We multiply by the multiplicative inverse of the divisor, or second rational expression.

EXAMPLE 9.14.   Divide $\dfrac{3x - 6}{x^2 + x} \div \dfrac{x^2 - 4}{x^2 + 2x + 1}$.

Solution.   First, we write the division as a multiplication by the multiplicative inverse of the second rational expression:

$$\frac{3x - 6}{x^2 + x} \div \frac{x^2 - 4}{x^2 + 2x + 1} = \frac{3x - 6}{x^2 + x} \cdot \frac{x^2 + 2x + 1}{x^2 - 4}.$$

To multiply, we factor the polynomials and divide out common factors as in the preceding examples:

$$\frac{3x - 6}{x^2 + x} \cdot \frac{x^2 + 2x + 1}{x^2 - 4} = \frac{3(x - 2)}{x(x + 1)} \cdot \frac{(x + 1)(x + 1)}{(x + 2)(x - 2)}$$

$$= \frac{3(x + 1)}{x(x + 2)}.$$

EXAMPLE 9.15.   Divide $\dfrac{x^3 - 5x^2 + 6x}{8 - x^3} \div \dfrac{x^2}{x^3 + 2x^2 + 4x}$.

Solution.   First, we write the division as a multiplication by the multiplicative inverse of the second rational expression:

$$\frac{x^3 - 5x^2 + 6x}{8 - x^3} \div \frac{x^2}{x^3 + 2x^2 + 4x} = \frac{x^3 - 5x^2 + 6x}{8 - x^3} \cdot \frac{x^3 + 2x^2 + 4x}{x^2}.$$

The numerator of the first expression has a common factor x, and then can be factored further:

$$x^3 - 5x^2 + 6x = x(x^2 - 5x + 6)$$

$$= x(x - 2)(x - 3).$$

The numerator of the second expression has a common factor, but cannot be factored further:

$$x^3 + 2x^2 + 4x = x(x^2 + 2x + 4).$$

The denominator of the first expression must be replaced by $(-1)(P - Q)$, and then factored:

$$8 - x^3 = (-1)(x^3 - 8)$$

$$= (-1)(x - 2)(x^2 + 2x + 4).$$

Thus the polynomials can be factored, and common factors divided out to give:

$$\frac{x^3 - 5x^2 + 6x}{8 - x^3} \div \frac{x^2}{x^3 + 2x^2 + 4x} = \frac{x^3 - 5x^2 + 6x}{8 - x^3} \cdot \frac{x^3 + 2x^2 + 4x}{x^2}$$

$$= \frac{x(x - 2)(x - 3)}{(-1)(x - 2)\ (x^2 + 2x + 4)} \cdot \frac{x(x^2 + 2x + 4)}{x \cdot x}$$

$$= \frac{x - 3}{-1}$$

$$= 3 - x.$$

EXAMPLE 9.16.   Divide $\dfrac{x}{x + y} \div \dfrac{x + y}{y}$.

Solution.   In this kind of example, there is a temptation to divide out the common factor $x + y$. However, the example is in the form of a division. When it is rewritten as a multiplication by the multiplicative inverse of the second rational expression, both factors $x + y$ will be in denominators and so cannot be divided out. The solution is

$$\frac{x}{x + y} \div \frac{x + y}{y} = \frac{x}{x + y} \cdot \frac{y}{x + y}$$

$$= \frac{xy}{(x + y)(x + y)}.$$

This expression may also be written in the form

$$\frac{xy}{(x + y)^2}.$$

The expression cannot be reduced.

Exercise 9.2

Multiply or divide as indicated:

1. $\dfrac{x}{3} \cdot \dfrac{x+3}{x+2}$

2. $\dfrac{x-4}{4} \cdot \dfrac{2}{x-2}$

3. $\dfrac{2x}{x+2} \cdot \dfrac{x+2}{4}$

4. $\dfrac{3x-6}{3} \cdot \dfrac{3x}{x-2}$

5. $\dfrac{x^2-3x}{3x} \cdot \dfrac{3}{x-3}$

6. $\dfrac{5xy+10x}{2xy+4x} \cdot \dfrac{2y+4}{5y+5}$

7. $\dfrac{x+3}{6x} \div \dfrac{x+3}{2x}$

8. $\dfrac{x}{3x-3} \div \dfrac{1}{x-1}$

9. $\dfrac{x}{x-2} \div \dfrac{x-2}{2}$

10. $\dfrac{2x}{xy+y} \div \dfrac{2x+2}{y}$

11. $\dfrac{x^2+x}{1-x^2} \cdot \dfrac{x-1}{1+x}$

12. $\dfrac{y}{y-x} \div \dfrac{x+y}{x^2-y^2}$

13. $\dfrac{x^2-x}{x^2+3x+2} \cdot \dfrac{4x+4}{x^2-3x+2}$

14. $\dfrac{2-6x}{3x^2-5x-2} \cdot \dfrac{2x^2-3x-2}{4-12x}$

15. $\dfrac{3x-6}{x^2+5x+6} \div \dfrac{x^2-5x+6}{x^2+2x}$

16. $\dfrac{x-2x^2}{x-x^2} \div \dfrac{2x^2+x-1}{x^2+x-2}$

17. $\dfrac{2x^2+3x-9}{3x^2+4x-4} \cdot \dfrac{3x^2-8x+4}{2x^2-9x+9}$

18. $\dfrac{9x^2-3x-2}{3x^2+7x-6} \div \dfrac{9x^2-1}{3x^2+8x-3}$

19. $\dfrac{x^2+x+xy+y}{xy+y} \div \dfrac{x^2+x-xy-y}{x^2+x}$

20. $\dfrac{x^2-4x-2xy+8y}{x^2-x-12} \cdot \dfrac{x^2-x-6}{x^2+2x-2xy-4y}$

21. $\dfrac{x^3-y^3}{x^2-y^2} \cdot \dfrac{2x+2y}{x^2+xy+y^2}$

22. $\dfrac{x^2-9}{3x+9} \div \dfrac{x^2-3x}{x^3+27}$

23. $\dfrac{4-x^2}{2x+4} \cdot \dfrac{4x+4}{x^2-x-2}$

24. $\dfrac{3x-x^2-3y+xy}{y^2-x^2} \div \dfrac{6-2x}{y^3+x^3}$

25. $\dfrac{x^2+ax-ay-xy}{x^3-a^3} \div \dfrac{x^2-y^2}{x^2-ax-ay+xy}$

26. $\dfrac{ax^2-ab^2}{ax+ab+bx+b^2} \cdot \dfrac{a^2-b^2}{ax-ab-bx+b^2}$

27. $\dfrac{x^2-1}{x^2-x-6} \cdot \dfrac{x^2-6x+9}{x^2-4x+4} \div \dfrac{x^2-3x-4}{x^2-4}$

28. $\dfrac{x^2-x}{2x+2y} \div \dfrac{x^2-2x+1}{x^2+2x+xy+2y} \cdot \dfrac{2x-2}{xy+2y}$

29. $\dfrac{x^2-a^2}{x^2+a^2} \div \dfrac{ax-a^2}{x^4-a^4} \cdot \dfrac{x-a}{x^2+2ax+a^2}$

30. $\dfrac{x^2-x-2}{x^4-y^4} \div \dfrac{x^2-1}{x^2+2x+xy+2y} \div \dfrac{x^2-4}{x^2-x-xy+y}$

Section
9.3

Rational Functions

A rational function is a function which involves one or more rational expressions. The most basic type of rational function is a function of the form

$$y = \frac{a}{x},$$

where $a \neq 0$. Observe that $x = 0$ gives an undefined y-value for this function. Many rational functions have one or more x-values which must be excluded because they lead to division by zero.

EXAMPLE 9.17.   Draw the graph of $y = \dfrac{2}{x}$.

Solution.   We use the $f(x)$ symbol, with

$$f(x) = \frac{2}{x}.$$

Observing that x cannot be zero, we start with positive values of x:

$$f(1) = \frac{2}{1} = 2, \text{ gives } (1, 2).$$

$$f(2) = \frac{2}{2} = 1, \text{ gives } (2, 1).$$

$$f(4) = \frac{2}{4} = \frac{1}{2}, \text{ gives } \left(4, \frac{1}{2}\right).$$

As x gets larger, the y-values are smaller and smaller positive fractions. To get y-values which are larger than 2, we must use positive x-values which are fractions less than 1. For example,

$$f\left(\frac{1}{2}\right) = \frac{2}{\frac{1}{2}} = \frac{2}{1} \cdot \frac{2}{1} = \frac{4}{1} = 4, \text{ gives } \left(\frac{1}{2}, 4\right).$$

Now, we repeat the process for negative values of x:

$$f(-1) = \frac{2}{-1} = -2, \text{ gives } (-1, -2).$$

$$f(-2) = \frac{2}{-2} = -1, \text{ gives } (-2, -1).$$

$$f(-4) = \frac{2}{-4} = -\frac{1}{2}, \text{ gives } \left(-4, -\frac{1}{2}\right).$$

Also,

$$f\left(-\frac{1}{2}\right) = \frac{2}{-\frac{1}{2}} = \frac{2}{1}\left(-\frac{2}{1}\right) = -\frac{4}{1} = -4, \text{ gives } \left(-\frac{1}{2}, -4\right).$$

Plotting all the points we have calculated, and connecting them by a smooth curve, we have the graph

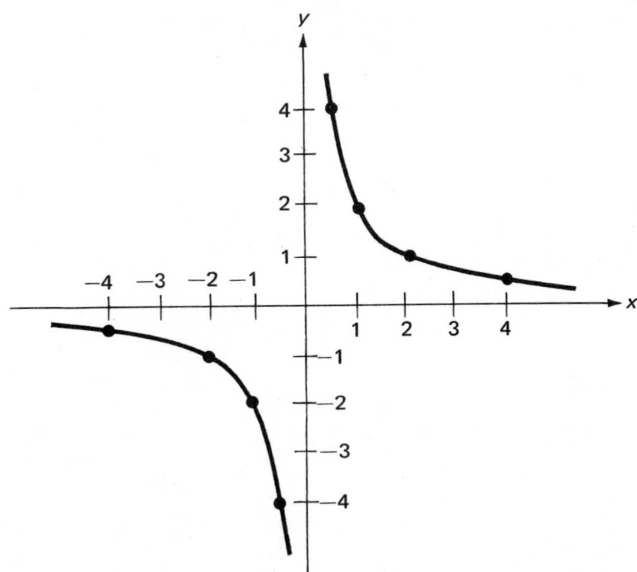

We have already observed that $y = \dfrac{a}{x}$ has no y-value if $x = 0$. Recall the definition of the domain of a function from Section 7.4. The domain of a function is the set of all x-values for which a y-value is defined. The domain of the function $y = \dfrac{a}{x}$ is all real numbers except zero because there is no y-value defined for $x = 0$. We may write this domain as $x \neq 0$.

Also, observe that the function $y = \dfrac{a}{x}$ never takes on the y-value $y = 0$. Recall that the range of a function is the set of all y-values of the function. The range of the function $y = \dfrac{a}{x}$ is all real numbers except zero, or $y \neq 0$.

Finally, we observe that the graph of $y = \dfrac{a}{x}$ gets closer and closer to the x-axis but never crosses it. Also, the graph gets closer and closer to the y-axis but never crosses it. We say the graph approaches the x- and y-axes **asymptotically**, and that the x-axis and y-axis are **asymptotes** of the graph. The graph has two branches which approach the asymptotes from opposite sides. When a is negative, the sides from which the graph approaches the asymptotes reverse.

Exercise 9.3

Draw the graph:

1. $y = \dfrac{1}{x}$ 2. $y = \dfrac{3}{x}$ 3. $y = \dfrac{-2}{x}$ 4. $y = \dfrac{-4}{x}$ 5. $y = \dfrac{-6}{x}$ 6. $y = \dfrac{-8}{x}$

7. $y = \dfrac{12}{x}$ 8. $y = \dfrac{-12}{x}$

Self-test

Reduce:

1. $\dfrac{x^2 + 8x + 15}{x^2 + 10x + 25}$

2. $\dfrac{2x - 4}{4 - x^2}$

Multiply or divide as indicated:

3. $\dfrac{3x + 3}{x^3 - 27} \div \dfrac{x + 1}{x^2 - 9}$

4. $\dfrac{x^2 - 4x + 4}{x + 2} \cdot \dfrac{x - y}{x^2 - 2x - xy + 2y}$

5. Draw the graph of $y = \dfrac{-10}{x}$.

1. _______________

2. _______________

3. _______________

4. _______________

Unit 10 Common Denominators

INTRODUCTION

Besides multiplying and dividing rational expressions, you must also be able to add and subtract rational expressions. To do this, you must know how to find the least common denominator. In this unit, you will learn how to find a least common multiple of two or more polynomials. You will then learn how to find the least common denominator, and to add or subtract rational expressions. Also, you will learn how to use the least common denominator to simplify combinations of rational expressions called complex fractions.

OBJECTIVES

When you have finished this unit you should be able to:

1. Find the least common multiple of two or more polynomials.
2. Add and subtract rational expressions, and reduce to lowest terms.
3. Simplify complex fractions.

Section
10.1

The Least Common Multiple

The **least common multiple** of two or more whole numbers is the smallest whole number divisible by each of the numbers. For example, the least common multiple of 2 and 3 is 6, because 6 is divisible by 2 and by 3, and there is no smaller whole number divisible by 2 and by 3. We will write LCM for least common multiple. The LCM of 2 and 3 is 6.

If the given numbers have common factors, then the LCM is a smaller number than the product of the given numbers. For example, the LCM of 6 and 8 is 24, not the product 48. It is true that 48 is divisible by both 6 and 8, but 48 is not the smallest whole number divisible by both 6 and 8. The smallest whole number divisible by both 6 and 8 is 24.

There is an easy way to find the LCM of two or more whole numbers. We factor each number completely:

$$6 = 2 \cdot 3$$

and

$$8 = 2 \cdot 2 \cdot 2.$$

Now, we take each factor the greatest number of times it appears in any of the numbers. Two appears once in 6 and three times in 8. The greatest number of times is three times. Therefore, the LCM has three factors of 2. Three appears once in 6 and no times in 8. The greatest number of times is once. Therefore, the LCM has one factor of 3. Thus the LCM of 6 and 8 is

$$2 \cdot 2 \cdot 2 \cdot 3 = 24.$$

It is possible for the LCM of two or more whole numbers to be one of the given numbers. For example, the LCM of 2 and 4 is 4. The smallest whole number divisible by 4 is clearly 4, and 4 is also divisible by 2.

We can apply the principles above to find the LCM of two or more polynomials.

EXAMPLE 10.1. Find the LCM of x and $x + 1$.

Solution.   Since neither x nor $x + 1$ can be factored, they cannot have any common factors. Therefore, the LCM of x and $x + 1$ is their product,

$$x(x + 1).$$

For most applications, it is easiest to leave the LCM of polynomials in factored form.

EXAMPLE 10.2. Find the LCM of $x^2 + 3x + 2$ and $x^2 - x - 2$.

Solution.   First we factor each of the polynomials:

$$x^2 + 3x + 2 = (x+1)(x+2)$$
$$x^2 - x - 2 = (x+1)(x-2).$$

Now, each factor, $x + 1$, $x + 2$, and $x - 2$, appears in any of the polynomials at most one time . Therefore, the LCM is

$$(x + 1)(x + 2)(x - 2).$$

EXAMPLE 10.3. Find the LCM of $x^2 + 2x + 1$ and $x + 1$.

Solution.   Factoring $x^2 + 2x + 1$, we have

$$x^2 + 2x + 1 = (x + 1)(x + 1),$$

which may also be written as

$$(x + 1)^2.$$

Therefore, the LCM has two factors of $x + 1$. Since there are no other factors in either polynomial, the LCM is

$$(x + 1)^2.$$

If we have more than two polynomials, we follow the same procedure.

EXAMPLE 10.4. Find the LCM of $x^2 + 6x + 9$, $x^2 - 9$, and $x^2 - 4x + 3$.

Solution.   Factoring each of the polynomials gives

$$x^2 + 6x + 9 = (x + 3)^2$$
$$x^2 - 9 = (x + 3)(x - 3)$$
$$x^2 - 4x + 3 = (x - 1)(x - 3).$$

The greatest number of times $x + 3$ appears in any of the polynomials is twice. Each of the factors $x - 3$ and $x - 1$ appears in any polynomial at most once. Thus the LCM is

$$(x + 3)^2(x - 3)(x - 1).$$

EXAMPLE 10.5.   Find the LCM of $x^2 + xy$, $2x + 2y$, and $x^2 - y^2$.

Solution.   Factoring each of the polynomials gives

$$x^2 + xy = x(x + y)$$

$$2x + 2y = 2(x + y)$$

$$x^2 - y^2 = (x + y)(x - y).$$

Each of the factors, x, 2, $x + y$, and $x - y$, appears in any polynomial at most once, so the LCM is

$$2x(x + y)(x - y).$$

It is usual to write constant factors first, then monomial factors, binomial factors, and so on.

For some examples, we need to use special kinds of factoring.

EXAMPLE 10.6.   Find the LCM of $x^3 - 8$ and $xy - 4x - 2y + 8$.

Solution.   Since $x^3 - 8$ is a difference of two cubes,

$$x^3 - 8 = (x - 2)(x^2 + 2x + 4).$$

We factor $xy - 4x - 2y + 8$ by grouping:

$$xy - 4x - 2y + 8 = x(y - 4) - 2(y - 4)$$

$$= (y - 4)(x - 2).$$

The LCM is

$$(x - 2)(y - 4)(x^2 + 2x + 4).$$

We may also use the fact that $Q - P = (-1)(P - Q)$ to find the LCM in some examples.

EXAMPLE 10.7.   Find the LCM of $x - 3$ and $3 - x$.

Solution.   It is a common error to write the LCM as the product $(x - 3)(3 - x)$. However, since

$$3 - x = (-1)(x - 3),$$

the LCM is

$$(-1)(x - 3).$$

Observe that the LCM is a first-degree polynomial, whereas the product of the two polynomials is second degree.

Exercise 10.1

Find the LCM:

1. $x^2 + x - 12$ and $x^2 - 5x + 6$

2. $2x^2 - 3x + 1$ and $2x^2 + 3x - 2$

3. $x^2 + 4x + 4$ and $x + 2$

4. $x^2 - 6x + 9$ and $2x - 6$

5. $x^2 - 10x + 25$, $x^2 - 25$, and $x^2 - 5x$

6. $4x + 16$, $x^2 + 8x + 16$, and $x^2 + 4x$

7. $x^2 - 4$, $x^2 + x - 6$, and $x^2 + 5x + 6$

8. $x^2 + 6x + 9$, $x^2 + 2x - 3$, $x^2 + 4x + 3$, and $x^2 - 1$

9. $x^2 - y^2$, $3x - 3y$, and $x^2 - 2xy + y^2$

10. $x^3 + y^3$, $x^2 + 2xy + y^2$, $x^2 + xy$, and $2x + 2y$

11. $x^3 - 27$ and $xy + 3x - 3y - 9$

12. $x^2 - y^2$ and $x^2 - x + xy - y$

13. $x - 2$ and $2 - x$

14. $6x - 4$ and $2 - 3x$

15. $x - 4$ and $16 - x^2$

16. $x^3 - y^3$, $y^2 - x^2$, and $2y - 2x$

Section 10.2   Addition and Subtraction

Recall from arithmetic that, in order to add or subtract fractions, we must use a **common denominator**. In multiplying fractions, and similarly in multiplying rational expressions, we simply multiply the numerators and multiply the denominators. It is a common error in arithmetic and in algebra to try to use a similar process to add fractions or to add rational expressions.

Suppose we want to add $\frac{1}{2} + \frac{1}{4}$. Were we simply to add the numerators and the denominators, we would have $\frac{2}{6}$ or $\frac{1}{3}$. This answer clearly is not correct since the correct sum must be more than $\frac{1}{2}$. Using a common denominator, we write

$$\frac{1}{2} + \frac{1}{4} = \frac{2}{4} + \frac{1}{4} = \frac{3}{4}$$

which is the correct sum. The sum of fractions is the sum of the numerators written over a common denominator.

We add rational expressions in a similar way.

EXAMPLE 10.8.   Add $\dfrac{x - 1}{x - 3} + \dfrac{x - 2}{x - 3}$.

Solution.   There is already a common denominator, $x - 3$, so we write the sum of the numerators over the common denominator:

$$\frac{x - 1}{x - 3} + \frac{x - 2}{x - 3} = \frac{(x - 1) + (x - 2)}{x - 3}$$

$$= \frac{x - 1 + x - 2}{x - 3}$$

$$= \frac{2x - 3}{x - 3}.$$

This rational expression cannot be reduced. Remember that you cannot divide out addends such as the x-terms or the constant terms.

To subtract fractions or rational expressions, we follow the same procedure. The difference of the numerators is written over a common denominator. However, when we are subtracting rational expressions, we must apply the rules for signs very carefully to the polynomials in the numerator.

EXAMPLE 10.9.   Subtract $\dfrac{x-1}{x-3} - \dfrac{x-2}{x-3}$.

Solution.   We write the difference of the numerators over the common denominator, being careful to use parentheses to indicate the separate numerators:

$$\frac{x-1}{x-3} - \frac{x-2}{x-3} = \frac{(x-1)-(x-2)}{x-3}.$$

Now, we remove the parentheses, being careful to apply the minus sign to both terms of the second polynomial:

$$\frac{(x-1)-(x-2)}{x-3} = \frac{x-1-x+2}{x-3}$$

$$= \frac{1}{x-3}.$$

Observe that the parentheses remind you to apply the minus sign to all terms of the second polynomial in the numerator. You should always use the parentheses to avoid errors in signs often made in subtracting rational expressions.

EXAMPLE 10.10.   Subtract $\dfrac{x^2+2x+1}{x+2} - \dfrac{1-2x-x^2}{x+2}$.

Solution.   We write the difference of the numerators over the common denominator, using parentheses, and then remove the parentheses:

$$\frac{x^2+2x+1}{x+2} - \frac{1-2x-x^2}{x+2} = \frac{(x^2+2x+1)-(1-2x-x^2)}{x+2}$$

$$= \frac{x^2+2x+1-1+2x+x^2}{x+2}$$

$$= \frac{2x^2+4x}{x+2}.$$

This expression can be reduced by factoring the polynomial in the numerator:

$$\frac{2x^2+4x}{x+2} = \frac{2x(x+2)}{x+2}$$

$$= 2x.$$

We factor and reduce the resulting rational expression whenever possible.

To add or subtract rational expressions where the denominators are not already the same, we find the **least common denominator**, LCD. The LCD of two or more rational expressions is simply the LCM of the denominators of the expressions.

EXAMPLE 10.11.   Add $\dfrac{x-1}{x} + \dfrac{x+2}{x+1}$.

Solution.   The LCD of the expressions is clearly $x(x+1)$. We write each expression with this denominator. To do this, we multiply the numerator and the denominator of the first expression by $x+1$:

$$\frac{x-1}{x} = \frac{(x-1)(x+1)}{x(x+1)},$$

and the numerator and the denominator of the second expression by x:

$$\frac{x+2}{x+1} = \frac{x(x+2)}{x(x+1)}.$$

Thus we have

$$\frac{x-1}{x} + \frac{x+2}{x+1} = \frac{(x-1)(x+1)}{x(x+1)} + \frac{x(x+2)}{x(x+1)}.$$

Now, writing the sum of the numerators over the common denominator,

$$\frac{(x-1)(x+1)}{x(x+1)} + \frac{x(x+2)}{x(x+1)} = \frac{(x-1)(x+1) + x(x+2)}{x(x+1)}.$$

Finally, we may multiply, remove parentheses, and simplify the numerator:

$$\frac{(x-1)(x+1) + x(x+2)}{x(x+1)} = \frac{(x^2-1) + (x^2+2x)}{x(x+1)}$$

$$= \frac{x^2 - 1 + x^2 + 2x}{x(x+1)}$$

$$= \frac{2x^2 + 2x - 1}{x(x+1)}.$$

The simplified numerator cannot be factored, and so the resulting rational expression cannot be reduced. Again, remember that you cannot divide out addends, so you cannot reduce the expression unless you can factor the numerator. As with the LCM of polynomials, the denominator may be left in factored form.

EXAMPLE 10.12.   Add $\dfrac{1}{x^2 + 5x + 6} + \dfrac{-2}{x^2 + 6x + 8}$.

Solution.   First we find the LCD:

$$x^2 + 5x + 6 = (x+2)(x+3)$$

$$x^2 + 6x + 8 = (x+2)(x+4),$$

therefore, the LCD is $(x+2)(x+3)(x+4)$. Now, we write each rational expression with the LCD:

$$\frac{1}{x^2 + 5x + 6} = \frac{1}{(x+2)(x+3)} = \frac{1(x+4)}{(x+2)(x+3)(x+4)},$$

and

$$\frac{-2}{x^2 + 6x + 8} = \frac{-2}{(x + 2)(x + 4)} = \frac{-2(x + 3)}{(x + 2)(x + 3)(x + 4)}.$$

Writing the sum of the resulting numerators over the common denominator, we have

$$\frac{1(x + 4)}{(x + 2)(x + 3)(x + 4)} + \frac{-2(x + 3)}{(x + 2)(x + 3)(x + 4)} = \frac{1(x + 4) + (-2)(x + 3)}{(x + 2)(x + 3)(x + 4)}.$$

Multiplying, removing parentheses, and simplifying the numerator,

$$\frac{1(x + 4) + (-2)(x + 3)}{(x + 2)(x + 3)(x + 4)} = \frac{(x + 4) + (-2x - 6)}{(x + 2)(x + 3)(x + 4)}$$

$$= \frac{x + 4 - 2x - 6}{(x + 2)(x + 3)(x + 4)}$$

$$= \frac{-x - 2}{(x + 2)(x + 3)(x + 4)}.$$

In this example, we do have a rational expression which can be reduced. We factor the numerator, and divide out the common factor:

$$\frac{-x - 2}{(x + 2)(x + 3)(x + 4)} = \frac{(-1)(x + 2)}{(x + 2)(x + 3)(x + 4)}$$

$$= \frac{-1}{(x + 3)(x + 4)}.$$

To subtract rational expressions where the denominators are not already the same, we follow essentially the same process. As before, we must remember to use parentheses carefully to be sure the signs of the second polynomial in the numerator are correct.

EXAMPLE 10.13.   Subtract $\dfrac{x + 1}{x^2 + 4x + 4} - \dfrac{x - 1}{x^2 - 4}.$

Solution.   Since

$$x^2 + 4x + 4 = (x + 2)^2$$

$$x^2 - 4 = (x + 2)(x - 2),$$

the LCD is $(x + 2)^2(x - 2)$. Then,

$$\frac{x + 1}{x^2 + 4x + 4} = \frac{x + 1}{(x + 2)^2} = \frac{(x + 1)(x - 2)}{(x + 2)^2(x - 2)}$$

and

$$\frac{x - 1}{x^2 - 4} = \frac{x - 1}{(x + 2)(x - 2)} = \frac{(x - 1)(x + 2)}{(x + 2)^2(x - 2)}.$$

Now we write the difference of the numerators over the common denominator:

$$\frac{(x + 1)(x - 2)}{(x + 2)^2(x - 2)} - \frac{(x - 1)(x + 2)}{(x + 2)^2(x - 2)} = \frac{(x + 1)(x - 2) - (x - 1)(x + 2)}{(x + 2)^2(x - 2)}.$$

We multiply and simplify the numerator, taking care to use parentheses to indicate the separate parts of the difference in the numerator:

$$\frac{(x + 1)(x - 2) - (x - 1)(x + 2)}{(x + 2)^2(x - 2)} = \frac{(x^2 - x - 2) - (x^2 + x - 2)}{(x + 2)^2(x - 2)}$$

$$= \frac{x^2 - x - 2 - x^2 - x + 2}{(x + 2)^2(x - 2)}$$

$$= \frac{-2x}{(x + 2)^2(x - 2)}.$$

This rational expression cannot be reduced.

Some examples may involve special types of factoring such as differences of cubes or factoring by grouping. We must always be sure to factor denominators completely, and to use the least common denominator. Using a larger common denominator may give a numerator larger than we know how to factor, so we will not know whether we can reduce the resulting rational expression.

EXAMPLE 10.14.   Subtract $\dfrac{x^2 + 4x + 6}{x^3 - 27} - \dfrac{1}{x - 3}$.

Solution.   Since

$$x^3 - 27 = (x - 3)(x^2 + 3x + 9)$$

the LCD is $(x - 3)(x^2 + 3x + 9)$. The first expression is already over the LCD. For the second expression,

$$\frac{1}{x - 3} = \frac{1(x^2 + 3x + 9)}{(x - 3)(x^2 + 3x + 9)}.$$

Then,

$$\frac{x^2 + 4x + 6}{x^3 - 27} - \frac{1}{x - 3} = \frac{x^2 + 4x + 6}{(x - 3)(x^2 + 3x + 9)} - \frac{1(x^2 + 3x + 9)}{(x - 3)(x^2 + 3x + 9)}$$

$$= \frac{(x^2 + 4x + 6) - 1(x^2 + 3x + 9)}{(x - 3)(x^2 + 3x + 9)}$$

$$= \frac{x^2 + 4x + 6 - x^2 - 3x - 9}{(x - 3)(x^2 + 3x + 9)}$$

$$= \frac{\cancel{x - 3}}{\cancel{(x - 3)}(x^2 + 3x + 9)}$$

$$= \frac{1}{x^2 + 3x + 9}.$$

In particular, we must watch for examples where we can use the replacement $Q - P = (-1)(P - Q)$.

EXAMPLE 10.15.   Add $\dfrac{y}{x - y} + \dfrac{x}{y - x}$.

Solution.   Using $Q - P = (-1)(P - Q)$, the LCD is $(-1)(x - y)$. Then we write the first expression as

$$\frac{y}{x - y} = \frac{(-1)y}{(-1)(x - y)}$$

and the second expression as

$$\frac{x}{y - x} = \frac{x}{(-1)(x - y)}.$$

Then we have the solution

$$\frac{y}{x - y} + \frac{x}{y - x} = \frac{(-1)y}{(-1)(x - y)} + \frac{x}{(-1)(x - y)}$$

$$= \frac{(-1)y + x}{(-1)(x - y)}$$

$$= \frac{-y + x}{(-1)(x - y)}$$

$$= \frac{x - y}{(-1)(x - y)}$$

$$= \frac{1}{-1}$$

$$= -1.$$

You must be careful not to divide out addends at any step of this solution. The expression can only be reduced by dividing out the common factor $x - y$, once it is derived.

Exercise 10.2

Add or subtract as indicated:

1. $\dfrac{x}{x + 3} + \dfrac{2x}{x + 3}$

2. $\dfrac{x - 1}{x + 4} + \dfrac{x - 3}{x + 4}$

3. $\dfrac{3x}{x + 1} - \dfrac{x}{x + 1}$

4. $\dfrac{x - 2}{x + 2} - \dfrac{x - 4}{x + 2}$

5. $\dfrac{2}{x^2} - \dfrac{3}{x}$

6. $\dfrac{1}{x} + \dfrac{4}{x^3}$

7. $\dfrac{x + 3}{x - 2} + \dfrac{x - 1}{x}$

8. $\dfrac{x - 2}{x} - \dfrac{x + 1}{x + 4}$

9. $\dfrac{x + 2}{x - 1} + \dfrac{x - 3}{x + 1}$

10. $\dfrac{x + 4}{x + 2} - \dfrac{x - 2}{x - 3}$

11. $\dfrac{-2}{x^2 - 2x - 3} + \dfrac{3}{x^2 - 9}$

12. $\dfrac{1}{x^2 - 5x + 6} + \dfrac{-2}{x^2 - 6x + 8}$

13. $\dfrac{2}{x^2 + 2x} - \dfrac{1}{x^2 + x}$

14. $\dfrac{8}{x^2 + 8x + 15} - \dfrac{4}{x^2 + 7x + 12}$

15. $\dfrac{3}{x - 3} - \dfrac{x}{3 - x}$

16. $\dfrac{y}{y-x} + \dfrac{x}{x-y}$

17. $\dfrac{4x}{16-x^2} + \dfrac{x}{x-4}$

18. $\dfrac{x}{x^3-1} + \dfrac{1}{1-x}$

19. $\dfrac{y^3}{y^3-x^3} + \dfrac{y}{x-y}$

20. $\dfrac{4}{x^3+8} - \dfrac{2}{4-x^2}$

21. $\dfrac{x}{x^2-1} + \dfrac{y}{x^2-x-xy+y}$

22. $\dfrac{3y}{xy+2x-3y-6} - \dfrac{2y}{xy+2x-2y-4}$

23. $\dfrac{1}{x-3} + \dfrac{1}{x^2-3x} - \dfrac{x-1}{x^2-6x+9}$

24. $\dfrac{1}{x} - \dfrac{2}{x^2+2x} - \dfrac{2}{x^2+4x+4}$

25. $\dfrac{1}{x^2-xy} + \dfrac{1}{xy+y^2} - \dfrac{2}{x^2-y^2}$

26. $\dfrac{2a}{x^2-a^2} + \dfrac{x}{ax+a^2} + \dfrac{x}{ax-a^2}$

<table><tr><td>Section
10.3</td><td></td></tr></table>

Complex Fractions

A **complex fraction** is a fraction which itself contains fractions. For example,

$$\frac{1 - \dfrac{2}{3}}{\dfrac{1}{9} + 2}$$

is a complex fraction.

One way to simplify a complex fraction is to combine the terms in the numerator, combine the terms in the denominator, and then divide the resulting fractions:

$$1 - \frac{2}{3} = \frac{3}{3} - \frac{2}{3} = \frac{1}{3}$$

$$\frac{1}{9} + 2 = \frac{1}{9} + \frac{18}{9} = \frac{19}{9}$$

and

$$\frac{1}{3} \div \frac{19}{9} = \frac{1}{3} \cdot \frac{9}{19} = \frac{3}{19}.$$

Thus the complex fraction reduces to $\frac{3}{19}$.

Although the method above can always be used, it is usually not the most efficient way of simplifying a complex fraction. In general, we will simplify complex fractions by multiplying the numerator and the denominator by the LCD of all the fractions in the complex fraction. For the example above, the LCD of $\frac{2}{3}$ and $\frac{1}{9}$ is 9, so we write

$$\frac{1 - \dfrac{2}{3}}{\dfrac{1}{9} + 2} = \frac{9\left(1 - \dfrac{2}{3}\right)}{9\left(\dfrac{1}{9} + 2\right)}$$

$$= \frac{9(1) - 9\left(\dfrac{2}{3}\right)}{9\left(\dfrac{1}{9}\right) + 9(2)}.$$

Observe that this process is the same as multiplying every term of the complex fraction by the LCD, so we will usually skip the first step. Thus

$$\frac{1 - \dfrac{2}{3}}{\dfrac{1}{9} + 2} = \frac{9(1) - 9\left(\dfrac{2}{3}\right)}{9\left(\dfrac{1}{9}\right) + 9(2)}$$

$$= \frac{9 - 6}{1 + 18}$$

$$= \frac{3}{19}.$$

Now we apply this method to complex fractions containing rational expressions.

EXAMPLE 10.16.   Simplify $\dfrac{1 - \dfrac{1}{x}}{1 - \dfrac{1}{x^2}}.$

Solution.   The LCD of all the rational expressions in the complex fraction is x^2. Multiplying each term of the complex fraction by x^2, we have

$$\frac{1 - \dfrac{1}{x}}{1 - \dfrac{1}{x^2}} = \frac{x^2(1) - x^2\left(\dfrac{1}{x}\right)}{x^2(1) - x^2\left(\dfrac{1}{x^2}\right)} = \frac{x^2 - x}{x^2 - 1}.$$

Of course, we always reduce the resulting rational expression if possible:

$$\frac{x^2 - x}{x^2 - 1} = \frac{x(x - 1)}{(x + 1)(x - 1)} = \frac{x}{x + 1}.$$

EXAMPLE 10.17.   Simplify $\dfrac{\dfrac{1}{x} + \dfrac{1}{y}}{\dfrac{1}{x^2} + \dfrac{1}{y^2}}.$

Solution.   The LCD of all the rational expressions is x^3y^2. Multiplying each rational expression by x^3y^2,

$$\frac{\dfrac{1}{x} + \dfrac{1}{y}}{\dfrac{1}{x^2} + \dfrac{1}{y^2}} = \frac{x^3y^2\left(\dfrac{1}{x}\right) + x^3y^2\left(\dfrac{1}{y}\right)}{x^3y^2\left(\dfrac{1}{x^2}\right) + x^3y^2\left(\dfrac{1}{y^2}\right)}.$$

$$= \frac{xy^2 + x^3y}{y^2 + x^2}$$

$$= \frac{xy(y + x)}{y^2 + x^2}.$$

Since $y^2 + x^2$ cannot be factored, the resulting rational expression cannot be reduced.

EXAMPLE 10.18.   Simplify $\dfrac{\dfrac{3}{x-3}+\dfrac{3}{x+3}}{\dfrac{x}{x^2-9}}$.

Solution.   The LCD is x^2-9, or $(x+3)(x-3)$. Multiplying each rational expression by $(x+3)(x-3)$, and reducing,

$$\frac{\dfrac{3}{x-3}+\dfrac{3}{x+3}}{\dfrac{x}{x^2-9}} = \frac{(x+3)(x-3)\left(\dfrac{3}{x-3}\right)+(x+3)(x-3)\left(\dfrac{3}{x+3}\right)}{(x+3)(x-3)\left(\dfrac{x}{x^2-9}\right)}$$

$$= \frac{(x+3)(3)+(x-3)(3)}{x}$$

$$= \frac{3x+9+3x-9}{x}$$

$$= \frac{6x}{x}$$

$$= 6.$$

Exercise 10.3

Simplify:

1. $\dfrac{5+\dfrac{1}{x}}{3+\dfrac{1}{x}}$ 2. $\dfrac{1-\dfrac{2}{x}}{1-\dfrac{4}{x}}$ 3. $\dfrac{2-\dfrac{2}{x}}{4-\dfrac{2}{x}}$ 4. $\dfrac{1+\dfrac{3}{x}}{3+\dfrac{9}{x}}$

5. $\dfrac{1-\dfrac{1}{x^2}}{\dfrac{1}{x}+\dfrac{1}{x^2}}$ 6. $\dfrac{1+\dfrac{1}{x}}{\dfrac{1}{x^2}-1}$ 7. $\dfrac{\dfrac{1}{xy}}{\dfrac{1}{x}+\dfrac{1}{y}}$ 8. $\dfrac{\dfrac{1}{x}+\dfrac{1}{y}}{\dfrac{1}{x}-\dfrac{1}{y}}$

9. $\dfrac{\dfrac{x}{y}-\dfrac{y}{x}}{\dfrac{1}{y}-\dfrac{1}{x}}$ 10. $\dfrac{\dfrac{1}{x}-\dfrac{1}{y}}{\dfrac{1}{x^2}-\dfrac{1}{y^2}}$ 11. $\dfrac{\dfrac{1}{x-2}-\dfrac{1}{x+2}}{\dfrac{4}{x^2-4}}$ 12. $\dfrac{\dfrac{25}{x^2-25}}{\dfrac{5}{x-5}-\dfrac{5}{x+5}}$

13. $\dfrac{\dfrac{1}{x-1}+\dfrac{1}{x+1}}{\dfrac{1}{x-1}-\dfrac{1}{x+1}}$ 14. $\dfrac{\dfrac{x+1}{x-1}-\dfrac{x-1}{x+1}}{\dfrac{x+1}{x-1}+\dfrac{x-1}{x+1}}$

Self-test

1. Find the LCM of $x^2 - 16$, $x^2 - 8x + 16$, and $4x + 16$.

1.___________________

Add or subtract as indicated:

2. $\dfrac{x - 4}{4 - x^2} + \dfrac{2}{x - 2}$

2.___________________

3. $\dfrac{6}{x + 5} - \dfrac{3}{x}$

3.___________________

4. $\dfrac{-2}{x^2 - 10x + 24} + \dfrac{1}{x^2 - 11x + 30}$

4.___________________

5. Simplify $\dfrac{1 - \dfrac{1}{x}}{\dfrac{1}{x^2} - \dfrac{1}{x^3}}$

5.___________________

Unit
11 Cumulative Review

INTRODUCTION

In this unit you should make certain you remember all the material in all of the preceding units.

OBJECTIVE

When you have finished this unit you should be able to demonstrate that you can fulfill every objective of each preceding unit.

To prepare for this unit you should review the Self-Tests for Units 1 through 10. Do each problem of each Self-Test over again. If you cannot do a problem, or even if you have the slightest difficulty, you should:

1. Find out from the answer section the Objective for the unit to which the problem relates.
2. Review all the material in the section which has the same number as the objective, and redo all the Exercises for the section.
3. Try the Self-Test for the unit again.

Repeat these steps until you can do each problem in each Self-Test for each of Units 1 through 10 easily and accurately.

Self-test

Solve for x:

1. $3x^2 + 4x = 4$

1. _______________

2. $2x^2 + 1 = 4x$

2. _______________

3. The width of a rectangle is 3 less than its length. Its area is $\frac{27}{4}$. Find the width and length.

3. _______________

4. Find the slope of the line through the points $(2, -3)$ and $(-4, -1)$.

4. _______________

5. Find the y-intercept, x-intercept, and draw the graph of $12x - 3y = 6$.

y-intercept_______________

x-intercept_______________

graph:

6. Find the y-intercept, x-intercepts, vertex, and draw the graph of $y = 12 - x - x^2$.

y-intercept________________________

x-intercepts________________________

vertex________________________

graph:

7. A ball is tossed into the air according to the formula $s = -16t^2 + 32t + 64$. How high does it rise?

7. ________________________

Perform the operation as indicated:

8. $\dfrac{27 - x^3}{3x + 9} \div \dfrac{x^2 - 9}{3}$

8. ________________________

9. $\dfrac{3}{2x - 8} - \dfrac{x - 2}{x^2 - 5x + 4}$

9. ________________________

10. Simplify $\dfrac{\dfrac{1}{y} - \dfrac{y}{x^2}}{\dfrac{1}{x} - \dfrac{x}{y^2}}$

10. ________________________

Unit 12

Integral Exponents

INTRODUCTION

You most likely have had some experience with exponents in previous courses. In fact, we have used exponents in several of the preceding units. Whenever we have spoken of an x^2-term we have used an exponent. In this unit you will learn the basic rules for working with exponents. You will use these rules to simplify expressions involving exponents which are positive, negative, or zero integers. In addition, you will learn to graph a basic type of function where one of the variables is in an exponent.

OBJECTIVES

When you have finished this unit you should be able to:

1. Use the rules for exponents to simplify expressions involving positive integral exponents.
2. Find the value of or simplify expressions containing zero exponents.
3. Find the value of or simplify expressions involving positive, zero, and negative integral exponents.
4. Draw the graph of a function of the form $y = a^x$.

Section 12.1 — Rules of Exponents

We have already used expressions such as x^2 in several units. Such an expression is called a **power** of x. The number 2 is called the **exponent**, and the letter x is called the **base**. In general, an expression of the form b^n is called the **nth power of b**. The exponent is n, and the base is b.

Of course you know from earlier courses and units that

$$x^2 = x \cdot x$$

or two factors of x. Similarly,

$$x^3 = x \cdot x \cdot x$$

or three factors of x,

$$x^4 = x \cdot x \cdot x \cdot x$$

or four factors of x, and so on.

> **Definition:** $\underbrace{b^n = b \cdot b \cdot b \cdot \ \ldots \ \cdot b}_{n \text{ factors}}$, or n factors of b.

This definition makes sense only for exponents n which are positive integers: 1, 2, 3, 4, and so on. In particular, observe that $b^1 = b$.

    We will now state five basic rules for exponents. We will not attempt to prove these rules, but we will illustrate each rule by an example. You are already familiar with Rule 1, Rule 2a, and Rule 3, which we reviewed in Unit 8.

Rules of Exponents: For m and n any positive integers:

1. $b^m \cdot b^n = b^{m+n}$.

2.   a. $\dfrac{b^m}{b^n} = b^{m-n}$ if $m > n$ (m larger than n), and $b \neq 0$.

     b. $\dfrac{b^m}{b^n} = \dfrac{1}{b^{n-m}}$ if $n > m$ (n larger than m), and $b \neq 0$.

     c. $\dfrac{b^m}{b^n} = 1$ if $m = n$, and $b \neq 0$.

3. $(b^m)^n = b^{mn}$.

4. $(ab)^n = a^n b^n$.

5. $\left(\dfrac{a}{b}\right)^n = \dfrac{a^n}{b^n}$, $b \neq 0$.

EXAMPLE 12.1.   Simplify $x^4 \cdot x^3$.

Solution.   Using Rule 1,

$$x^4 \cdot x^3 = x^{4+3}$$
$$= x^7.$$

Observe that $x^4 \cdot x^3 = (x \cdot x \cdot x \cdot x)(x \cdot x \cdot x)$, or seven factors of x. Although such an example is not a proof of the rule, the example should help you to see why the rule works. To multiply two powers with the same base, we add the exponents.

    It is important to remember that in using Rule 1, although we add the exponents, we are actually multiplying the powers. There are no rules for adding or subtracting powers. Expressions of the form $b^m + b^n$ or $b^m - b^n$ usually cannot be simplified.

EXAMPLE 12.2.   Simplify $x^3 + x^2$.

Solution.   Since x^3 and x^2 are not like terms, the expression cannot be simplified.

EXAMPLE 12.3.   Simplify:

    a. $\dfrac{x^5}{x^2}$ b. $\dfrac{x}{x^6}$ c. $\dfrac{x^3}{x^3}$

Solutions.   a.   Using Rule 2a,

$$\frac{x^5}{x^2} = x^{5-2}$$

$$= x^3.$$

Observe that $\dfrac{x^5}{x^2} = \dfrac{x \cdot x \cdot x \cdot x \cdot x}{x \cdot x} = x \cdot x \cdot x$, or three factors of x.

b.   Using Rule 2b,

$$\frac{x}{x^6} = \frac{1}{x^{6-1}}$$

$$= \frac{1}{x^5}.$$

Observe that $\dfrac{x}{x^6} = \dfrac{x}{x \cdot x \cdot x \cdot x \cdot x \cdot x} = \dfrac{1}{x \cdot x \cdot x \cdot x \cdot x}$, leaving five factors of x in the denominator. To divide powers with the same base but different exponents, we subtract the smaller exponent from the larger. The result is written in the numerator if the larger exponent is in the numerator, with the denominator understood to be 1. The result is written in the denominator if the larger exponent is in the denominator and the numerator is 1.

c.   Using Rule 2c,

$$\frac{x^3}{x^3} = 1.$$

Observe that $\dfrac{x^3}{x^3} = \dfrac{x \cdot x \cdot x}{x \cdot x \cdot x} = \dfrac{1}{1} = 1.$

EXAMPLE 12.4.   Simplify $(x^3)^4$.

Solution.   Using Rule 3,

$$(x^3)^4 = x^{3 \cdot 4}$$

$$= x^{12}.$$

Observe that, extending Rule 1 to a product of four powers, $(x^3)^4 = x^3 \cdot x^3 \cdot x^3 \cdot x^3 = x^{3+3+3+3}$ $= x^{12}$. Alternatively, $x^3 \cdot x^3 \cdot x^3 \cdot x^3 = (x \cdot x \cdot x)(x \cdot x \cdot x)(x \cdot x \cdot x)(x \cdot x \cdot x)$, which is indeed twelve factors of x.

EXAMPLE 12.5.   Simplify $(xy)^3$.

Solution.   Using Rule 4,

$$(xy)^3 = x^3 y^3.$$

Observe that $(xy)^3 = (xy)(xy)(xy) = (x \cdot x \cdot x)(y \cdot y \cdot y)$, or the product of three factors of x and three factors of y.

EXAMPLE 12.6. Simplify $\left(\dfrac{x}{y}\right)^5$.

Solution.   Using Rule 5,

$$\left(\frac{x}{y}\right)^5 = \frac{x^5}{y^5}.$$

Observe that $\left(\dfrac{x}{y}\right)^5 = \dfrac{x}{y} \cdot \dfrac{x}{y} \cdot \dfrac{x}{y} \cdot \dfrac{x}{y} \cdot \dfrac{x}{y} = \dfrac{x \cdot x \cdot x \cdot x \cdot x}{y \cdot y \cdot y \cdot y \cdot y}$, or the quotient of five factors of x and five factors of y.

We will often encounter problems in which two or more of the rules for exponents are involved.

EXAMPLE 12.7.   Simplify $(pq^2)^3$.

Solution.   First, we use Rule 4, with $a = p$ and $b = q^2$:

$$(pq^2)^3 = p^3(q^2)^3.$$

The second factor now can be simplified using Rule 3:

$$p^3(q^2)^3 = p^3q^6.$$

Thus we have

$$(pq^2)^3 = p^3q^6.$$

We will usually do problems of this type in one step.

You must be careful to distinguish between examples of the form $(pq^2)^3$ and examples of the form $p(q^2)^3$. In the second example, the parentheses indicate that the exponent 3 applies only to the factor q^2, and not to the factor p. Thus

$$p(q^2)^3 = pq^6.$$

In the next two examples, the placement of parentheses is also important.

EXAMPLE 12.8.   Simplify $\left(\dfrac{r^5st^2}{rs^3t^2}\right)^4$.

Solution.   First working inside the parentheses, we apply the different parts of Rule 2 to each pair of powers with the same base. Using Rule 2a,

$$\frac{r^5}{r} = r^4, \text{ or } \frac{r^4}{1}.$$

Rule 2b gives

$$\frac{s}{s^3} = \frac{1}{s^2},$$

and Rule 2c gives

$$\frac{t^2}{t^2} = 1.$$

Therefore,

$$\left(\frac{r^5st^2}{rs^3t^2}\right)^4 = \left(\frac{r^4}{s^2}\right)^4.$$

Now, applying Rule 5,

$$\left(\frac{r^4}{s^2}\right)^4 = \frac{(r^4)^4}{(s^2)^4}.$$

Finally, by Rule 3,

$$\frac{(r^4)^4}{(s^2)^4} = \frac{r^{16}}{s^8}.$$

Usually, you can apply Rules 5 and 3 in the same step. Thus we can write

$$\left(\frac{r^5st^2}{rs^3t^2}\right)^4 = \left(\frac{r^4}{s^2}\right)^4$$

$$= \frac{r^{16}}{s^8}.$$

EXAMPLE 12.9.   Simplify $\dfrac{(r^2s)^3}{(rs^2)^2}$.

Solution.   In this example we cannot begin inside the parentheses because a different exponent applies to each set of parentheses. Therefore, we use Rule 3 first:

$$\frac{(r^2s)^3}{(rs^2)^2} = \frac{r^6s^3}{r^2s^4}$$

$$= \frac{r^4}{s}.$$

Exercise 12.1

Simplify:

1. x^3x^5

2. $\dfrac{x^6}{x^2}$

3. $\dfrac{x^2}{x^4}$

4. $\dfrac{x^2}{x^2}$

5. $\dfrac{x^{10}}{x^9}$

6. $x^{10}x^5$

7. $x^{10} + x^5$

8. $x^3 - x^4$

9. $(xy)^3$

10. $(x^3)^6$

11. $(x^5)^2$

12. $\left(\dfrac{x}{y}\right)^8$

13. $(p^2q^3)^4$

14. $(p^5q^6)^2$

15. $p^3(q^3)^5$

16. $\left(\dfrac{p^2}{q^4}\right)^3$

17. $\left(\dfrac{rs^3}{st^3}\right)^2$

18. $\left(\dfrac{r^2s^3t^4}{r^4s^3t^2}\right)^3$

19. $\dfrac{(rs^3)^5}{(rs)^4}$

20. $\dfrac{(rst)^3}{(r^2s^3t^4)^2}$

<table>
<tr><td>Section
12.2</td><td># The Zero Exponent</td></tr>
</table>

The definition of a power given in Section 12.1 does not include expressions such as x^0. Such an expression would mean we have no factors of x. A common mistake is to assume that in this case the expression itself is zero. However, if we think of an expression such as x^2 as

$$x^2 = x \cdot x = 1 \cdot x \cdot x,$$

then no factors of x would indicate that the expression x^0 is

$$x^0 = 1.$$

It is easy to show that this is indeed the case.

Consider Rule 2c,

$$\frac{b^m}{b^n} = 1$$

if $m = n$ and $b \neq 0$; that is,

$$\frac{b^m}{b^m} = 1.$$

Now, let us apply Rule 2a to $\dfrac{b^m}{b^m}$:

$$\frac{b^m}{b^m} = b^{m-m} = b^0.$$

Thus we have $\dfrac{b^m}{b^m} = 1$ and $\dfrac{b^m}{b^m} = b^0$, and so $b^0 = 1$.

Zero Exponent Rule: $b^0 = 1, b \neq 0.$

The rule does not hold for $b = 0$, because in this case we cannot divide by b^m. However, the rule is true for any other base b, regardless of whether b is a number, variable, or expression involving constants and variables.

EXAMPLE 12.10. Simplify:

a. 7^0 b. $(pq)^0$ c. $\left(\dfrac{x^2yz}{xy}\right)^0$

Solutions.   Using the rule for zero exponents,

a. $7^0 = 1$

b. $(pq)^0 = 1$

c. $\left(\dfrac{x^2yz}{xy}\right)^0 = 1$

We must be more careful when the zero exponent does not apply to the entire expression. As in Section 12.1, we distinguish between cases by use of parentheses.

EXAMPLE 12.11.   Simplify:

a. $(2x)^0$ b. $2x^0$

Solutions.

a.   In this case, the zero exponent does apply to the entire expression, therefore we have

$$(2x)^0 = 1.$$

b.   In this case, the zero exponent applies only to the factor x, therefore we have

$$2x^0 = 2(1) = 2.$$

When zero exponents appear within an expression containing other exponents, we set all powers with a zero exponent equal to 1. Then the remaining expression can be simplified using the rules for exponents in Section 12.1.

EXAMPLE 12.12.   Simplify $\left(\dfrac{a^3bc^0}{ab}\right)^3$.

Solution.   Since $c^0 = 1$, we have

$$\left(\frac{a^3bc^0}{ab}\right)^3 = \left(\frac{a^3b}{ab}\right)^3$$

$$= (a^2)^3$$

$$= a^6.$$

EXAMPLE 12.13.   Simplify $\dfrac{(p^2qr^2)^0}{(pq^0r)^2}$.

Solution.   Since $q^0 = 1$ and also $(p^2qr^2)^0 = 1$, we have

$$\frac{(p^2qr^2)^0}{(pq^0r)^2} = \frac{1}{(pr)^2} = \frac{1}{p^2r^2}.$$

Exercise 12.2

Simplify:

1. 2^0

2. $(-10)^0$

3. x^0

4. $(xy)^0$

5. x^0y

6. $10x^0$

7. r^2st^0

8. $(rs^3t^2)^0$

9. $\dfrac{(a^2bc)^2}{(a^3b^2c)^0}$

10. $\dfrac{(ab^2c)^0}{(a^2bc^0)^3}$

11. $\left(\dfrac{m^3n^0}{m^0n^2}\right)^3$

12. $\left(\dfrac{k^3mn^2}{k^2m^3n}\right)^0$

13. $\left(\dfrac{p^2q^0r^2}{pr}\right)^4$

14. $\left(\dfrac{p^3q^3r^0}{p^2q^3r^2}\right)^5$

15. $(x^2y + xy^2)^0$

16. $x^2y^0 + x^0y^2$

<table>
<tr><td>Section
12.3</td><td><h1 style="text-align:center">Negative Exponents</h1></td></tr>
</table>

The definition of a power given in Section 12.1 does not include expressions b^n where n is a negative integer: $-1, -2, -3$, and so on. We can find a meaning for negative exponents using the rules for positive and zero exponents.

Suppose we have an expression b^n where n is a positive integer. Then b^{-n} is a power where the exponent is a negative integer. Using Rule 1 and the zero exponent rule:

$$b^n \cdot b^{-n} = b^{n+(-n)}$$
$$= b^{n-n}$$
$$= b^0$$
$$= 1,$$

and so

$$b^n \cdot b^{-n} = 1.$$

Dividing first by one factor of the left-hand side of this equation, and then by the other factor, we get two forms of the rule for negative exponents.

Negative Exponent Rule:

1. $b^{-n} = \dfrac{1}{b^n}$

2. $b^n = \dfrac{1}{b^{-n}}$

EXAMPLE 12.14.   Find the value of:

a. 3^{-2} b. $\dfrac{1}{4^{-3}}$ c. $\left(\dfrac{1}{2}\right)^{-5}$

Solutions.   a.   Using part 1 of the negative exponent rule:

$$3^{-2} = \frac{1}{3^2} = \frac{1}{9}.$$

b.   Using part 2 of the negative exponent rule:

$$\frac{1}{4^{-3}} = 4^3 = 64.$$

c.   Again, we use part 1 of the negative exponent rule:

$$\left(\frac{1}{2}\right)^{-5} = \frac{1}{\left(\frac{1}{2}\right)^5} = \frac{1}{\frac{1}{32}} = \frac{1}{1} \cdot \frac{32}{1} = 32.$$

In simplifying expressions containing negative exponents, it is often easiest to use the negative exponent rule to change all exponents to positive exponents. We may then use the rules in Section 12.1.

EXAMPLE 12.15.   Simplify $\dfrac{a^{-3}b^{-2}}{c^{-4}}$.

Solution.

$$\frac{a^{-3}b^{-2}}{c^{-4}} = \frac{\dfrac{1}{a^3} \cdot \dfrac{1}{b^2}}{\dfrac{1}{c^4}}$$

$$= \frac{\dfrac{1}{a^3 b^2}}{\dfrac{1}{c^4}}$$

$$= \frac{1}{a^3 b^2} \cdot \frac{c^4}{1}$$

$$= \frac{c^4}{a^3 b^2}.$$

Problems such as Example 12.15 can be solved more quickly if we observe that a negative exponent becomes positive when the power it applies to is moved from numerator to denominator or from denominator to numerator. Thus a^{-3} and b^{-2} move to the denominator and become a^3 and b^2, and c^{-4} moves to the numerator and becomes c^4. In one step we have the solution

$$\frac{a^{-3}b^{-2}}{c^{-4}} = \frac{c^4}{a^3 b^2}.$$

EXAMPLE 12.16.   Simplify $\dfrac{p^4 r^{-3}}{s^{-6}}$.

Solution.   Using the observation above,

$$\frac{p^4 r^{-3}}{s^{-6}} = \frac{p^4 s^6}{r^3}.$$

In this example, p^4 remains unchanged since this power already has a positive exponent.

In simplifying more complicated expressions, it is usually easiest to proceed in the following order:

1.   Set all powers with zero exponents equal to 1.
2.   Remove all parentheses using Rules 3, 4, and 5.
3.   Make all negative exponents positive.
4.   Combine all powers which have the same base using Rules 1 and 2.

EXAMPLE 12.17.   Simplify $\left(\dfrac{x^{-3}y^4 z^0}{x^4 y^{-3} z^{-2}} \right)^{-3}$.

Solution.   First, we set $z^0 = 1$:

$$\left(\frac{x^{-3}y^4z^0}{x^4y^{-3}z^{-2}} \right)^{-3} = \left(\frac{x^{-3}y^4}{x^4y^{-3}z^{-2}} \right)^{-3}.$$

Second, we remove the parentheses by multiplying each exponent inside the parentheses by -3:

$$\left(\frac{x^{-3}y^4}{x^4y^{-3}z^{-2}} \right)^{-3} = \frac{x^9y^{-12}}{x^{-12}y^9z^6}.$$

Third, we make all negative exponents positive:

$$\frac{x^9y^{-12}}{x^{-12}y^9z^6} = \frac{x^9x^{12}}{y^9y^{12}z^6}.$$

Finally, we combine the powers of x and the powers of y:

$$\frac{x^9x^{12}}{y^9y^{12}z^6} = \frac{x^{21}}{y^{21}z^6}.$$

Thus the solution can be written:

$$\left(\frac{x^{-3}y^4z^0}{x^4y^{-3}z^{-2}} \right)^{-3} = \left(\frac{x^{-3}y^4}{x^4y^{-3}z^{-2}} \right)^{-3}$$

$$= \frac{x^9y^{-12}}{x^{-12}y^9z^6}$$

$$= \frac{x^9x^{12}}{y^9y^{12}z^6}$$

$$= \frac{x^{21}}{y^{21}z^6}.$$

The steps can be done in other orders. For instance, the powers of x and the powers of y can be combined before dealing with the exponent -3. However, other orders are usually longer and more complicated.

EXAMPLE 12.18.   Simplify $\dfrac{(x^0y^{-2}z^2)^3}{(x^{-3}y^0z^{-3})^{-2}}$.

Solution.

$$\frac{(x^0y^{-2}z^2)^3}{(x^{-3}y^0z^{-3})^{-2}} = \frac{(y^{-2}z^2)^3}{(x^{-3}z^{-3})^{-2}}$$

$$= \frac{y^{-6}z^6}{x^6z^6}$$

$$= \frac{z^6}{x^6y^6z^6}$$

$$= \frac{1}{x^6y^6}.$$

In Section 8.3, we saw that $(P + Q)^2$ is not the same as $P^2 + Q^2$. It is important to remember that for any integer n other than 1, $(P + Q)^n$ *cannot be written as* $P^n + Q^n$. When n is negative, we must treat the entire expression $(P + Q)^n$ as we would a single term.

EXAMPLE 12.19.   Simplify $(x + y)^{-1}$.

Solution.   We treat the entire expression as a single term:

$$(x + y)^{-1} = \frac{1}{(x + y)^1}$$

$$= \frac{1}{x + y}.$$

If negative exponents apply to separate terms of an expression, we must apply the negative exponent rule to each separate term.

EXAMPLE 12.20.   Simplify $\dfrac{x^{-2} + y^{-2}}{(xy)^{-2}}$.

Solution.   We apply the negative exponent rule separately to x^{-2}, y^{-2}, and $(xy)^{-2}$:

$$x^{-2} = \frac{1}{x^2}$$

$$y^{-2} = \frac{1}{y^2}$$

$$(xy)^{-2} = \frac{1}{(xy)^2}$$

$$= \frac{1}{x^2 y^2}.$$

Therefore,

$$\frac{x^{-2} + y^{-2}}{(xy)^{-2}} = \frac{\dfrac{1}{x^2} + \dfrac{1}{y^2}}{\dfrac{1}{(xy)^2}}$$

$$= \frac{\dfrac{1}{x^2} + \dfrac{1}{y^2}}{\dfrac{1}{x^2 y^2}}.$$

Now, we must simplify the complex fraction. We multiply by the LCD, $x^2 y^2$:

$$\frac{\dfrac{1}{x^2} + \dfrac{1}{y^2}}{\dfrac{1}{x^2 y^2}} = \frac{x^2 y^2 \left(\dfrac{1}{x^2}\right) + x^2 y^2 \left(\dfrac{1}{y^2}\right)}{x^2 y^2 \left(\dfrac{1}{x^2 y^2}\right)} = y^2 + x^2.$$

EXAMPLE 12.21. Simplify $\dfrac{x^{-1} - y^{-1}}{x - y}$.

Solution.   We apply the negative exponent rule separately to x^{-1} and y^{-1}. Then we simplify the complex fraction. We observe that the resulting fraction can be reduced:

$$\frac{x^{-1} - y^{-1}}{x - y} = \frac{\dfrac{1}{x} - \dfrac{1}{y}}{x - y}$$

$$= \frac{xy\left(\dfrac{1}{x}\right) - xy\left(\dfrac{1}{y}\right)}{xy(x - y)}$$

$$= \frac{y - x}{xy(x - y)}$$

$$= \frac{(-1)(x - y)}{xy(x - y)}$$

$$= \frac{-1}{xy}.$$

Exercise 12.3

Find the value of:

1. 6^{-2}

2. $\dfrac{1}{5^{-3}}$

3. $\left(\dfrac{1}{3}\right)^{-4}$

4. $\left(\dfrac{1}{4}\right)^{-2}$

Simplify:

5. $\dfrac{a^{-3}}{b^{-2}}$

6. $\dfrac{p^{-4}}{r^{-3}}$

7. $\dfrac{m^5}{n^{-4}}$

8. $\dfrac{x^{-1}}{y}$

9. $\dfrac{p^{-1}q^0}{r^{-1}}$

10. $\dfrac{(pq)^0}{r^{-2}s^{-2}}$

11. $\dfrac{(mn)^{-3}}{k^0}$

12. $\dfrac{h^{-3}k^0}{h^0k^{-2}}$

13. $\left(\dfrac{a^{-3}b^0}{a^{-1}b^{-2}}\right)^{-2}$

14. $\left(\dfrac{p^{-4}q^{-8}}{p^{-4}q^{-2}}\right)^{-3}$

15. $\left(\dfrac{x^{-3}y^2}{x^3y^{-4}}\right)^{-2}$

16. $\left(\dfrac{p^{-2}q^0r^{-1}}{p^0q^{-4}r^{-4}}\right)^3$

17. $\dfrac{(p^0qr^2)^2}{(p^2q^{-3}r^0)^{-2}}$

18. $\dfrac{(x^2y^3z)^{-4}}{(x^{-2}y^{-3}z^{-1})^{-1}}$

19. $x(y+z)^{-1}$

20. $\left(\dfrac{x+y}{z}\right)^{-1}$

21. $\dfrac{x^{-1} + y^{-1}}{x^{-1} - y^{-1}}$

22. $\dfrac{xy^{-1} - x^{-1}y}{x - y}$

23. $\dfrac{x^{-2} - y^{-2}}{x - y}$

24. $\dfrac{x^{-1} - y^{-1}}{(x-y)^{-1}}$

<table><tr><td>Section
12.4</td><td><h1 align="center">Exponential Functions</h1></td></tr></table>

An **exponential function** is a function where a variable is in the exponent of a power. A basic exponential function has the form

$$y = a^x,$$

where $a > 0$ and $a \neq 1$.

EXAMPLE 12.22. Draw the graph of $y = 2^x$.

Solution.   We will use

$$f(x) = 2^x$$

to find several points of the graph:

$$f(0) = 2^0 = 1, \text{ gives } (0, 1).$$

$$f(1) = 2^1 = 2, \text{ gives } (1, 2).$$

$$f(2) = 2^2 = 4, \text{ gives } (2, 4).$$

$$f(-1) = 2^{-1} = \frac{1}{2^1} = \frac{1}{2}, \text{ gives } \left(-1, \frac{1}{2}\right).$$

$$f(-2) = 2^{-2} = \frac{1}{2^2} = \frac{1}{4}, \text{ gives } \left(-2, \frac{1}{4}\right).$$

Plotting these points, and drawing a smooth curve, we have the graph

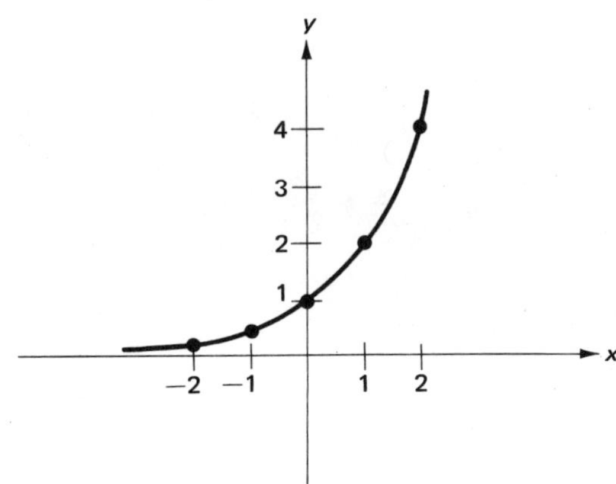

For now, we can only find points of $y = a^x$ where x is an integer. We must assume that points exist where x is a fraction such as $\frac{1}{2}$, or x is an irrational number such as $\sqrt{3}$.

In Unit 7 we introduced the concepts of domain and range of a function. Although we only know how to find values of y for integral values of x, the domain of the exponential function is all real numbers. Observe however, that all the values of y are positive. Since the y-values are always positive, the range of the function is only positive real numbers. We write the range as the numbers $y > 0$.

Finally, we observe that the graph gets closer and closer to the x-axis as we go to the left. We say the graph approaches the x-axis **asymptotically**, or that the x-axis is an **asymptote** of the function. In some cases, the graph of $y = a^x$ will reverse, and the graph will approach the x-axis asymptotically as we go to the right.

Exercise 12.4

Draw the graph:

1. $y = 3^x$ 2. $y = 4^x$ 3. $y = 5^x$ 4. $y = \left(\frac{1}{2}\right)^x$ 5. $y = \left(\frac{1}{3}\right)^x$ 6. $y = \left(\frac{1}{4}\right)^x$

7. $y = 10^x$ (Use a scale of 10, 20, 30, . . . , on the y-axis.)

8. $y = \left(\frac{1}{10}\right)^x$ (Use a scale of 10, 20, 30, . . . , on the y-axis.)

Self-test

1. Find the value of:

    a. $(-8)^0$

    b. $\left(\dfrac{1}{5}\right)^{-2}$

1a. ______________________

1b. ______________________

Simplify:

2. $\dfrac{(a^2b^3)^4}{(ab^2)^3}$

2. ______________________

3. $\left(\dfrac{p^0q^2r^{-1}}{p^{-4}q^{-2}r^0}\right)^{-3}$

3. ______________________

4. $\dfrac{xy^{-2} + x^{-2}y}{x + y}$

4. ______________________________

5. Draw the graph of $y = \left(\dfrac{1}{5}\right)^x$.

Rational Exponents

INTRODUCTION

In previous units we have used rational numbers, irrational numbers, and real numbers. However, we have not yet used these names to classify types of numbers. In this unit you will review briefly the meaning of the terms rational, irrational, and real number. Then you will learn how exponents which are rational numbers can be defined. You will learn how to apply the rules for exponents to rational exponents, and also how to use rational exponents to simplify radicals.

OBJECTIVES

When you have finished this unit you should be able to:

1. Find the value of a power involving rational exponents where the value is a rational number.
2. Simplify expressions involving rational exponents.
3. Use rational exponents to simplify expressions involving radicals.

Section
13.1

Rational Exponents and Radicals

So far we have tried to avoid using the names "real number," "rational number," and "irrational number." Although you have probably seen these names before, we have been using the numbers without naming them. They are numbers such as $\frac{1}{2}$ or $\sqrt{3}$, or even 1 or 0. They have been with us since antiquity (with the possible exception of zero, the origin of which is still disputed by mathematics historians), and we could not do even simple arithmetic without them.

While the types of numbers named above are simple to use, to define them is difficult and was done successfully only a little over a hundred years ago. The process was begun by Karl Weierstrass (1815–1897), and the definitions used now are based on the "Dedekind cut," given by Richard Dedekind (1831–1916) in 1872. The Dedekind cut is far too complicated to attempt to explain here. However, we can stop briefly to give a simple description of the names "real number," "rational number," and "irrational number."

A **real number** may be described as a number which can be written as or approximated by a decimal. The number $\frac{1}{2}$ is a real number because it can be written as the decimal 0.5. The numbers $\sqrt{3}$ and π are real numbers because they can be approximated by decimals. One and zero are real numbers because they can be written as 1.0 and 0.0. All real numbers can be plotted on a number line such as the x-axis or y-axis of a coordinate system, and real numbers are the only type of number that can be plotted on a number line.

A **rational number** is a real number which can be written exactly as a decimal. The numbers 0.5, 1.0, and 0.0 are rational numbers. The rational number is so named because it can be written as a **ratio** $\dfrac{a}{b}$. The number 0.5 can be written as the ratio $\frac{1}{2}$, 1.0 as $\frac{1}{1}$, and 0.0 as $\frac{0}{1}$. Many rational numbers have decimal representations which are unending. For example, $\frac{1}{3} = 0.3333\ldots$. However, we still have an exact decimal because we know all the digits are 3.

An **irrational number** is a real number which can be approximated by a decimal but not written exactly as a decimal. The number $\sqrt{3}$ is approximately 1.732050808, but not exactly. The number π is approximately 3.141592654, but not exactly. The decimal representations of these numbers are unending, and we do not know what the rest of the digits are. Although by using computers, these numbers can be approximated to thousands of places, an irrational number still has digits we do not know. The term "irrational" means "not rational"; that is, a number which cannot be written as a ratio.

You should be familiar with the term **radical**, which means a square root, cube root, fourth root, or in general, an nth root. A radical which is a real number can be either rational or irrational. The radical $\sqrt{3}$ is irrational. Radicals such as $\sqrt{4}$ or $\sqrt[3]{27}$ are rational.

A general expression for a radical, rational or irrational, is

$$\sqrt[n]{a}\,,$$

called **the nth root of a**. In this expression, n is the **index** of the radical and a is the **radicand**.

Definition: $\sqrt[n]{a} = x$ if $a = x^n$, where n is an integer and $n > 1$. When $n = 2$, we simply write $\sqrt{a}$, and the index 2 is understood.

A radical is a rational number if we can find a rational number x which fulfills the definition. For example,

$$\sqrt{4} = 2 \text{ because } 4 = 2^2.$$

$$\sqrt[3]{27} = 3 \text{ because } 27 = 3^3.$$

However, $\sqrt{3}$ is not rational because there is no rational number x such that $3 = x^2$.

It is possible for a radical to be a number which is not a real number. For example, $\sqrt{-4}$ is not a real number because there is no real number x such that $-4 = x^2$. We will consider this kind of number in Unit 15. However, $\sqrt[3]{-27}$ is a real number:

$$\sqrt[3]{-27} = -3 \text{ because } -27 = (-3)^3.$$

If a radical has an even index and a negative radicand, then the radical is not a real number. If the index is odd, however, then the radical is a real number.

We are now ready to turn to rational exponents. Consider the expression $a^{\frac{1}{2}}$. We may write

$$a^{\frac{1}{2}} \cdot a^{\frac{1}{2}} = a^{\frac{1}{2}+\frac{1}{2}}$$

$$= a^1$$

$$= a.$$

or

$$\left(a^{\frac{1}{2}}\right)^2 = a,$$

if a is positive. But then $a^{\frac{1}{2}} = \sqrt{a}$ since, using the definition,

$$\sqrt{a} \ = \ a^{\frac{1}{2}} \text{ because } a \ = \ \left(a^{\frac{1}{2}}\right)^2,$$

if a is positive. Similarly,

$$\sqrt[3]{a} \ = \ a^{\frac{1}{3}} \text{ because } a \ = \ \left(a^{\frac{1}{3}}\right)^3,$$

where a can be positive or negative. In general,

$$\sqrt[n]{a} \ = \ a^{\frac{1}{n}} \text{ because } a \ = \ \left(a^{\frac{1}{n}}\right)^n,$$

where n is an integer, $n > 1$, and a must be positive when n is even.

EXAMPLE 13.1.   Find the value of:

$$\text{a.} \quad 36^{\frac{1}{2}} \qquad \text{b.} \quad \left(-\frac{1}{125}\right)^{\frac{1}{3}} \qquad \text{c.} \quad \left(-\frac{1}{16}\right)^{\frac{1}{4}}$$

Solutions.

a. $36^{\frac{1}{2}} = \sqrt{36} = 6.$

b. $\left(-\dfrac{1}{125}\right)^{\frac{1}{3}} = \sqrt[3]{-\dfrac{1}{125}} = -\dfrac{1}{5}.$

c. $\left(-\dfrac{1}{16}\right)^{\frac{1}{4}}$ has a negative radicand $-\dfrac{1}{16}$ with an even index 4, so there is no real number solution.

A rational exponent is an exponent of the form $\dfrac{m}{n}$, where m and n are integers and $n > 1$. If $n = 1$, the exponent is an integral exponent, which we studied in the preceding unit. If $m = 1$, the exponent is of the form $\dfrac{1}{n}$, and follows the rule above:

$$a^{\frac{1}{n}} = \sqrt[n]{a} \, ,$$

where a must be positive when n is even.

Now, suppose m and n are integers, n is positive, and neither m nor n is 1. Then we may write

$$a^{\frac{m}{n}} \ = \ a^{m\left(\frac{1}{n}\right)}$$

$$= \ (a^m)^{\frac{1}{n}}$$

$$= \sqrt[n]{a^m} \, .$$

Alternatively, we may write

$$a^{\frac{m}{n}} \ = \ a^{\left(\frac{1}{n}\right)m}$$

$$= \ \left(a^{\frac{1}{n}}\right)^m$$

$$= \ \left(\sqrt[n]{a}\right)^m.$$

Thus we have the following rule for rational exponents:

Rational Exponent Rule:

1. $a^{\frac{m}{n}} = \sqrt[n]{a^m}$

2. $a^{\frac{m}{n}} = (\sqrt[n]{a})^m$

where m and n are integers, $n > 1$, and a must be positive when n is even.

EXAMPLE 13.2.   Find the value of:

$$\text{a.} \quad \left(\frac{1}{16}\right)^{\frac{3}{4}} \qquad \text{b.} \quad (-8)^{\frac{2}{3}} \qquad \text{c.} \quad (-16)^{\frac{3}{2}}$$

Solutions.   We use part 2 of the rational exponent rule:

$$\text{a.} \quad \left(\frac{1}{16}\right)^{\frac{3}{4}} = \left(\sqrt[4]{\frac{1}{16}}\right)^3 = \left(\frac{1}{2}\right)^3 = \frac{1}{8}.$$

$$\text{b.} \quad (-8)^{\frac{2}{3}} = (\sqrt[3]{-8})^2 = (-2)^2 = 4.$$

c. $(-16)^{\frac{3}{2}}$ has a negative radicand -16 with an even index 2, so there is no real number solution.

Of course, we may also have negative rational exponents. In the case of a negative rational exponent, we apply the negative exponent rule and then the rational exponent rule.

EXAMPLE 13.3.   Find the value of:

$$\text{a.} \quad 64^{-\frac{1}{3}} \qquad \text{b.} \quad \left(\frac{1}{25}\right)^{-\frac{3}{2}} \qquad \text{c.} \quad \left(-\frac{8}{27}\right)^{-\frac{2}{3}}$$

Solution.   Again using part 2 of the rational exponent rule, after applying the negative exponent rule:

$$\text{a.} \quad 64^{-\frac{1}{3}} = \frac{1}{64^{\frac{1}{3}}} = \frac{1}{\sqrt[3]{64}} = \frac{1}{4}.$$

$$\text{b.} \quad \left(\frac{1}{25}\right)^{-\frac{3}{2}} = \frac{1}{\left(\frac{1}{25}\right)^{\frac{3}{2}}} = \frac{1}{\left(\sqrt{\frac{1}{25}}\right)^3} = \frac{1}{\left(\frac{1}{5}\right)^3} = \frac{1}{\frac{1}{125}} = 125.$$

$$\text{c.} \quad \left(-\frac{8}{27}\right)^{-\frac{2}{3}} = \frac{1}{\left(-\frac{8}{27}\right)^{\frac{2}{3}}} = \frac{1}{\left(\sqrt[3]{-\frac{8}{27}}\right)^2} = \frac{1}{\left(-\frac{2}{3}\right)^2} = \frac{1}{\frac{4}{9}} = \frac{9}{4}.$$

Exercise 13.1

Find the value of:

1. $25^{\frac{1}{2}}$

2. $8^{\frac{1}{3}}$

3. $(-32)^{\frac{1}{5}}$

4. $(-16)^{\frac{1}{4}}$

5. $\left(\frac{1}{216}\right)^{\frac{1}{3}}$

6. $\left(-\frac{1}{243}\right)^{\frac{1}{5}}$

7. $\left(-\frac{4}{25}\right)^{\frac{1}{2}}$

8. $\left(-\frac{8}{125}\right)^{\frac{1}{3}}$

9. $16^{\frac{3}{2}}$ 10. $(-27)^{\frac{2}{3}}$ 11. $\left(\dfrac{1}{25}\right)^{\frac{3}{2}}$ 12. $\left(-\dfrac{1}{16}\right)^{\frac{3}{4}}$

13. $\left(-\dfrac{64}{125}\right)^{\frac{2}{3}}$ 14. $\left(-\dfrac{32}{243}\right)^{\frac{3}{5}}$ 15. $\left(-\dfrac{1}{8}\right)^{\frac{5}{3}}$ 16. $\left(\dfrac{125}{64}\right)^{\frac{4}{3}}$

17. $\left(\dfrac{256}{81}\right)^{\frac{5}{4}}$ 18. $\left(-\dfrac{729}{64}\right)^{\frac{5}{6}}$ 19. $100^{-\frac{1}{2}}$ 20. $(-216)^{-\frac{1}{3}}$

21. $\left(-\dfrac{1}{243}\right)^{-\frac{3}{5}}$ 22. $\left(\dfrac{1}{125}\right)^{-\frac{4}{3}}$ 23. $\left(-\dfrac{27}{64}\right)^{-\frac{5}{3}}$ 24. $\left(\dfrac{625}{1296}\right)^{-\frac{3}{4}}$

Section 13.2   Expressions Involving Rational Exponents

We can simplify expressions containing rational exponents in exactly the same way as we did expressions containing integral exponents in Unit 12. The rules for exponents stated in Unit 12 may also be used for rational exponents. We may use the zero exponent rule in the same way as in Unit 12. We have already applied the negative exponent rule to rational exponents.

In each of the following examples, just one of the five basic rules for exponents is applied to rational exponents.

EXAMPLE 13.4. Simplify $x^{\frac{1}{3}} \cdot x^{\frac{1}{2}}$.

Solution. Using Rule 1,

$$x^{\frac{1}{3}} \cdot x^{\frac{1}{2}} = x^{\frac{1}{3}+\frac{1}{2}}$$

$$= x^{\frac{5}{6}}.$$

EXAMPLE 13.5. Simplify $\dfrac{x^{\frac{2}{3}}}{x^{\frac{4}{5}}}$.

Solution. Since $\frac{4}{5}$ is larger than $\frac{2}{3}$, we use Rule 2b:

$$\frac{x^{\frac{2}{3}}}{x^{\frac{4}{5}}} = \frac{1}{x^{\frac{4}{5}-\frac{2}{3}}}$$

$$= \frac{1}{x^{\frac{2}{15}}}.$$

EXAMPLE 13.6. Simplify $\left(x^{\frac{3}{2}}\right)^{\frac{2}{5}}$.

Solution. Using Rule 3,

$$\left(x^{\frac{3}{2}}\right)^{\frac{2}{5}} = x^{\frac{3}{2}\cdot\frac{2}{5}}$$

$$= x^{\frac{3}{5}}.$$

In the next two examples, Rules 4 and 5 are used in combination with Rule 3.

EXAMPLE 13.7.   Simplify $\left(x^{\frac{1}{2}}y^{\frac{3}{2}}\right)^{\frac{1}{3}}$.

Solution.   Using Rule 4,

$$\left(x^{\frac{1}{2}}y^{\frac{3}{2}}\right)^{\frac{1}{3}} = x^{\frac{1}{2}\cdot\frac{1}{3}}y^{\frac{3}{2}\cdot\frac{1}{3}}$$

$$= x^{\frac{1}{6}}y^{\frac{1}{2}}.$$

EXAMPLE 13.8.   Simplify $\left(\dfrac{x^{\frac{3}{4}}}{y^{\frac{5}{8}}}\right)^{4}$.

Solution.   Using Rule 5,

$$\left(\frac{x^{\frac{3}{4}}}{y^{\frac{5}{8}}}\right)^{4} = \frac{x^{\frac{3}{4}\cdot 4}}{y^{\frac{5}{8}\cdot 4}}$$

$$= \frac{x^{3}}{y^{\frac{5}{2}}}.$$

We may have expressions which involve several of the five rules. The same order of solution that we used in Unit 12 is usually the most efficient. We remove all parentheses first, and then combine all powers which have the same base.

EXAMPLE 13.9.   Simplify $\left(\dfrac{x^{\frac{1}{2}}yz^{\frac{1}{3}}}{xy^{\frac{1}{4}}z}\right)^{6}$.

Solution.   Using Rules 4, 5, and 3, we remove the parentheses and multiply each exponent by 6:

$$\left(\frac{x^{\frac{1}{2}}yz^{\frac{1}{3}}}{xy^{\frac{1}{4}}z}\right)^{6} = \frac{x^{\frac{1}{2}\cdot 6}y^{6}z^{\frac{1}{3}\cdot 6}}{x^{6}y^{\frac{1}{4}\cdot 6}z^{6}}$$

$$= \frac{x^{3}y^{6}z^{2}}{x^{6}y^{\frac{3}{2}}z^{6}}.$$

Now, we combine the powers involving x, the powers involving y, and the powers involving z:

$$\frac{x^{3}y^{6}z^{2}}{x^{6}y^{\frac{3}{2}}z^{6}} = \frac{y^{\frac{9}{2}}}{x^{3}z^{4}}.$$

In this example it would have been possible to combine powers with the same base first. However, this order of solution often results in more difficult arithmetic involving fractions.

EXAMPLE 13.10.   Simplify $\dfrac{\left(x^{\frac{3}{4}}y^{\frac{2}{3}}\right)^{\frac{1}{2}}}{\left(x^{\frac{3}{2}}y^{\frac{4}{3}}\right)^{\frac{1}{3}}}$.

Solution.   First, removing the parentheses,

$$\frac{\left(x^{\frac{3}{4}}y^{\frac{2}{3}}\right)^{\frac{1}{2}}}{\left(x^{\frac{3}{2}}y^{\frac{4}{3}}\right)^{\frac{1}{3}}} = \frac{x^{\frac{3}{4}\cdot\frac{1}{2}}y^{\frac{2}{3}\cdot\frac{1}{2}}}{x^{\frac{3}{2}\cdot\frac{1}{3}}y^{\frac{4}{3}\cdot\frac{1}{3}}}$$

$$= \frac{x^{\frac{3}{8}}y^{\frac{1}{3}}}{x^{\frac{1}{2}}y^{\frac{4}{9}}}.$$

Combining the powers involving x and the powers involving y,

$$\frac{x^{\frac{3}{8}}y^{\frac{1}{3}}}{x^{\frac{1}{2}}y^{\frac{4}{9}}} = \frac{1}{x^{\frac{1}{8}}y^{\frac{1}{9}}}.$$

In this example we could not combine powers with the same base first because different exponents apply to the two sets of parentheses.

If an expression contains zero exponents, we set all parts with zero exponents equal to 1.

EXAMPLE 13.11.   Simplify $\dfrac{\left(x^{\frac{1}{2}}yz^{\frac{1}{2}}\right)^{0}}{\left(x^{\frac{3}{2}}y^{0}z^{\frac{5}{2}}\right)^{2}}.$

Solution.   First we set all parts with zero exponents equal to 1:

$$\frac{\left(x^{\frac{1}{2}}yz^{\frac{1}{2}}\right)^{0}}{\left(x^{\frac{3}{2}}y^{0}z^{\frac{5}{2}}\right)^{2}} = \frac{1}{\left(x^{\frac{3}{2}}z^{\frac{5}{2}}\right)^{2}}.$$

Now, we simplify the resulting expression as in the preceding example:

$$\frac{1}{\left(x^{\frac{3}{2}}z^{\frac{5}{2}}\right)^{2}} = \frac{1}{x^{\frac{3}{2}\cdot 2}z^{\frac{5}{2}\cdot 2}}$$

$$= \frac{1}{x^{3}z^{5}}.$$

If an expression contains negative exponents, we make them positive before combining powers with the same base.

EXAMPLE 13.12.   Simplify $\left[\dfrac{x^{-\frac{1}{4}}y^{\frac{1}{4}}z^{0}}{x^{-\frac{1}{2}}y^{-\frac{1}{3}}z^{\frac{1}{2}}}\right]^{-\frac{4}{3}}.$

Solution.

$$\left(\frac{x^{-\frac{1}{4}}y^{\frac{1}{4}}z^{0}}{x^{-\frac{1}{2}}y^{-\frac{1}{3}}z^{\frac{1}{2}}}\right)^{-\frac{4}{3}} = \left(\frac{x^{-\frac{1}{4}}y^{\frac{1}{4}}}{x^{-\frac{1}{2}}y^{-\frac{1}{3}}z^{\frac{1}{2}}}\right)^{-\frac{4}{3}}$$

$$= \frac{x^{\left(-\frac{1}{4}\right)\left(-\frac{4}{3}\right)}y^{\left(\frac{1}{4}\right)\left(-\frac{4}{3}\right)}}{x^{\left(-\frac{1}{2}\right)\left(-\frac{4}{3}\right)}y^{\left(-\frac{1}{3}\right)\left(-\frac{4}{3}\right)}z^{\left(\frac{1}{2}\right)\left(-\frac{4}{3}\right)}}$$

$$= \frac{x^{\frac{1}{3}}y^{-\frac{1}{3}}}{x^{\frac{2}{3}}y^{\frac{4}{9}}z^{-\frac{2}{3}}}$$

$$= \frac{x^{\frac{1}{3}}z^{\frac{2}{3}}}{x^{\frac{2}{3}}y^{\frac{4}{9}}y^{\frac{1}{3}}}$$

$$= \frac{z^{\frac{2}{3}}}{x^{\frac{1}{3}}y^{\frac{7}{9}}}\,.$$

Exercise 13.2

Simplify:

1. $x^{\frac{1}{4}}\cdot x^{\frac{2}{3}}$

2. $\dfrac{x^{\frac{5}{2}}}{x^{\frac{5}{3}}}$

3. $\left(x^{\frac{4}{3}}\right)^{\frac{3}{2}}$

4. $\left(x^{\frac{1}{2}}y^{\frac{2}{3}}\right)^{3}$

5. $\left(\dfrac{x^{\frac{2}{3}}}{y^{\frac{3}{4}}}\right)^{\frac{1}{6}}$

6. $\left(x^{\frac{3}{2}}y^{\frac{4}{3}}\right)^{0}$

7. $\left(\dfrac{p^{\frac{5}{6}}q}{pq^{\frac{5}{3}}}\right)^{\frac{3}{5}}$

8. $\dfrac{\left(p^{2}q^{\frac{2}{3}}\right)^{\frac{1}{2}}}{\left(p^{\frac{1}{3}}q^{3}\right)^{\frac{1}{3}}}$

9. $\dfrac{\left(a^{\frac{3}{2}}b^{\frac{4}{3}}c^{\frac{5}{4}}\right)^{2}}{\left(a^{\frac{1}{2}}b^{\frac{2}{3}}c^{\frac{3}{4}}\right)^{3}}$

10. $\left(\dfrac{a^{\frac{5}{3}}b^{0}c^{\frac{5}{4}}}{a^{\frac{10}{9}}b^{\frac{1}{2}}c^{\frac{15}{4}}}\right)^{\frac{1}{5}}$

11. $\dfrac{\left(p^{-\frac{2}{3}}q^{\frac{1}{2}}\right)^{\frac{1}{2}}}{\left(p^{\frac{4}{3}}q^{-\frac{2}{3}}\right)^{\frac{3}{2}}}$

12. $\dfrac{\left(p^{-\frac{1}{5}}q^{\frac{1}{3}}\right)^{0}}{\left(p^{\frac{1}{10}}q^{\frac{1}{15}}\right)^{-3}}$

13. $\left(\dfrac{p^{\frac{4}{3}}q^{-\frac{2}{9}}r^{\frac{8}{9}}}{p^{\frac{4}{9}}q^{\frac{4}{3}}r^{-\frac{2}{9}}}\right)^{-\frac{1}{2}}$

14. $\left(\dfrac{p^{-\frac{1}{2}}q^{\frac{1}{2}}r^{-1}}{pq^{-\frac{1}{8}}r^{\frac{1}{4}}}\right)^{-4}$

15. $\dfrac{\left(x^{0}y^{-\frac{1}{4}}z^{-\frac{1}{3}}\right)^{-4}}{\left(xy^{-\frac{1}{3}}z^{-\frac{1}{4}}\right)^{-3}}$

16. $\left(\dfrac{x^{\frac{2}{3}}y^{\frac{1}{3}}z^{0}}{x^{0}y^{-\frac{5}{6}}z^{-\frac{1}{6}}}\right)^{-\frac{2}{3}}$

Section 13.3

Expressions Involving Radicals

Recall from Section 13.1 the rational exponent rule:

1. $a^{\frac{m}{n}} = \sqrt[n]{a^{m}}$

2. $a^{\frac{m}{n}} = \left(\sqrt[n]{a}\right)^{m}$

where a must be positive when n is even. In this section we will assume that any radicand containing variables is positive.

If we have an expression containing a radical, in either form of the right-hand side of the rational exponent rule, we may be able to use the rule to simplify the radical.

EXAMPLE 13.13.   Simplify $\sqrt[6]{a^3}$.

Solution.   Using part 1 of the rational exponent rule,

$$\sqrt[6]{a^3} = a^{\frac{3}{6}}.$$

We reduce the rational exponent,

$$a^{\frac{3}{6}} = a^{\frac{1}{2}},$$

and then write the expression as a radical using the rational exponent rule again:

$$a^{\frac{1}{2}} = \sqrt{a} .$$

Thus we can reduce the radical:

$$\sqrt[6]{a^3} = a^{\frac{3}{6}}$$

$$= a^{\frac{1}{2}}$$

$$= \sqrt{a} .$$

EXAMPLE 13.14.   Simplify $\sqrt[4]{a^8 b^6 c^2}$

Solution.   Again using part 1 of the rational exponent rule,

$$\sqrt[4]{a^8 b^6 c^2} = (a^8 b^6 c^2)^{\frac{1}{4}}$$

$$= a^{\frac{8}{4}} b^{\frac{6}{4}} c^{\frac{2}{4}}$$

$$= a^2 b^{\frac{3}{2}} c^{\frac{1}{2}}.$$

To return to a radical expression, we write

$$a^2 b^{\frac{3}{2}} c^{\frac{1}{2}} = a^2 (b^3 c)^{\frac{1}{2}}$$

$$= a^2 \sqrt{b^3 c} .$$

If there is an exponent outside of the radical, part 2 of the rational exponent rule applies.

EXAMPLE 13.15.   Simplify $\left(\sqrt[6]{x^2 y z^3} \right)^2$.

Solution.   Using part 2 of the rational exponent rule,

$$\left(\sqrt[6]{x^2 y z^3} \right)^2 = (x^2 y z^3)^{\frac{2}{6}}.$$

Reducing and continuing with the rules for exponents,

$$\left(x^2yz^3\right)^{\frac{2}{6}} = \left(x^2yz^3\right)^{\frac{1}{3}}$$

$$= x^{2\cdot\frac{1}{3}}y^{\frac{1}{3}}z^{3\cdot\frac{1}{3}}$$

$$= x^{\frac{2}{3}}y^{\frac{1}{3}}z.$$

Returning to a rational expression,

$$x^{\frac{2}{3}}y^{\frac{1}{3}}z = \left(x^2y\right)^{\frac{1}{3}}z$$

$$= \sqrt[3]{x^2y}\ z,$$

$$= z\sqrt[3]{x^2y}\ .$$

It is conventional to write any factors not inside the radical in front of the radical. This convention lets us avoid any ambiguity with respect to whether a factor such as z is inside or outside of the radical.

The index of a radical must be an integer greater than 1. Other exponents, however, can be any rational number. We use the rules for exponents applied to rational exponents, as in the preceding section.

EXAMPLE 13.16.   Simplify $\left(\sqrt{xy}\ \right)^{\frac{1}{3}}$.

Solution.   Although we cannot apply either part of the rational exponent rule directly, we know that

$$\sqrt{xy} = (xy)^{\frac{1}{2}},$$

and so

$$\left(\sqrt{xy}\ \right)^{\frac{1}{3}} = \left[(xy)^{\frac{1}{2}}\right]^{\frac{1}{3}}.$$

Continuing, using the rules for exponents applied to rational exponents,

$$\left[(xy)^{\frac{1}{2}}\right]^{\frac{1}{3}} = (xy)^{\frac{1}{6}}.$$

Returning to radical form,

$$(xy)^{\frac{1}{6}} = \sqrt[6]{xy}\ .$$

Exercise 13.3

Simplify:

1. $\sqrt[6]{a^2}$ 2. $\sqrt[8]{a^4}$ 3. $\sqrt[3]{a^6}$ 4. $\sqrt[4]{a^4}$ 5. $\sqrt[4]{p^2q^4}$ 6. $\sqrt[6]{p^{12}q^2}$

7. $\left(\sqrt[4]{xy}\ \right)^2$ 8. $\left(\sqrt[12]{x^6y^4z^2}\ \right)^2$ 9. $\left(\sqrt{xy}\ \right)^{\frac{1}{4}}$ 10. $\left(\sqrt[3]{x^3y^2}\ \right)^{\frac{1}{4}}$ 11. $\sqrt{\sqrt{a}}$ 12. $\sqrt[4]{\sqrt[3]{a^2}}$

Self-test

1. Find the value of:

   a. $9^{-\frac{5}{2}}$

   b. $(-32)^{\frac{2}{5}}$

Simplify:

2. $x^{\frac{1}{2}} \cdot x^{-\frac{1}{3}}$

3. $\left(\dfrac{p^{\frac{2}{3}} q^{\frac{2}{9}}}{pq^{\frac{1}{3}}} \right)^{3}$

1a. ______________________

1b. ______________________

2. ______________________

3. ______________________

4. $$\dfrac{\left(x^{-\frac{1}{2}}y^{\frac{1}{2}}z\right)^{\frac{1}{3}}}{\left(x^{-\frac{1}{3}}y^{-\frac{1}{3}}z^{0}\right)^{-\frac{1}{2}}}$$

4. __________________

5. Simplify, and write your answer in radical form:
$$\sqrt[12]{a^{8}b^{12}}$$

5. __________________

Radical Expressions

INTRODUCTION

In Unit 13, rational exponents were used to simplify a specific type of radical. Using a similar method, rules for radicals can be derived from some of the rules for exponents. In this unit, you will learn the rules for radicals and use them to simplify radicals. You can use the rules to add and subtract expressions containing radicals and to multiply certain expressions. You will also learn a technique called rationalizing the denominator or numerator of a rational expression containing radicals. Finally in this unit, you will learn to draw the graph of a basic type of function involving radicals.

OBJECTIVES

When you have finished this unit you should be able to:

1. Use the rules for radicals to simplify radicals by removing factors and by rationalizing denominators.
2. Add and subtract expressions containing radicals.
3. Multiply an expression containing radicals by a single radical term, and multiply two two-term expressions containing radicals.
4. Rationalize a one-term or two-term denominator or numerator containing square roots.
5. Draw the graph of a function of the form $y = \sqrt{x} - a$ or $y = -\sqrt{x} - a$.

Simplifying Radicals

Suppose we have an expression of the form $\sqrt{a}\,\sqrt{b}$, where a and b are positive. If we write the radicals in the form of rational exponents,

$$\sqrt{a}\,\sqrt{b} = a^{\frac{1}{2}}b^{\frac{1}{2}},$$

then, using Rule 4 for exponents,

$$a^{\frac{1}{2}}b^{\frac{1}{2}} = (ab)^{\frac{1}{2}}.$$

Returning to radical form,

$$(ab)^{\frac{1}{2}} = \sqrt{ab}.$$

Therefore,

$$\sqrt{a}\,\sqrt{b} = \sqrt{ab}\,.$$

We can apply the method above to any index n. In general, if a and b are positive when n is even,

$$\sqrt[n]{a}\,\sqrt[n]{b} = a^{\frac{1}{n}}b^{\frac{1}{n}}$$

$$= (ab)^{\frac{1}{n}}$$

$$= \sqrt[n]{ab}\,.$$

Similarly, using Rule 5 for exponents,

$$\frac{\sqrt[n]{a}}{\sqrt[n]{b}} = \frac{a^{\frac{1}{n}}}{b^{\frac{1}{n}}}$$

$$= \left(\frac{a}{b}\right)^{\frac{1}{n}}$$

$$= \sqrt[n]{\frac{a}{b}}\,.$$

Rules for Radicals:

1. $\quad \sqrt[n]{a}\,\sqrt[n]{b} = \sqrt[n]{ab}$

2. $\quad \dfrac{\sqrt[n]{a}}{\sqrt[n]{b}} = \sqrt[n]{\dfrac{a}{b}}$

where a and b must be positive when n is even.

Observe that these rules are for multiplication and division of radicals. As with exponents, there are no rules for addition and subtraction. It is a common error to write $\sqrt[n]{a} + \sqrt[n]{b}$ as $\sqrt[n]{a + b}$, and vice versa. We can easily see that the two expressions are not equal. For example, $\sqrt{9} + \sqrt{16} = 3 + 4 = 7$, but $\sqrt{9 + 16} = \sqrt{25} = 5$. Therefore, $\sqrt{9} + \sqrt{16}$ is not the same as $\sqrt{9 + 16}$.

We can use the rules for radicals to do two different kinds of simplification of radicals. First, we can remove factors from a radical. In Unit 4, we simplified a radical such as $\sqrt{50}$ by writing it as $5\sqrt{2}$. This simplification really uses Rule 1 for radicals. We are actually writing

$$\sqrt{50} = \sqrt{25 \cdot 2}$$

$$= \sqrt{25}\,\sqrt{2}$$

$$= 5\sqrt{2}\,.$$

The second step uses Rule 1 for radicals. The same method works for radicals with index greater than 2.

EXAMPLE 14.1.   Simplify $\sqrt[4]{162}$.

Solution.   Using Rule 1,

$$\sqrt[4]{162} = \sqrt[4]{81 \cdot 2}$$

$$= \sqrt[4]{81}\ \sqrt[4]{2}$$

$$= 3\sqrt[4]{2}\ .$$

When we encounter variables in a radical, it will be useful to know that, if a is positive when n is even,

$$\sqrt[n]{a^n} = a,$$

since $\sqrt[n]{a^n} = (a^n)^{\frac{1}{n}} = a^{n(\frac{1}{n})} = a^1 = a$. Thus, if a is positive when n is even,

$$\sqrt{a^2} = a$$

$$\sqrt[3]{a^3} = a$$

$$\sqrt[4]{a^4} = a$$

and so on. Throughout this unit, we will assume that all variables are positive.

EXAMPLE 14.2.   Simplify $\sqrt{x^3}$.

Solution.   Using Rule 1,

$$\sqrt{x^3} = \sqrt{x^2 \cdot x}$$

$$= \sqrt{x^2}\ \sqrt{x}\ .$$

But, since $\sqrt{x^2} = x$,

$$\sqrt{x^2}\ \sqrt{x} = x\sqrt{x}\ .$$

You will usually be able to skip the second step, which shows the actual application of Rule 1. Thus

$$\sqrt{x^3} = \sqrt{x^2 \cdot x}$$

$$= x\sqrt{x}\ .$$

EXAMPLE 14.3.   Simplify $\sqrt[3]{24x^4y^6}$.

Solution.

$$\sqrt[3]{24x^4y^6} = \sqrt[3]{8 \cdot 3x^3xy^3y^3}$$

$$= 2xy^2\sqrt[3]{3x}\ .$$

Using Rule 2 for radicals, we can simplify radicals where the radicand is a quotient.

EXAMPLE 14.4.   Simplify $\sqrt[3]{\dfrac{27x^5}{y^9}}$.

Solution.   Using Rule 2,

$$\sqrt[3]{\frac{27x^5}{y^9}} = \frac{\sqrt[3]{27x^5}}{\sqrt[3]{y^9}} .$$

Then, Rule 1 gives

$$\sqrt[3]{27x^5} = \sqrt[3]{27x^3x^2}$$

$$= 3x\sqrt[3]{x^2} ,$$

and

$$\sqrt[3]{y^9} = \sqrt[3]{y^3y^3y^3}$$

$$= y \cdot y \cdot y$$

$$= y^3.$$

Therefore,

$$\frac{\sqrt[3]{27x^5}}{\sqrt[3]{y^9}} = \frac{3x\sqrt[3]{x^2}}{y^3} .$$

Again, you will probably be able to skip the actual application of Rule 2 in the first step, and write

$$\sqrt[3]{\frac{27x^5}{y^9}} = \sqrt[3]{\frac{27x^3x^2}{y^3y^3y^3}}$$

$$= \frac{3x\sqrt[3]{x^2}}{y \cdot y \cdot y}$$

$$= \frac{3x\sqrt[3]{x^2}}{y^3} .$$

If a radical has been simplified, and the radicand still is a quotient, it is often considered conventional to **rationalize the denominator**. For example, we often see $\sqrt{\dfrac{1}{2}}$ written as $\dfrac{\sqrt{2}}{2}$.

The second form is considered to be **rationalized** since there are no radicals in the denominator. The two forms are equal because

$$\sqrt{\frac{1}{2}} = \sqrt{\frac{1 \cdot 2}{2 \cdot 2}} = \sqrt{\frac{2}{4}} = \frac{\sqrt{2}}{2} .$$

EXAMPLE 14.5.   Simplify, and rationalize the denominator of $\sqrt{\dfrac{64x^2y}{z^3}}$.

Solution.   Using Rules 1 and 2,

$$\sqrt{\frac{64x^2y}{z^3}} = \frac{8x}{z}\sqrt{\frac{y}{z}} \ .$$

Now, to rationalize the denominator, we multiply the numerator and the denominator of the radicand by z in order to create the radical $\sqrt{z^2}$ in the denominator:

$$\frac{8x}{z}\sqrt{\frac{y}{z}} = \frac{8x}{z}\sqrt{\frac{yz}{z^2}} \ .$$

Then, applying Rule 2 again,

$$\frac{8x}{z}\sqrt{\frac{yz}{z^2}} = \frac{8x\sqrt{yz}}{z \cdot z}$$

$$= \frac{8x\sqrt{yz}}{z^2} \ ,$$

and the denominator is rationalized.

EXAMPLE 14.6. Simplify, and rationalize the denominator of $\sqrt[3]{\dfrac{54x^2y}{2xy^2}}$.

Solution.   First, we reduce the radicand by dividing out any common factors:

$$\sqrt[3]{\frac{54x^2y}{2xy^2}} = \sqrt[3]{\frac{27x}{y}} = 3\sqrt[3]{\frac{x}{y}} \ .$$

Now, to rationalize the denominator, we need y^3 to create the radical $\sqrt[3]{y^3}$. Therefore, we multiply the numerator and the denominator of the radicand by y^2:

$$3\sqrt[3]{\frac{x}{y}} = 3\sqrt[3]{\frac{xy^2}{y^3}} = \frac{3}{y}\sqrt[3]{xy^2}.$$

An answer of this type can be written as either

$$\frac{3}{y}\sqrt[3]{xy^2}$$

or

$$\frac{3\sqrt[3]{xy^2}}{y} \ .$$

We will also encounter radical expressions where the denominator is a separate radical. For such expressions, the process of rationalizing the denominator follows the same principle.

EXAMPLE 14.7.   Rationalize the denominator of $\dfrac{x}{\sqrt[4]{x^3yz^2}}$.

Solution.   We need to create the radical $\sqrt[4]{x^4y^4z^4}$ in the denominator. However, we cannot simply multiply under the radical because the denominator is a separate radical. Therefore, we multiply the numerator and the denominator by $\sqrt[4]{xy^3z^2}$:

$$\frac{x}{\sqrt[4]{x^3yz^2}} = \frac{x\,\sqrt[4]{xy^3z^2}}{\sqrt[4]{x^3yz^2}\,\sqrt[4]{xy^3z^2}}$$

$$= \frac{x\,\sqrt[4]{xy^3z^2}}{\sqrt[4]{x^4y^4z^4}}$$

$$= \frac{x\,\sqrt[4]{xy^3z^2}}{xyz}$$

$$= \frac{\sqrt[4]{xy^3z^2}}{yz}$$

$$= \frac{1}{yz}\sqrt[4]{xy^3z^2}\;.$$

Exercise 14.1

Simplify, and rationalize the denominator if necessary:

1. $\sqrt[3]{54}$

2. $\sqrt[4]{80}$

3. $\sqrt[5]{128}$

4. $\sqrt[3]{500}$

5. $\sqrt[3]{x^5}$

6. $\sqrt[4]{x^9}$

7. $\sqrt{12x^2y^3}$

8. $\sqrt[3]{64x^9y^8}$

9. $\sqrt[4]{32x^4y^8}$

10. $\sqrt[3]{125x^6y^{12}}$

11. $\sqrt[3]{40x^4y^5z^6}$

12. $\sqrt{200x^5y^7z^9}$

13. $\sqrt{\dfrac{16x^3}{y^2}}$

14. $\sqrt[4]{\dfrac{2x^5y^9}{z^8}}$

15. $\sqrt[3]{\dfrac{x^2}{y^3z^6}}$

16. $\sqrt[5]{\dfrac{64x^5}{y^{10}z^{15}}}$

17. $\sqrt[4]{\dfrac{48x^{10}}{3x^5}}$

18. $\sqrt[3]{\dfrac{250x^4y^9}{2x^7y^6}}$

19. $\sqrt{\dfrac{64x^3}{y}}$

20. $\sqrt[3]{\dfrac{54x^4}{y^2}}$

21. $\sqrt[3]{\dfrac{250x^5z^6}{y^4}}$

22. $\sqrt{\dfrac{72y^3z^5}{x^7}}$

23. $\sqrt[6]{\dfrac{x^2y^3}{z^4u^3}}$

24. $\sqrt[4]{\dfrac{81x^5y^{10}}{z^5u^{10}}}$

25. $\sqrt[4]{\dfrac{32x^3y^4}{2x^2y^5}}$

26. $\sqrt[5]{\dfrac{x^{12}y^6z^9}{x^3yz^{15}}}$

27. $\dfrac{1}{\sqrt[3]{9x^2y}}$

28. $\dfrac{1}{\sqrt[4]{8xyz^3}}$

29. $\dfrac{2x}{\sqrt[4]{16x^3y^6}}$

30. $\dfrac{5xy}{\sqrt[3]{5x^5y^7}}$

Addition and Subtraction

Radicals may be combined by addition or subtraction only if they are **like radicals**. Like radicals have both the same indices and the same radicands. Recall from Section 14.1 that, for example, $\sqrt{2} + \sqrt{3}$ cannot be combined. These radicals have different radicands. Similarly, a sum with different indices cannot be combined; for example, $\sqrt{2} + \sqrt[3]{2}$ cannot be combined. However, if the indices and the radicands are the same, radicals can be added just like other kinds of like terms. For example,

$$\sqrt[3]{2} + 3\sqrt[3]{2} = (1 + 3)\sqrt[3]{2}$$

$$= 4\sqrt[3]{2} .$$

Usually we skip the middle step and simply write

$$\sqrt[3]{2} + 3\sqrt[3]{2} = 4\sqrt[3]{2} .$$

EXAMPLE 14.8. Add $\sqrt{xy} + 2\sqrt{xz} + 3\sqrt{xz} + 4\sqrt{xy}$.

Solution. $\sqrt{xy}$ and $4\sqrt{xy}$ have like radicals, so

$$\sqrt{xy} + 4\sqrt{xy} = 5\sqrt{xy} .$$

Also, $2\sqrt{xz}$ and $3\sqrt{xz}$ have like radicals, so

$$2\sqrt{xz} + 3\sqrt{xz} = 5\sqrt{xz} .$$

Therefore,

$$\sqrt{xy} + 2\sqrt{xz} + 3\sqrt{xz} + 4\sqrt{xy} = 5\sqrt{xy} + 5\sqrt{xz} .$$

The remaining sum does not have like radicals and cannot be combined further.

To subtract radicals, we subtract the coefficients of the radicals if the radicals are like radicals. For example,

$$6\sqrt{3} - 2\sqrt{3} = 4\sqrt{3} .$$

EXAMPLE 14.9. Subtract $5\sqrt[3]{xy^2} - \sqrt[3]{xy} - 3\sqrt[3]{xy^2}$.

Solution. $5\sqrt[3]{xy^2}$ and $3\sqrt[3]{xy^2}$ have like radicals, so

$$5\sqrt[3]{xy^2} - 3\sqrt[3]{xy^2} = 2\sqrt[3]{xy^2} .$$

Therefore,

$$5\sqrt[3]{xy^2} - \sqrt[3]{xy} - 3\sqrt[3]{xy^2} = 2\sqrt[3]{xy^2} - \sqrt[3]{xy} .$$

The remaining radicals are not like radicals, and cannot be combined.

Sometimes radicals which are not alike can be transformed into like radicals by the methods of Section 14.1.

EXAMPLE 14.10.   Combine $2x\sqrt{50x} + x\sqrt{72x} - 5\sqrt{2x^3}$.

Solution.   Simplifying each radical, we have

$$2x\sqrt{50x} = 2x\sqrt{25(2x)} = 2x(5)\sqrt{2x} = 10x\sqrt{2x}$$

$$x\sqrt{72x} = x\sqrt{36(2x)} = x(6)\sqrt{2x} = 6x\sqrt{2x}$$

$$5\sqrt{2x^3} = 5\sqrt{x^2(2x)} = 5x\sqrt{2x} .$$

Thus

$$2x\sqrt{50x} + x\sqrt{72x} - 5\sqrt{2x^3} = 10x\sqrt{2x} + 6x\sqrt{2x} - 5x\sqrt{2x}$$

$$= 11x\sqrt{2x} .$$

EXAMPLE 14.11.   Combine $\dfrac{2}{y}\sqrt{9x^3y} - \sqrt{\dfrac{9x^3}{y}} + \dfrac{2x\sqrt{x}}{\sqrt{y}}$.

Solution.   We simplify each radical, including rationalizing the denominator:

$$\frac{2}{y}\sqrt{9x^3y} = \frac{2}{y}(3x)\sqrt{xy} = \frac{6x}{y}\sqrt{xy}$$

$$\sqrt{\frac{9x^3}{y}} = 3x\sqrt{\frac{x}{y}} = 3x\sqrt{\frac{xy}{y^2}} = \frac{3x}{y}\sqrt{xy}$$

$$\frac{2x\sqrt{x}}{\sqrt{y}} = \frac{2x\sqrt{x}\ \sqrt{y}}{\sqrt{y}\ \sqrt{y}} = \frac{2x\sqrt{xy}}{\sqrt{y^2}} = \frac{2x}{y}\sqrt{xy} .$$

Thus the combination becomes

$$\frac{2}{y}\sqrt{9x^3y} - \sqrt{\frac{9x^3}{y}} + \frac{2x\sqrt{x}}{\sqrt{y}} = \frac{6x}{y}\sqrt{xy} - \frac{3x}{y}\sqrt{xy} + \frac{2x}{y}\sqrt{xy} .$$

Then, since the radicals are all like radicals, and also the other variables in each term are alike, we have like terms. Combining the like terms, we have

$$\frac{6x}{y}\sqrt{xy} - \frac{3x}{y}\sqrt{xy} + \frac{2x}{y}\sqrt{xy} = \frac{5x}{y}\sqrt{xy} .$$

It is possible to have like radicals but not like terms.

EXAMPLE 14.12.   Combine $x\sqrt[3]{27x} - 2\sqrt[3]{x^4} - x\sqrt[3]{x^4}$.

Solution.   Since

$$x\sqrt[3]{27x} = 3x\sqrt[3]{x}$$

$$2\sqrt[3]{x^4} = 2x\sqrt[3]{x}$$

$$x\sqrt[3]{x^4} = x^2\sqrt[3]{x},$$

we have

$$x\sqrt[3]{27x} - 2\sqrt[3]{x^4} - x\sqrt[3]{x^4} = 3x\sqrt[3]{x} - 2x\sqrt[3]{x} - x^2\sqrt[3]{x}$$

$$= x\sqrt[3]{x} - x^2\sqrt[3]{x}.$$

Although the two terms have like radicals, x and x^2 are not like terms and therefore the expression cannot be combined further.

Exercise 14.2

Add or subtract as indicated:

1. $2\sqrt{3} + 3\sqrt{3}$

2. $4\sqrt{2} - \sqrt{2}$

3. $5\sqrt{xy} + 6\sqrt{xy} - 10\sqrt{xy}$

4. $2\sqrt[3]{x} - 3\sqrt[3]{x} + 4\sqrt[3]{x}$

5. $2\sqrt{xy} - 3\sqrt{xy} + \sqrt{x}$

6. $2\sqrt{x} - \sqrt[3]{x} + 2\sqrt[3]{x}$

7. $\sqrt[3]{x^2} + 2\sqrt[3]{x} - 5\sqrt[3]{x^2} + 3\sqrt[3]{x}$

8. $10\sqrt{xy} - 6\sqrt[3]{xy} - 4\sqrt{xy} - 2\sqrt[3]{xy}$

9. $2\sqrt{12} - 3\sqrt{3}$

10. $4\sqrt[3]{54} + 3\sqrt[3]{128}$

11. $2\sqrt{72} - 2\sqrt{50} + \sqrt{200}$

12. $3\sqrt{125} + \sqrt{180} - 6\sqrt{80}$

13. $2\sqrt{x^3} + x\sqrt{x} - 4x\sqrt{x}$

14. $x\sqrt[3]{27x} + 3\sqrt[3]{8x^4} - 4\sqrt[3]{x^4}$

15. $y\sqrt[3]{x^4y^2} + x\sqrt[3]{xy^5} + \sqrt[3]{8x^4y^5}$

16. $6xy\sqrt{18x^3y^3} - 2x\sqrt{50x^3y^5} - 2\sqrt{72x^5y^5}$

17. $\sqrt[3]{128xy^3} + x\sqrt[3]{250y^3} + y\sqrt[3]{54x}$

18. $x\sqrt[4]{16xy} + 2\sqrt[4]{x^5y} - 2\sqrt[4]{xy^5}$

19. $\sqrt{27} - \sqrt{\dfrac{16}{3}} - \dfrac{2}{\sqrt{3}}$

20. $\dfrac{1}{\sqrt{2}} - 2\sqrt{18} + \sqrt{\dfrac{9}{2}}$

21. $\dfrac{2}{x}\sqrt{9x} + 2\sqrt{\dfrac{4}{x}} - \dfrac{2}{\sqrt{x}}$

22. $\sqrt{\dfrac{75x}{y}} - \dfrac{1}{y}\sqrt{48xy} + \dfrac{6x}{\sqrt{12xy}}$

23. $4x\sqrt{\dfrac{x}{y}} + \dfrac{2}{y}\sqrt{25xy} + \sqrt{\dfrac{36x^3}{y}}$

24. $2x\sqrt{\dfrac{x^3}{5}} - \dfrac{1}{5}\sqrt{45x^3} - \dfrac{3x^2}{\sqrt{5x}}$

<table>
<tr><td>Section
14.3</td><td><h1 style="text-align:center">Multiplication</h1></td></tr>
</table>

Radicals may be multiplied if they have the same indices, even though they may not have the same radicands. We use Rule 1 for radicals: $\sqrt[n]{a}\ \sqrt[n]{b} = \sqrt[n]{ab}$.

EXAMPLE 14.13.   Multiply $\sqrt[4]{2x^2y^3}\ \sqrt[4]{2x^3y^5}$.

Solution.   Using Rule 1,

$$\sqrt[4]{2x^2y^3}\ \sqrt[4]{2x^3y^5} = \sqrt[4]{(2x^2y^3)(2x^3y^5)}$$

$$= \sqrt[4]{4x^5y^8}$$

$$= xy^2\sqrt[4]{4x}\ .$$

Observe that in the last step we have simplified the radical using the methods of Section 14.1. We will always simplify the resulting radical when possible.

To multiply a single-term radical expression and an expression containing more than one term, we use the usual rule for multiplying a multi-term expression by a single term. Using the distributive property, each term of the multi-term expression is multiplied by the single term. (We do not call the multi-term expression a polynomial, or the single-term expression a monomial, because in their strict definition these do not contain radicals. Polynomials and monomials contain only positive integral powers of the variables.)

EXAMPLE 14.14.   Multiply $\sqrt{2x}\,(\sqrt{2x} + x\sqrt{2} - 2\sqrt{x}\,)$.

Solution.   Multiplying each term of the multi-term expression by the single term, we have

$$\sqrt{2x}\,(\sqrt{2x} + x\sqrt{2} - 2\sqrt{x}\,) = \sqrt{2x}\ \sqrt{2x} + \sqrt{2x}\,(x\sqrt{2}\,) - \sqrt{2x}\,(2\sqrt{x}\,).$$

Then, we multiply each product of radicals:

$$\sqrt{2x}\ \sqrt{2x} + \sqrt{2x}\,(x\sqrt{2}\,) - \sqrt{2x}\,(2\sqrt{x}\,) = \sqrt{4x^2} + x\sqrt{4x} - 2\sqrt{2x^2}\ .$$

Simplifying each radical,

$$\sqrt{4x^2} + x\sqrt{4x} - 2\sqrt{2x^2} = 2x + 2x\sqrt{x} - 2x\sqrt{2}\ .$$

Since there are no like radicals in the final expression, none of the terms can be combined and the expression cannot be simplified further.

To multiply radical expressions where each has two terms, we use the same rule as to multiply binomials. (Again, we do not call such expressions binomials in their strict definition.)

EXAMPLE 14.15.   Multiply $(\sqrt{2x} - \sqrt{5x}\,)(\sqrt{8x} - \sqrt{5x}\,)$.

Solution.   We multiply the expressions, following the rule for binomials:

$$(\sqrt{2x} - \sqrt{5x})(\sqrt{8x} - \sqrt{5x}) = \sqrt{2x}\sqrt{8x} - \sqrt{2x}\sqrt{5x} - \sqrt{5x}\sqrt{8x} + \sqrt{5x}\sqrt{5x}.$$

Now, we multiply each product of radicals, and simplify:

$$\sqrt{2x}\sqrt{8x} - \sqrt{2x}\sqrt{5x} - \sqrt{5x}\sqrt{8x} + \sqrt{5x}\sqrt{5x} = \sqrt{16x^2} - \sqrt{10x^2} - \sqrt{40x^2} + \sqrt{25x^2}$$

$$= 4x - x\sqrt{10} - 2x\sqrt{10} + 5x$$

$$= 9x - 3x\sqrt{10}.$$

EXAMPLE 14.16.   Multiply $(\sqrt{x} + 2\sqrt[3]{x^2})(2\sqrt{x} - \sqrt[3]{x^2})$.

Solution.   We follow the same procedure as in the example above:

$$\left(\sqrt{x} + 2\sqrt[3]{x^2}\right)\left(2\sqrt{x} - \sqrt[3]{x^2}\right) = \sqrt{x}\,(2\sqrt{x}) - \sqrt{x}\,\sqrt[3]{x^2} + 2\sqrt[3]{x^2}\,(2\sqrt{x}) - 2\sqrt[3]{x^2}\,\sqrt[3]{x^2}$$

$$= 2\sqrt{x^2} - \sqrt{x}\,\sqrt[3]{x^2} + 4\sqrt{x}\,\sqrt[3]{x^2} - 2\sqrt[3]{x^4}$$

$$= 2x - \sqrt{x}\,\sqrt[3]{x^2} + 4\sqrt{x}\,\sqrt[3]{x^2} - 2x\sqrt[3]{x}.$$

Since $\sqrt{x}$ and $\sqrt[3]{x^2}$ do not have the same index, they cannot be multiplied using Rule 1. However, the terms $\sqrt{x}\,\sqrt[3]{x^2}$ and $4\sqrt{x}\,\sqrt[3]{x^2}$ contain all like radicals and can be combined:

$$2x - \sqrt{x}\,\sqrt[3]{x^2} + 4\sqrt{x}\,\sqrt[3]{x^2} - 2x\sqrt[3]{x} = 2x + 3\sqrt{x}\,\sqrt[3]{x^2} - 2x\sqrt[3]{x}.$$

Since there are no other like terms, the expression cannot be simplified further.

Exercise 14.3

Multiply:

1. $\sqrt{12}\,\sqrt{24}$

2. $\sqrt[3]{8x^2}\,\sqrt[3]{8x^2}$

3. $\sqrt[3]{3xy^2}\,\sqrt[3]{9x^2y}$

4. $\sqrt[4]{2x^3y}\,\sqrt[4]{2x^3y^2}$

5. $\sqrt{10x^3y^5z}\,\sqrt{20xy^4z^3}$

6. $\sqrt[3]{50x^4y^4z^3}\,\sqrt[3]{15x^6y^4z^4}$

7. $\sqrt{3x}\,(x\sqrt{x} - 2\sqrt{3x})$

8. $\sqrt[3]{16xy}\,(\sqrt[3]{4xy^2} + \sqrt[3]{4x^2y})$

9. $2\sqrt[3]{5x}\,(\sqrt[3]{25x} + \sqrt[3]{15x^3} - \sqrt[3]{150x^5})$

10. $2\sqrt[4]{4x^3y^2}\,(3\sqrt[4]{4x^3y^2} - 2\sqrt[4]{8x^6y^6} - 2\sqrt[4]{16x^8y^4})$

11. $(\sqrt{2} + \sqrt{12})(\sqrt{2} + \sqrt{6})$

12. $(\sqrt{5} - \sqrt{12})(\sqrt{6} + \sqrt{15})$

13. $(3\sqrt{2} + 2\sqrt{10})(3\sqrt{8} - 2\sqrt{10})$

14. $(3\sqrt{5} + \sqrt{7})(\sqrt{7} + 3\sqrt{5})$

15. $(2\sqrt{6} + 2\sqrt{2})(2\sqrt{6} - 2\sqrt{2})$

16. $(\sqrt{15} + 4)(\sqrt{15} - 4)$

17. $(\sqrt{3x} - \sqrt{6x})(\sqrt{3x} - \sqrt{24x})$

18. $(\sqrt{2xy} - \sqrt{7x})(\sqrt{2xy} + \sqrt{7x})$

19. $(2\sqrt{xy} - 4\sqrt{x})(3\sqrt{y} - 6)$

20. $(\sqrt[3]{xy} + \sqrt[3]{x^2y})(\sqrt[3]{xy} + \sqrt[3]{xy^2})$

21. $(\sqrt{2} + 2\sqrt[3]{4})(2\sqrt{2} + \sqrt[3]{4})$

22. $(\sqrt[3]{9} - 3\sqrt{3})(3\sqrt[3]{9} + \sqrt{3})$

23. $(\sqrt{x} + \sqrt[5]{x^3})(\sqrt{x} - \sqrt[5]{x^3})$

24. $(2\sqrt{xy} - 3\sqrt[3]{xy^2})(3\sqrt{xy} - 2\sqrt[3]{x^2y})$

<table>
<tr><td>Section
14.4</td><td><h1>Rationalizing Denominators and Numerators</h1></td></tr>
</table>

Division of radical expressions is actually the same as rationalizing the denominator. In Section 14.1 we used the example

$$\sqrt{\frac{1}{2}} = \sqrt{\frac{1 \cdot 2}{2 \cdot 2}} = \frac{\sqrt{2}}{2}.$$

Since $\sqrt{\dfrac{1}{2}} = \dfrac{1}{\sqrt{2}}$, this example is in fact the division of 1 by $\sqrt{2}$.

When the denominator of an expression consists of a single radical, we can proceed just as in Section 14.1.

EXAMPLE 14.17. Rationalize the denominator of $\dfrac{\sqrt{3}}{\sqrt{6xy}}$.

Solution. We multiply the numerator and the denominator by $\sqrt{6xy}$:

$$\frac{\sqrt{3}}{\sqrt{6xy}} = \frac{\sqrt{3}\,\sqrt{6xy}}{\sqrt{6xy}\,\sqrt{6xy}}$$

$$= \frac{\sqrt{18xy}}{6xy}.$$

Then, simplifying the numerator, and reducing,

$$\frac{\sqrt{18xy}}{6xy} = \frac{3\sqrt{2xy}}{6xy}$$

$$= \frac{\sqrt{2xy}}{2xy}.$$

Observe that you cannot divide out factors such as 2 or x or y because, in each case, one factor is under the radical but the other is not. For example, you cannot divide out $\sqrt{2}$ and 2 since these are not the same number.

If the numerator has more than one term, we simply use multiplication of a multi-term expression by a single term.

EXAMPLE 4.18.   Rationalize the denominator of $\dfrac{2\sqrt{2xy} - 2\sqrt{x} - 2\sqrt{y}}{\sqrt{2xy}}$.

Solution.   Multiplying the numerator and the denominator by $\sqrt{2xy}$, simplifying, and reducing, gives

$$\frac{2\sqrt{2xy} - 2\sqrt{x} - 2\sqrt{y}}{\sqrt{2xy}} = \frac{\left(2\sqrt{2xy} - 2\sqrt{x} - 2\sqrt{y}\,\right)\sqrt{2xy}}{\sqrt{2xy}\ \sqrt{2xy}}$$

$$= \frac{2\sqrt{2xy}\ \sqrt{2xy} - 2\sqrt{x}\ \sqrt{2xy} - 2\sqrt{y}\ \sqrt{2xy}}{2xy}$$

$$= \frac{2\sqrt{4x^2y^2} - 2\sqrt{2x^2y} - 2\sqrt{2xy^2}}{2xy}$$

$$= \frac{4xy - 2x\sqrt{2y} - 2y\sqrt{2x}}{2xy}$$

$$= \frac{2xy - x\sqrt{2y} - y\sqrt{2x}}{xy}.$$

You may divide out 2 because it is a common factor. However, you must be careful in this kind of example not to divide out addends involving x or y.

In Exercise 14.3, there were some problems which involved two two-term expressions which were the sum and difference of the same two radical expressions. For example,

$$(2\sqrt{6} + 2\sqrt{2}\,)(2\sqrt{6} - 2\sqrt{2}\,) = 2\sqrt{6}\,(2\sqrt{6}\,) - 2\sqrt{6}\,(2\sqrt{2}\,) + 2\sqrt{2}\,(2\sqrt{6}\,) - 2\sqrt{2}\,(2\sqrt{2}\,)$$

$$= 4(6) - 4\sqrt{12} + 4\sqrt{12} - 4(2)$$

$$= 24 - 8$$

$$= 16.$$

Observe that, in this kind of example, there are no radicals in the answer. In general, for two positive numbers a and b,

$$(\sqrt{a} + \sqrt{b}\,)(\sqrt{a} - \sqrt{b}\,) = \sqrt{a}\ \sqrt{a} - \sqrt{a}\ \sqrt{b} + \sqrt{b}\ \sqrt{a} - \sqrt{b}\ \sqrt{b}$$

$$= a - \sqrt{ab} + \sqrt{ab} - b$$

$$= a - b.$$

The expression $a - b$ does not involve any radicals. The expressions $\sqrt{a} + \sqrt{b}$ and $\sqrt{a} - \sqrt{b}$ are called **conjugates**. The product of conjugates of this form never contains radicals.

Using conjugates, we can rationalize denominators with two terms, one or both of which are square roots.

EXAMPLE 14.19. Rationalize the denominator of $\dfrac{\sqrt{5}}{\sqrt{10}-\sqrt{5}}$.

Solution. Remember not to divide out addends such as $\sqrt{5}$. We multiply the numerator and the denominator by the conjugate of $\sqrt{10}-\sqrt{5}$, which is $\sqrt{10}+\sqrt{5}$:

$$\frac{\sqrt{5}}{\sqrt{10}-\sqrt{5}}=\frac{\sqrt{5}\,(\sqrt{10}+\sqrt{5}\,)}{(\sqrt{10}-\sqrt{5}\,)(\sqrt{10}+\sqrt{5}\,)}$$

$$=\frac{\sqrt{50}+5}{10+\sqrt{50}-\sqrt{50}-5}$$

$$=\frac{5\sqrt{2}+5}{5}.$$

Since 5 is a common factor, it may be divided out from each term:

$$\frac{5\sqrt{2}+5}{5}=\sqrt{2}+1.$$

If the numerator also has two terms, the process is essentially the same.

EXAMPLE 14.20. Rationalize the denominator of $\dfrac{3\sqrt{2}-\sqrt{3}}{2\sqrt{3}+\sqrt{5}}$.

Solution. We multiply the numerator and the denominator by the conjugate of $2\sqrt{3}+\sqrt{5}$, which is $2\sqrt{3}-\sqrt{5}$:

$$\frac{3\sqrt{2}-\sqrt{3}}{2\sqrt{3}+\sqrt{5}}=\frac{(3\sqrt{2}-\sqrt{3}\,)(2\sqrt{3}-\sqrt{5}\,)}{(2\sqrt{3}+\sqrt{5}\,)(2\sqrt{3}-\sqrt{5}\,)}$$

$$=\frac{6\sqrt{6}-3\sqrt{10}-2\sqrt{9}+\sqrt{15}}{4\sqrt{9}-2\sqrt{15}+2\sqrt{15}-5}$$

$$=\frac{6\sqrt{6}-3\sqrt{10}-6+\sqrt{15}}{12-5}$$

$$=\frac{6\sqrt{6}-3\sqrt{10}-6+\sqrt{15}}{7}.$$

This expression cannot be reduced further, either by simplifying radicals, or by dividing out common factors.

EXAMPLE 14.21. Rationalize the denominator of $\dfrac{3\sqrt{3}-8}{2\sqrt{3}-4}$.

Solution.   Multiplying by the conjugate of $2\sqrt{3} - 4$, which is $2\sqrt{3} + 4$,

$$\frac{3\sqrt{3} - 8}{2\sqrt{3} - 4} = \frac{(3\sqrt{3} - 8)(2\sqrt{3} + 4)}{(2\sqrt{3} - 4)(2\sqrt{3} + 4)}$$

$$= \frac{6(3) + 12\sqrt{3} - 16\sqrt{3} - 32}{4(3) + 8\sqrt{3} - 8\sqrt{3} - 16}$$

$$= \frac{18 - 4\sqrt{3} - 32}{12 - 16}$$

$$= \frac{-14 - 4\sqrt{3}}{-4}$$

$$= \frac{7 + 2\sqrt{3}}{2}.$$

In this example, we had an expression where we could combine terms and reduce. As usual, we reduce the resulting expression whenever possible.

There are times when it is useful to rationalize the numerator of an expression. This process is exactly like rationalizing the denominator, except we multiply the numerator and the denominator by the conjugate of the numerator.

EXAMPLE 14.22.   Rationalize the numerator of $\dfrac{\sqrt{x} + 2}{2\sqrt{x} + 1}$.

Solution.   We multiply the numerator and the denominator by the conjugate of the numerator, $\sqrt{x} + 2$, which is $\sqrt{x} - 2$:

$$\frac{\sqrt{x} + 2}{2\sqrt{x} + 1} = \frac{(\sqrt{x} + 2)(\sqrt{x} - 2)}{(2\sqrt{x} + 1)(\sqrt{x} - 2)}$$

$$= \frac{x - 2\sqrt{x} + 2\sqrt{x} - 4}{2x - 4\sqrt{x} + \sqrt{x} - 2}$$

$$= \frac{x - 4}{2x - 3\sqrt{x} - 2}.$$

When the numerator is rationalized, no radicals appear in the numerator.

Exercise 14.4

Rationalize the denominator:

1. $\dfrac{\sqrt{5x}}{\sqrt{5y}}$

2. $\dfrac{4x}{\sqrt{6x}}$

3. $\dfrac{\sqrt[3]{4x^2y}}{\sqrt[3]{2xy^2}}$

4. $\dfrac{\sqrt[3]{3xyz}}{\sqrt[3]{9x^2yz^2}}$

5. $\dfrac{\sqrt{3x} - 3}{\sqrt{3x}}$

6. $\dfrac{2\sqrt{x} + 2\sqrt{y} - \sqrt{2xy}}{2\sqrt{2xy}}$

7. $\dfrac{\sqrt{xy} + \sqrt{10y}}{\sqrt{10xy}}$

8. $\dfrac{x\sqrt{y} + y\sqrt{x} + xy}{\sqrt{xy}}$

9. $\dfrac{\sqrt{3}}{\sqrt{15} - \sqrt{3}}$ 10. $\dfrac{2}{2\sqrt{2} + 3}$ 11. $\dfrac{\sqrt{x}}{\sqrt{x} + 1}$ 12. $\dfrac{2x}{\sqrt{3x} + \sqrt{5x}}$

13. $\dfrac{\sqrt{5} + \sqrt{3}}{\sqrt{3} - \sqrt{2}}$ 14. $\dfrac{3\sqrt{2} - \sqrt{3}}{2\sqrt{3} + \sqrt{6}}$ 15. $\dfrac{\sqrt{5} - \sqrt{7}}{\sqrt{5} + \sqrt{7}}$ 16. $\dfrac{2\sqrt{6} + 3\sqrt{3}}{3\sqrt{3} - 2\sqrt{6}}$

17. $\dfrac{\sqrt{5} - \sqrt{3}}{\sqrt{15} + 4}$ 18. $\dfrac{3\sqrt{6} - 6}{\sqrt{6} - 4}$ 19. $\dfrac{\sqrt{x} + \sqrt{3}}{\sqrt{x} - \sqrt{3}}$ 20. $\dfrac{x + 2\sqrt{x}}{2x + \sqrt{x}}$

Rationalize the numerator:

21. $\dfrac{\sqrt{10} + \sqrt{6}}{\sqrt{10} - \sqrt{6}}$ 22. $\dfrac{2\sqrt{5} - 5}{10}$ 23. $\dfrac{\sqrt{x} + \sqrt{y}}{y\sqrt{x} + x\sqrt{y}}$ 24. $\dfrac{\sqrt{x} - \sqrt{a}}{x - a}$

Section 14.5

Functions Involving Radicals

It is useful to know about a basic type of function involving radicals. This type of function has the form

$$y = \sqrt{x - a}$$

or

$$y = -\sqrt{x - a}\,.$$

EXAMPLE 14.23.   Draw the graph of $y = \sqrt{x - 2}$.

Solution.   We use

$$f(x) = \sqrt{x - 2}\,.$$

Observe that, in order to have real number values, $x - 2$ must not be negative. Thus x must be greater than 2, or equal to 2. Starting with $x = 2$, and then choosing values of x greater than 2 which will give whole number values for y, we calculate some points of the graph:

$$f(2) = \sqrt{2 - 2} = \sqrt{0} = 0, \text{ gives } (2, 0).$$

$$f(3) = \sqrt{3 - 2} = \sqrt{1} = 1, \text{ gives } (3, 1).$$

$$f(6) = \sqrt{6 - 2} = \sqrt{4} = 2, \text{ gives } (6, 2).$$

$$f(11) = \sqrt{11 - 2} = \sqrt{9} = 3, \text{ gives } (11, 3).$$

Plotting these points and drawing a smooth curve, we have the graph

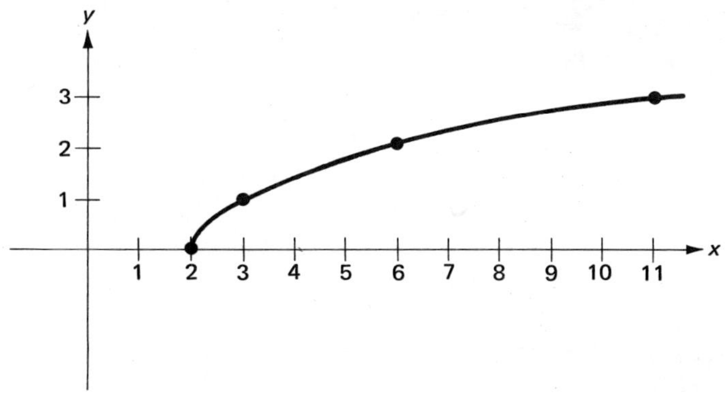

Values of x between the plotted points will give either rational y-values which are fractions, or irrational y-values which can be approximated by decimals.

The domain of this function is all real numbers greater than or equal to 2, which we write $x \geqslant 2$. Different values of a in the function $y = \sqrt{x - a}$ will give different domains. All the y-values are positive or zero, so the range is $y \geqslant 0$.

If we change the function to $y = -\sqrt{x - a}$, we change the range to negative real numbers or zero, or $y \leqslant 0$.

EXAMPLE 14.24.   Draw the graph of $y = -\sqrt{x - 2}$.

Solution.   We use

$$f(x) = -\sqrt{x - 2}.$$

Since the value of a is again 2, we start with $x = 2$:

$$f(2) = -\sqrt{2 - 2} = -\sqrt{0} = 0, \text{ gives } (2, 0).$$

$$f(3) = -\sqrt{3 - 2} = -\sqrt{1} = -1, \text{ gives } (3, -1).$$

$$f(6) = -\sqrt{6 - 2} = -\sqrt{4} = -2, \text{ gives } (6, -2).$$

$$f(11) = -\sqrt{11 - 2} = -\sqrt{9} = -3, \text{ gives } (11, -3).$$

Plotting these points and drawing a smooth curve, we have the graph

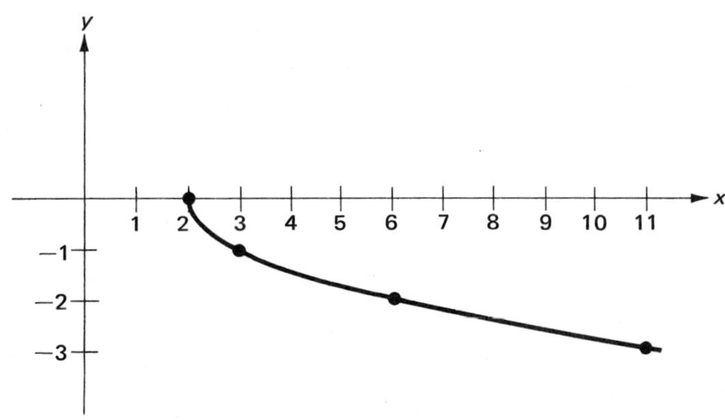

The graph of $y = -\sqrt{x-2}$ is the mirror image of the graph of $y = \sqrt{x-2}$. Both have the domain $x \geqslant 2$. But $y = \sqrt{x-2}$ has the range $y \geqslant 0$ and $y = -\sqrt{x-2}$ has the range $y \leqslant 0$.

Exercise 14.5

Draw the graph:

1. $y = \sqrt{x-1}$ 2. $y = -\sqrt{x-1}$ 3. $y = \sqrt{x+1}$ 4. $y = \sqrt{x+2}$

5. $y = \sqrt{x-3}$ 6. $y = -\sqrt{x-3}$ 7. $y = \sqrt{x+3}$ 8. $y = -\sqrt{x+4}$

Self-test

1. Simplify, and rationalize the denominator of $\sqrt[4]{\dfrac{16x^6}{yz^3}}$.

1. ______________

Perform the operation as indicated:

2. $2\sqrt{6x}\,(x\sqrt{2x} - 5\sqrt{6x}\,)$

2. ______________

3. $\sqrt{5x^5} + 4x\sqrt{20x^3} - x^2\sqrt{80x}$

3. ______________

4. Rationalize the denominator of $\dfrac{\sqrt{5} + \sqrt{15}}{\sqrt{15} - \sqrt{5}}$.

4. ______________

5. Draw the graph of $y = \sqrt{x} + 5$.

Unit 15 — Imaginary Numbers

INTRODUCTION

In the preceding units, it has been assumed that all radicands in radical expressions are positive. Otherwise, the radicals would not represent real numbers. However, it is quite possible for a radicand to be negative. For example, the quadratic formula can give the square root of a negative number just as well as a positive number. In this unit, you will learn the definition of imaginary numbers involving square roots of negative numbers. You will learn how to add, subtract, multiply and divide imaginary numbers. Then, you will use square roots of negative numbers to solve quadratic equations which have imaginary number solutions.

OBJECTIVES

When you have finished this unit you should be able to:

1. Write the square root of a negative number as a pure imaginary number.
2. Add and subtract imaginary numbers.
3. Multiply pure imaginary numbers, an imaginary by a pure imaginary number, or two imaginary numbers.
4. Divide by a pure imaginary number, or an imaginary number.
5. Solve quadratic equations where the solutions are imaginary numbers.

Section 15.1 — Pure Imaginary Numbers

Consider the number $\sqrt{-1}$. Suppose $x = \sqrt{-1}$. Then $x^2 = -1$; but this is not possible because the square of a number is always zero or positive. That is, the square of a **real number** is always zero or positive. We need to define a new kind of number represented by $\sqrt{-1}$.

Definition:

$$i = \sqrt{-1}$$

and

$$i^2 = -1$$

The letter i is used to stand for $\sqrt{-1}$. Any number of the form bi, where b is a real number, is called a **pure imaginary number**. These are some examples of pure imaginary numbers:

$$2i, \quad -i, \quad \frac{1}{2}i, \quad i\sqrt{2}.$$

Observe that, in the last number, we have written i in front of b. This order is conventional when b is a square root, because then it is clear that i is not under the radical.

EXAMPLE 15.1. Write $\sqrt{-4}$ as a pure imaginary number.

Solution.   Since

$$\sqrt{-4} = \sqrt{(-1)4}$$
$$= 2\sqrt{-1},$$

and since $i = \sqrt{-1}$, we have

$$2\sqrt{-1} = 2i.$$

EXAMPLE 15.2. Write $\sqrt{-10}$ as a pure imaginary number.

Solution.
$$\sqrt{-10} = \sqrt{(-1)10}$$
$$= \sqrt{-1}\,\sqrt{10}$$
$$= i\sqrt{10}.$$

Since $\sqrt{10}$ cannot be simplified, the pure imaginary number $i\sqrt{10}$ cannot be simplified further.

EXAMPLE 15.3. Write $\sqrt{-12}$ as a pure imaginary number.

Solution.
$$\sqrt{-12} = \sqrt{(-1)12}$$
$$= \sqrt{-1}\,\sqrt{12}$$
$$= i\sqrt{12}.$$

But $\sqrt{12} = 2\sqrt{3}$, therefore

$$i\sqrt{12} = 2i\sqrt{3}.$$

EXAMPLE 15.4. Write $-\sqrt{-3}$ as a pure imaginary number.

Solution.   It is a common error to make the two negatives into a positive. However,

$$\sqrt{-3} = i\sqrt{3},$$

therefore

$$-\sqrt{-3} = -i\sqrt{3}.$$

The two negatives have no effect on one another. The negative under the square root simply makes the number a pure imaginary number.

Real numbers and pure imaginary numbers can be added to make **complex numbers**. For example, $3 + 2i$ is a complex number. It has a **real part** 3, and a **pure imaginary part** $2i$. In general, a complex number is a number of the form $a + bi$, where a is the real part and bi is the pure imaginary part.

If $b = 0$, then $bi = 0$, and we have a number $a + 0 = a$. In this case, the complex number is a real number. If $b \neq 0$, but $a = 0$, then the number is $0 + bi = bi$. In this case, the complex number is a pure imaginary number. If neither $a = 0$ nor $b = 0$, then $a + bi$ is called simply an **imaginary number**. For example,

$3 + 0i = 3$ is a real number.
$0 + 2i = 2i$ is a pure imaginary number.
$3 + 2i$ is an imaginary number.

Real numbers, pure imaginary numbers, and imaginary numbers, all are complex numbers.

EXAMPLE 15.5.   Write $12 + \sqrt{-25}$ as an imaginary number.

Solution.   First, we write $\sqrt{-25}$ as a pure imaginary number:

$$\sqrt{-25} = \sqrt{(-1)25}$$

$$= 5\sqrt{-1}$$

$$= 5i.$$

Then, we write the imaginary number

$$12 + \sqrt{-25} = 12 + 5i.$$

Observe that the negative under the square root creates an imaginary number, but in no way affects the sign between the real and pure imaginary parts of the imaginary number.

EXAMPLE 15.6.   Write $\sqrt{3} - \sqrt{-48}$ as an imaginary number.

Solution.   The pure imaginary part is

$$\sqrt{-48} = \sqrt{(-1)48}$$

$$= \sqrt{-1}\,\sqrt{48}$$

$$= i\sqrt{48}$$

$$= 4i\sqrt{3}.$$

Thus the imaginary number is

$$\sqrt{3} - \sqrt{-48} = \sqrt{3} - 4i\sqrt{3}.$$

The name "imaginary number" is the result of an unfortunate twist of historical fate. Imaginary numbers are no more "imaginary," in the sense of artificial, than real numbers. The symbol $2i$ is a symbol made up to stand for a numerical concept, just like the symbol -2 or the symbol $\sqrt{2}$ are made up to stand for numerical concepts.

Exercise 15.1

Write as a pure imaginary number:

1. $\sqrt{-36}$ 2. $-\sqrt{-9}$ 3. $\sqrt{-7}$ 4. $-\sqrt{-11}$

5. $\sqrt{-18}$ 6. $\sqrt{-27}$ 7. $\sqrt{-80}$ 8. $\sqrt{-98}$

9. $-\sqrt{-50}$ 10. $-\sqrt{-45}$ 11. $\sqrt{-\frac{1}{4}}$ 12. $\sqrt{-\frac{9}{25}}$

13. $-\sqrt{-\frac{16}{9}}$ 14. $-\sqrt{-\frac{1}{16}}$ 15. $\sqrt{-\frac{5}{4}}$ 16. $-\sqrt{-\frac{8}{9}}$

Write as an imaginary number:

17. $9 + \sqrt{-81}$ 18. $4 - \sqrt{-64}$ 19. $10 - \sqrt{-20}$ 20. $\sqrt{3} + \sqrt{-72}$

21. $\frac{3}{4} - \sqrt{-\frac{9}{4}}$ 22. $\frac{2}{3} + \sqrt{-\frac{2}{9}}$

Section 15.2	Addition and Subtraction

To add imaginary numbers we combine like terms. In many cases, the real parts of imaginary numbers are like terms, and the pure imaginary parts are like terms.

EXAMPLE 15.7.   Add $(4 + 2i) + (6 - 4i)$.

Solution.   First, we remove the parentheses:

$$(4 + 2i) + (6 - 4i) = 4 + 2i + 6 - 4i.$$

Then, we combine the real parts, which are like terms, and the pure imaginary parts, which are like terms:

$$4 + 2i + 6 - 4i = 10 - 2i.$$

To subtract imaginary numbers, we follow the same procedure, but we must be careful in removing the parentheses.

EXAMPLE 15.8.   Subtract $(3 - i) - (2 - 5i)$.

Solution.   We remove the parentheses, being careful with the signs of the second number:

$$(3 - i) - (2 - 5i) = 3 - i - 2 + 5i.$$

Now, we can combine the real parts and the pure imaginary parts:

$$3 - i - 2 + 5i = 1 + 4i.$$

In the examples above, the sum or difference of imaginary numbers is also an imaginary number. It is possible, in special cases, for the sum or difference to be a real number or a pure imaginary number.

EXAMPLE 15.9.   Add $(5 - 2i) + (1 + 2i)$.

Solution.   Removing parentheses, and combining like terms,

$$(5 - 2i) + (1 + 2i) = 5 - 2i + 1 + 2i$$
$$= 6 + 0i$$
$$= 6,$$

which is a real number.

EXAMPLE 15.10.   Subtract $(4 - 5i) - (4 + 3i)$.

Solution.   Carefully removing parentheses, and combining like terms,

$$(4 - 5i) - (4 + 3i) = 4 - 5i - 4 - 3i$$
$$= 0 - 8i$$
$$= -8i,$$

which is a pure imaginary number.

It is possible that the real parts, or the pure imaginary parts, or both, will not be like terms.

EXAMPLE 15.11.   Add $(\sqrt{2} + i\sqrt{3}) + (\sqrt{2} + i\sqrt{5})$.

Solution. $$(\sqrt{2} + i\sqrt{3}) + (\sqrt{2} + i\sqrt{5}) = \sqrt{2} + i\sqrt{3} + \sqrt{2} + i\sqrt{5}$$
$$= 2\sqrt{2} + i\sqrt{3} + i\sqrt{5}.$$

Since $\sqrt{3} + \sqrt{5}$ cannot be combined, $i\sqrt{3} + i\sqrt{5}$ cannot be combined, and the addition cannot be carried further. The pure imaginary part can be written as one term by factoring out the common factor i:

$$2\sqrt{2} + i\sqrt{3} + i\sqrt{5} = 2\sqrt{2} + (\sqrt{3} + \sqrt{5})i.$$

An imaginary number is often written in this form.

If the pure imaginary parts are not already written in terms of i, this should be done first.

EXAMPLE 15.12.   Subtract $(3 - \sqrt{-4}) - (2 - \sqrt{-9})$.

Solution.   Writing the numbers in imaginary number form,

$$(3 - \sqrt{-4}) - (2 - \sqrt{-9}) = (3 - 2i) - (2 - 3i).$$

Now, continuing as before, we have

$$(3 - 2i) - (2 - 3i) = 3 - 2i - 2 + 3i$$
$$= 1 + i.$$

Observe that we would not have been able to combine the pure imaginary parts were they in the form $\sqrt{-4}$ and $\sqrt{-9}$. It is important always to write imaginary and pure imaginary numbers in terms of i.

Exercise 15.2

Add or subtract as indicated:

1. $2i + 3i$

2. $4i - i$

3. $(1 + 2i) + (2 + 4i)$

4. $(5 - 2i) + (2 - i)$

5. $(8 + 6i) - (5 + 2i)$

6. $(4 - i) - (5 - 3i)$

7. $(10 - 8i) - (4 + 4i)$

8. $(2 + 3i) + (3 - 6i)$

9. $(-5 + 6i) + (3 - 2i)$

10. $(-6 - 8i) - (-1 + 2i)$

11. $(6 + 8i) - (6 + 4i)$

12. $(3 + i) + (-3 - 2i)$

13. $(7 + 5i) - (3 + 5i)$

14. $(4 + 5i) + (4 - 5i)$

15. $(2 + i) - (3 - 4i) + (6 - 8i)$

16. $(10 - 5i) - (3 + 4i) - (7 - 2i)$

17. $(4 - \sqrt{-25}) + (4 - \sqrt{-81})$

18. $(6 - \sqrt{-36}) - (-3 - \sqrt{-49})$

19. $(\sqrt{2} + \sqrt{-8}) + (\sqrt{2} + \sqrt{-18})$

20. $(\sqrt{12} + \sqrt{-20}) + (\sqrt{3} - \sqrt{-12})$

Section 15.3

Multiplication

To multiply imaginary numbers, we recall that $i^2 = -1$. Thus, for example,

$$(3i)(4i) = 12i^2 = 12(-1) = -12.$$

It is very important to write pure imaginary numbers in terms of i before multiplying. If a and b are both negative, then $\sqrt{a}\,\sqrt{b}$ is not equal to $\sqrt{ab}$.

EXAMPLE 15.13.   Multiply $\sqrt{-5}\,\sqrt{-5}$.

Solution.   It is a common error to write $\sqrt{(-5)(-5)}$, and then $\sqrt{25} = 5$. However, since both radicands are negative, we must write the numbers as pure imaginary numbers in terms of i:

$$\sqrt{-5}\,\sqrt{-5} = (i\sqrt{5})(i\sqrt{5}).$$

Now, we have

$$(i\sqrt{5})(i\sqrt{5}) = i^2(\sqrt{5}\,\sqrt{5})$$

$$= (-1)(5)$$

$$= -5.$$

The product is -5, not 5, because $i^2 = -1$.

To multiply an imaginary number by a pure imaginary, we follow the usual procedure for multiplying by a single term, remembering that $i^2 = -1$.

EXAMPLE 15.14.   Multiply $3i(2 + 6i)$.

Solution.   Multiplying each term of the imaginary number by $3i$,

$$3i(2 + 6i) = 3i(2) + (3i)(6i)$$
$$= 6i + 18i^2.$$

Now, since $i^2 = -1$, we write

$$6i + 18i^2 = 6i + 18(-1)$$
$$= 6i - 18.$$

EXAMPLE 15.15.   Multiply $2i(5 - 4i)$.

Solution.   Multiplying each term by $2i$, and using $i^2 = -1$,

$$2i(5 - 4i) = 2i(5) - (2i)(4i)$$
$$= 10i - 8i^2$$
$$= 10i - 8(-1)$$
$$= 10i + 8.$$

When the real part of an imaginary number is positive, we usually write the real part first:

$$10i + 8 = 8 + 10i.$$

We multiply two imaginary numbers in the same way we multiply two binomials. Again, it is important to use $i^2 = -1$.

EXAMPLE 15.16.   Multiply $(5 - 2i)(3 - 6i)$.

Solution.   First, we multiply the different parts of the imaginary numbers as if they were binomials:

$$(5 - 2i)(3 - 6i) = 15 - 30i - 6i + 12i^2.$$

We may combine like terms:

$$15 - 30i - 6i + 12i^2 = 15 - 36i + 12i^2.$$

A common error is to stop at this point. However, using $i^2 = -1$, we can combine more terms:

$$15 - 36i + 12i^2 = 15 - 36i + 12(-1)$$
$$= 15 - 36i - 12$$
$$= 3 - 36i.$$

The product of two imaginary numbers always can be reduced by using $i^2 = -1$. However, as with addition and subtraction, we may not always have like terms.

EXAMPLE 15.17.   Multiply $(\sqrt{3} + i\sqrt{2})(\sqrt{3} + i\sqrt{5})$.

Solution.   Multiplying the different parts, and using $i^2 = -1$, we have

$$(\sqrt{3} + i\sqrt{2})(\sqrt{3} + i\sqrt{5}) = 3 + i\sqrt{15} + i\sqrt{6} + i^2\sqrt{10}$$

$$= 3 + i\sqrt{15} + i\sqrt{6} + (-1)\sqrt{10}$$

$$= 3 + i\sqrt{15} + i\sqrt{6} - \sqrt{10}$$

$$= 3 - \sqrt{10} + (\sqrt{15} + \sqrt{6})i.$$

Although the terms cannot be combined, the resulting number is an imaginary number. The real part is $3 - \sqrt{10}$, and the pure imaginary part is $(\sqrt{15} + \sqrt{6})i$.

It is possible for the product of two imaginary numbers to be a real number or a pure imaginary number.

EXAMPLE 15.18.   Multiply $(2 + 3i)(2 - 3i)$.

Solution.

$$(2 + 3i)(2 - 3i) = 4 - 6i + 6i - 9i^2$$

$$= 4 - 9i^2$$

$$= 4 - 9(-1)$$

$$= 4 + 9$$

$$= 13.$$

Exercise 15.3

Multiply:

1. $(7i)(2i)$

2. $(4i)i$

3. $\sqrt{-4}\,\sqrt{-4}$

4. $\sqrt{-6}\,\sqrt{-6}$

5. $\sqrt{-2}\,\sqrt{-8}$

6. $\sqrt{-3}\,\sqrt{-6}$

7. $8i(3 + 2i)$

8. $2i(6 - 4i)$

9. $-i(4 + i)$

10. $-5i(5 - 2i)$

11. $(2 + 3i)(4 + i)$

12. $(3 + 5i)(2 + 3i)$

13. $(6 - 2i)(3 + 4i)$

14. $(2 - 5i)(6 - 3i)$

15. $(5 + 4i)(5 - 4i)$

16. $(2 - i)(2 + i)$

17. $(3 + 2i)(2 + 3i)$

18. $(4 - 2i)(2 - 4i)$

19. $(2 + i\sqrt{2})(3 + i\sqrt{2})$ 20. $(\sqrt{3} + 2i)(2\sqrt{3} - i)$

21. $(\sqrt{3} - i\sqrt{5})(\sqrt{3} + i)$ 22. $(\sqrt{2} + i\sqrt{6})(\sqrt{2} - i\sqrt{6})$

23. $(2 + \sqrt{-3})(3 + \sqrt{-5})$ 24. $(\sqrt{3} + \sqrt{-6})(\sqrt{6} + \sqrt{-3})$

<table>
<tr><td>Section
15.4</td><td><h1>Division</h1></td></tr>
</table>

Division of imaginary numbers is usually written in the form of a fraction. The division is accomplished by making the denominator of the fraction a real number. If the denominator is a pure imaginary number, we need only multiply the numerator and the denominator by i.

EXAMPLE 15.19.   Divide $\dfrac{5}{3i}$.

Solution.   Multiplying the numerator and the denominator by i,

$$\frac{5}{3i} = \frac{5i}{3i^2}.$$

Recalling that $i^2 = -1$,

$$\frac{5i}{3i^2} = \frac{5i}{3(-1)}$$

$$= \frac{5i}{-3}$$

$$= -\frac{5i}{3}.$$

The denominator is a real number. The entire quotient is a pure imaginary number which may be written

$$-\frac{5i}{3}$$

or

$$-\frac{5}{3}i.$$

EXAMPLE 15.20.   Divide $\dfrac{3 + 10i}{2i}$.

Solution.   Again, we begin by multiplying the numerator and the denominator by i:

$$\frac{3 + 10i}{2i} = \frac{(3 + 10i)i}{2i^2}.$$

Now, we must multiply the expression in the numerator:

$$\frac{(3 + 10i)i}{2i^2} = \frac{3i + 10i^2}{2i^2}.$$

Using $i^2 = -1$ in both the numerator and the denominator,

$$\frac{3i + 10i^2}{2i^2} = \frac{3i + 10(-1)}{2(-1)}$$

$$= \frac{3i - 10}{-2}$$

$$= -\frac{3i - 10}{2}.$$

Since $-(3i - 10) = 10 - 3i$, a good form for the resulting imaginary number is

$$\frac{10 - 3i}{2}.$$

Another common form is

$$5 - \frac{3}{2}i,$$

which shows the real part, 5, and the pure imaginary part, $-\frac{3}{2}i$.

EXAMPLE 15.21. Divide $\dfrac{6 - 4i}{5i}$.

Solution.

$$\frac{6 - 4i}{5i} = \frac{(6 - 4i)i}{5i^2}$$

$$= \frac{6i - 4i^2}{5i^2}$$

$$= \frac{6i - 4(-1)}{5(-1)}$$

$$= \frac{6i + 4}{-5}$$

$$= -\frac{6i + 4}{5}.$$

This imaginary number can be written in the form

$$-\frac{4 + 6i}{5}$$

or

$$\frac{-4 - 6i}{5}$$

or

$$-\frac{4}{5} - \frac{6}{5}i.$$

If a division is written as a fraction and the denominator is an imaginary number, we multiply the numerator and the denominator of the fraction by the **conjugate** of the denominator. The conjugate of an imaginary number $a + bi$ is $a - bi$, and vice versa. The product of two conjugate imaginary numbers always is a real number:

$$(a + bi)(a - bi) = a^2 - abi + abi - b^2i^2$$

$$= a^2 - b^2i^2$$

$$= a^2 - b^2(-1)$$

$$= a^2 + b^2.$$

Since a and b are real numbers, and no i appears, $a^2 + b^2$ is a real number.

EXAMPLE 15.22.   Divide $\dfrac{2 + 3i}{2 + i}$.

Solution.   The conjugate of $2 + i$ is $2 - i$, so we multiply the numerator and the denominator by $2 - i$:

$$\frac{2 + 3i}{2 + i} = \frac{(2 + 3i)(2 - i)}{(2 + i)(2 - i)}.$$

Multiplying the imaginary numbers in the numerator and denominator, we have

$$\frac{(2 + 3i)(2 - i)}{(2 + i)(2 - i)} = \frac{4 - 2i + 6i - 3i^2}{4 - 2i + 2i - i^2}$$

$$= \frac{4 + 4i - 3i^2}{4 - i^2}.$$

Then, using $i^2 = -1$,

$$\frac{4 + 4i - 3i^2}{4 - i^2} = \frac{4 + 4i - 3(-1)}{4 - (-1)}$$

$$= \frac{4 + 4i + 3}{4 + 1}$$

$$= \frac{7 + 4i}{5}.$$

The denominator is a real number, and we may write the resulting imaginary number in the form

$$\frac{7 + 4i}{5}$$

or

$$\frac{7}{5} + \frac{4}{5}i.$$

EXAMPLE 15.23.   Divide $\dfrac{3 - 4i}{4 - 2i}$.

Solution.   We multiply the numerator and the denominator by $4 + 2i$, the conjugate of the denominator:

$$\frac{3 - 4i}{4 - 2i} = \frac{(3 - 4i)(4 + 2i)}{(4 - 2i)(4 + 2i)}.$$

Multiplying the imaginary numbers, combining like terms, and using $i^2 = -1$, we have

$$\frac{(3 - 4i)(4 + 2i)}{(4 - 2i)(4 + 2i)} = \frac{12 + 6i - 16i - 8i^2}{16 + 8i - 8i - 4i^2}$$

$$= \frac{12 - 10i - 8i^2}{16 - 4i^2}$$

$$= \frac{12 - 10i - 8(-1)}{16 - 4(-1)}$$

$$= \frac{12 - 10i + 8}{16 + 4}$$

$$= \frac{20 - 10i}{20}.$$

We reduce the number by dividing each term by 10:

$$\frac{20 - 10i}{20} = \frac{2 - i}{2}.$$

This number may be written

$$\frac{2 - i}{2}$$

or

$$1 - \frac{1}{2}i.$$

Exercise 15.4

Divide:

1. $\dfrac{5}{6i}$

2. $\dfrac{-2}{3i}$

3. $\dfrac{6}{8i}$

4. $\dfrac{25}{5i}$

5. $\dfrac{3}{\sqrt{-9}}$

6. $\dfrac{\sqrt{2}}{\sqrt{-16}}$

7. $\dfrac{5 + 8i}{2i}$

8. $\dfrac{3 - 4i}{9i}$

9. $\dfrac{-2 + 8i}{5i}$

10. $\dfrac{1 + i}{i}$

11. $\dfrac{5i}{2 + 4i}$

12. $\dfrac{-5i}{2 - i}$

13. $\dfrac{6 + 3i}{2 - i}$

14. $\dfrac{4 + 5i}{2 - 3i}$

15. $\dfrac{2 - 3i}{1 - 4i}$

16. $\dfrac{3 - i}{4 + 3i}$

17. $\dfrac{5 - 10i}{6 - 2i}$

18. $\dfrac{7 - 5i}{3 + 3i}$

19. $\dfrac{3 + i}{3 - i}$

20. $\dfrac{1 - 2i}{1 + 2i}$

21. $\dfrac{4 + 3i}{3 - 4i}$

22. $\dfrac{1 - 7i}{7 + i}$

23. $\dfrac{2\sqrt{5} + i}{1 + i\sqrt{5}}$

24. $\dfrac{2\sqrt{2} + i}{\sqrt{2} - i}$

<table><tr><td>

Section
15.5

</td><td>

Imaginary Solutions

</td></tr></table>

In Unit 4, we solved equations such as $x^2 - 4 = 0$ by writing

$$x^2 - 4 = 0$$
$$x^2 = 4$$
$$x = \pm\sqrt{4}$$
$$x = \pm 2.$$

If we encountered an equation such as $x^2 + 4 = 0$, however, we had to write "no real number solution." Using pure imaginary numbers, now we can find solutions to such equations.

EXAMPLE 15.24. Solve $x^2 + 4 = 0$.

Solution. Following the method of Unit 4,

$$x^2 + 4 = 0$$
$$x^2 = -4$$
$$x = \pm\sqrt{-4}.$$

Since we can write $\sqrt{-4}$ as the pure imaginary number $2i$, the solutions are

$$x = \pm\sqrt{-4}$$
$$x = \pm 2i.$$

The solutions are $x = 2i$ and $x = -2i$.

Recall that the quadratic equation in standard form,

$$ax^2 + bx + c = 0,$$

can be solved by the quadratic formula,

$$x = \frac{-b \pm \sqrt{b^2 - 4ac}}{2a}.$$

Again, using the quadratic formula, there were equations for which we could only write "no real number solution." The solutions to such equations are imaginary numbers.

EXAMPLE 15.25. Solve $2x^2 - 3x + 2 = 0$.

Solution. In the quadratic formula, $a = 2$, $b = -3$, and $c = 2$:

$$x = \frac{-b \pm \sqrt{b^2 - 4ac}}{2a}$$

$$= \frac{-(-3) \pm \sqrt{(-3)^2 - 4(2)(2)}}{2(2)}$$

$$= \frac{3 \pm \sqrt{9 - 16}}{4}$$

$$= \frac{3 \pm \sqrt{-7}}{4}$$

$$= \frac{3 \pm i\sqrt{7}}{4}.$$

The solutions are the imaginary numbers $\dfrac{3 + i\sqrt{7}}{4}$ and $\dfrac{3 - i\sqrt{7}}{4}$. Imaginary solutions to quadratic equations always will be conjugates of one another.

EXAMPLE 15.26. Solve $x^2 + 2x + 4 = 0$.

Solution.   In the quadratic formula, $a = 1$, $b = 2$, and $c = 4$:

$$x = \frac{-b \pm \sqrt{b^2 - 4ac}}{2a}$$

$$= \frac{-2 \pm \sqrt{2^2 - 4(1)(4)}}{2(1)}$$

$$= \frac{-2 \pm \sqrt{4 - 16}}{2}$$

$$= \frac{-2 \pm \sqrt{-12}}{2}$$

$$= \frac{-2 \pm i\sqrt{12}}{2}.$$

Of course, we reduce imaginary number solutions whenever possible:

$$\frac{-2 \pm i\sqrt{12}}{2} = \frac{-2 \pm 2i\sqrt{3}}{2}$$

$$= -1 \pm i\sqrt{3}.$$

The solutions are $-1 + i\sqrt{3}$ and $-1 - i\sqrt{3}$.

We should realize that imaginary number solutions are no less "real" than real number solutions. Imaginary numbers have applications ranging from theory of electricity to a geometry that might turn out to describe the shape of the universe. However, we cannot plot imaginary numbers on the number line.

Suppose a parabola has the equation $y = x^2 + 2x + 4$. Then its y-intercept is $(0, 4)$, and its x-intercepts are given by the equation $0 = x^2 + 2x + 4$, which we have just found to have imaginary number solutions. The vertex is $(-1, 3)$, and the graph of the parabola is

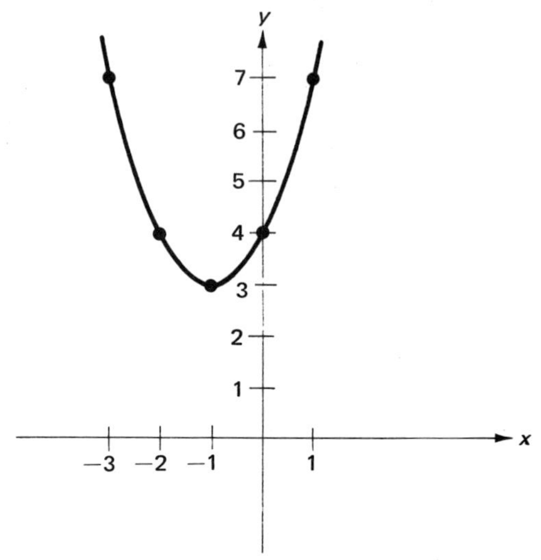

The graph does not cross the x-axis because its x-intercepts are imaginary numbers.

Exercise 15.5

Solve:

1. $x^2+1=0$

2. $x^2+25=0$

3. $x^2+45=0$

4. $4x^2+48=0$

5. $x^2+x+1=0$

6. $2x^2-5x+5=0$

7. $x^2-2x+2=0$

8. $x^2+2x+26=0$

9. $9=2x-2x^2$

10. $3x-4=3x^2$

Self-test

1. Write $-\sqrt{-90}$ as a pure imaginary number.

1. _________________

Perform the operation as indicated:

2. $(5 - i)(4 - 6i)$

2. _________________

3. $(5 + 2i) - (3 - 4i)$

3. _________________

4. $\dfrac{-5i}{4 - 3i}$

4. _________________

5. Solve $x^2 + 2x + 15 = 0$.

5. _________________

Cumulative Review

INTRODUCTION

In this unit you should make certain you remember all the material in all of the preceding units.

OBJECTIVE

When you have finished this unit you should be able to demonstrate that you can fulfill every objective of each preceding unit.

To prepare for this unit you should review the Self-Tests for Units 1 through 15. Do each problem of each Self-Test over again. If you cannot do a problem, or even if you have the slightest difficulty, you should:

1. Find out from the answer section the Objective for the unit to which the problem relates.
2. Review all the material in the section which has the same number as the objective, and redo all the Exercises for the section.
3. Try the Self-Test for the unit again.

Repeat these steps until you can do each problem in each Self-Test for each of Units 1 through 15 easily and accurately.

Solve for x:

1. $3x^2 = 10x + 8$

1. ____________________

2. $x^2 - 6x = -13$

2. ____________________

3. An airplane leaves an airport traveling due east at an average rate of 160 miles per hour. Thirty minutes later, another airplane leaves traveling due west at an average rate of 120 miles per hour. How long after the second plane leaves will they be 150 miles apart?

3. ____________________

4. Find the slope of the line and the equation of the line through the points $(-3, 8)$ and $(6, -4)$.

slope ____________________

equation ____________________

5. Find the y-intercept, x-intercepts, vertex, and draw the graph of $y = x^2 + 2x + 3$.

y-intercept ____________

x-intercepts ____________

vertex ____________

graph:

6. Use rules of exponents to simplify $\left(\dfrac{x^2 y^5 z^0}{x^4 y z^4} \right)^{-\frac{1}{2}}$

6. ____________

Perform the operation as indicated:

7. $\dfrac{2x - 10}{x} \cdot \dfrac{x^2}{25 - x^2}$

7. ____________

8. $3x \sqrt[3]{192x^2} - 2\sqrt[3]{375x^5}$

8. ____________

9. $\dfrac{1 + 3i}{2 - 4i}$

9. _______________

10. Draw the graph of $y = 6^x$.

Equations with Rational Expressions

INTRODUCTION

In the last several units, you have learned how to use a number of tools of algebra. Now you can apply these tools to solving equations and solving applied problems. In this unit you will learn how to solve equations involving rational expressions. You can then solve applied problems where the equations involve rational expressions.

OBJECTIVES

When you have finished this unit you should be able to:

1. Solve equations involving rational expressions.
2. Solve applied problems using equations involving rational expressions.

Solving Equations with Rational Expressions

Recall from Unit 10 how to find the least common multiple (LCM) of two or more polynomials, or the least common denominator (LCD) of two or more rational expressions. To add or subtract rational expressions, we rewrite each rational expression so that its denominator is the LCD. Then we write the sum or difference of the numerators over the LCD. For example, to add

$$\frac{1}{x} + \frac{1}{x + 1},$$

we use the LCD = $x(x + 1)$ and write

$$\frac{1}{x} + \frac{1}{x + 1} = \frac{1(x + 1)}{x(x + 1)} + \frac{1(x)}{x(x + 1)}$$

$$= \frac{x + 1 + x}{x(x + 1)}$$

$$= \frac{2x + 1}{x(x + 1)}.$$

It is important to remember that, to add or subtract, we must write a rational expression where the denominator is the LCD.

Now we turn to solving equations involving rational expressions. In Unit 2 we saw that we can multiply both sides of an equation by the same number, if the number is not zero. Similarly, we can multiply both sides of an equation by the same expression, if the expression is not zero. In particular, if we have an equation involving rational expressions, we can multiply both sides of the equation by the LCD of all the rational expressions. This method will remove all the denominators from the equation. Again, it is important to remember that we must have an equation to use this method. In adding or subtracting rational expressions the denominators cannot be removed.

EXAMPLE 17.1.   Solve $\dfrac{1}{x} + 3 = \dfrac{2}{x}$.

Solution.   The LCD is x. Multiplying both sides of the equation by x, we have

$$\frac{1}{x} + 3 = \frac{2}{x}$$

$$x\left(\frac{1}{x} + 3\right) = x\left(\frac{2}{x}\right)$$

$$x\left(\frac{1}{x}\right) + x(3) = x\left(\frac{2}{x}\right).$$

Observe that multiplying both sides of the equation by the LCD is the same as multiplying each term by the LCD. We will usually skip the first step and begin by multiplying each term by the LCD. Then, dividing out common factors in each term,

$$x\left(\frac{1}{x}\right) + x(3) = x\left(\frac{2}{x}\right)$$

$$1 + 3x = 2.$$

Finally, we solve the resulting linear equation:

$$1 + 3x = 2$$

$$3x = 1$$

$$x = \frac{1}{3}.$$

Since x is not zero, we have not multiplied by zero. Therefore, $\frac{1}{3}$ is the solution. To check, we substitute $\frac{1}{3}$ for x in the equation, and calculate each side:

$$\frac{1}{x} + 3 = \frac{2}{x}$$

$$\frac{1}{\frac{1}{3}} + 3 \stackrel{?}{=} \frac{2}{\frac{1}{3}}$$

$$\frac{1}{1} \cdot \frac{3}{1} + 3 \stackrel{?}{=} \frac{2}{1} \cdot \frac{3}{1}$$

$$3 + 3 = 6.$$

EXAMPLE 17.2.   Solve $\dfrac{x}{x+1} = \dfrac{5}{3} - \dfrac{2}{x}$.

Solution.   The LCD is $3x(x+1)$. We multiply each term by the LCD:

$$\frac{x}{x+1} = \frac{5}{3} - \frac{2}{x}$$

$$3x(x+1)\left(\frac{x}{x+1}\right) = 3x(x+1)\left(\frac{5}{3}\right) - 3x(x+1)\left(\frac{2}{x}\right).$$

Dividing out common factors in each term,

$$3x(x) = x(x+1)(5) - 3(x+1)(2).$$

Now we may remove parentheses in each expression and simplify. We must be careful with the parentheses because of the minus on the right-hand side:

$$3x(x) = 5x(x+1) - 6(x+1)$$

$$3x^2 = 5x^2 + 5x - 6x - 6.$$

The resulting equation is a quadratic equation. In this example, we can collect the terms on the right-hand side and solve by factoring:

$$0 = 2x^2 - x - 6$$

$$0 = (x-2)(2x+3)$$

$$x - 2 = 0 \text{ and } 2x + 3 = 0$$

$$x = 2 \text{ and } x = -\frac{3}{2}.$$

Since neither $x = 2$ nor $x = -\frac{3}{2}$ makes the LCD zero, both 2 and $-\frac{3}{2}$ are solutions. To check each solution, we first substitute $x = 2$:

$$\frac{x}{x+1} = \frac{5}{3} - \frac{2}{x}$$

$$\frac{2}{2+1} \overset{?}{=} \frac{5}{3} - \frac{2}{2}$$

$$\frac{2}{3} = \frac{5}{3} - 1.$$

Now we substitute $x = -\frac{3}{2}$:

$$\frac{x}{x+1} = \frac{5}{3} - \frac{2}{x}$$

$$\frac{-\dfrac{3}{2}}{-\dfrac{3}{2}+1} \overset{?}{=} \frac{5}{3} - \frac{2}{-\dfrac{3}{2}}$$

$$\frac{-\dfrac{3}{2}}{-\dfrac{1}{2}} \overset{?}{=} \frac{5}{3} - \frac{2}{-\dfrac{3}{2}}$$

$$\left(-\frac{3}{2}\right)\left(-\frac{2}{1}\right) \overset{?}{=} \frac{5}{3} - \frac{2}{1}\left(-\frac{2}{3}\right)$$

$$3 \overset{?}{=} \frac{5}{3} + \frac{4}{3}$$

$$3 = \frac{9}{3}.$$

When we derive a quadratic equation from an equation involving rational expressions, we may not be able to solve the quadratic equation by factoring. In this case, we must use the quadratic formula.

EXAMPLE 17.3. Solve $\dfrac{2x}{x - 2} = 3 + \dfrac{3x}{x + 2}$.

Solution. The LCD is $(x - 2)(x + 2)$. We multiply each term by the LCD, divide out common factors, and simplify:

$$\frac{2x}{x - 2} = 3 + \frac{3x}{x + 2}$$

$$(x - 2)(x + 2)\left(\frac{2x}{x - 2}\right) = (x - 2)(x + 2)(3) + (x - 2)(x + 2)\left(\frac{3x}{x + 2}\right)$$

$$(x + 2)(2x) = (x^2 - 4)(3) + (x - 2)(3x)$$

$$2x^2 + 4x = 3x^2 - 12 + 3x^2 - 6x$$

$$0 = 4x^2 - 10x - 12.$$

Since there is a constant factor of 2, we may divide each term by 2:

$$0 = 2x^2 - 5x - 6.$$

This quadratic equation cannot be solved by factoring, so we use the quadratic formula:

$$x = \frac{-b \pm \sqrt{b^2 - 4ac}}{2a}$$

$$= \frac{-(-5) \pm \sqrt{(-5)^2 - 4(2)(-6)}}{2(2)}$$

$$= \frac{5 \pm \sqrt{25 + 48}}{4}$$

$$= \frac{5 \pm \sqrt{73}}{4}.$$

The solutions are $\dfrac{5 + \sqrt{73}}{4}$ and $\dfrac{5 - \sqrt{73}}{4}$.

It is possible to check these solutions by substituting and using the techniques for radical expressions in Unit 14. We will show this method for $\dfrac{5+\sqrt{73}}{4}$:

$$\frac{2x}{x-2} = 3 + \frac{3x}{x+2}$$

$$\frac{2\left(\dfrac{5+\sqrt{73}}{4}\right)}{\dfrac{5+\sqrt{73}}{4}-2} \overset{?}{=} 3 + \frac{3\left(\dfrac{5+\sqrt{73}}{4}\right)}{\dfrac{5+\sqrt{73}}{4}+2}$$

$$\frac{\dfrac{5+\sqrt{73}}{2}}{\dfrac{5+\sqrt{73}}{4}-2} \overset{?}{=} 3 + \frac{\dfrac{15+3\sqrt{73}}{4}}{\dfrac{5+\sqrt{73}}{4}+2}$$

Multiplying the numerator and denominator of each fraction by 4,

$$\frac{2(5+\sqrt{73})}{5+\sqrt{73}-8} \overset{?}{=} 3 + \frac{15+3\sqrt{73}}{5+\sqrt{73}+8}$$

$$\frac{10+2\sqrt{73}}{-3+\sqrt{73}} \overset{?}{=} 3 + \frac{15+3\sqrt{73}}{13+\sqrt{73}}$$

Now, we rationalize the denominator of each fraction:

$$\frac{(10+2\sqrt{73})(-3-\sqrt{73})}{(-3+\sqrt{73})(-3-\sqrt{73})} \overset{?}{=} 3 + \frac{(15+3\sqrt{73})(13-\sqrt{73})}{(13+\sqrt{73})(13-\sqrt{73})}$$

$$\frac{-30-10\sqrt{73}-6\sqrt{73}-2(73)}{9-73} \overset{?}{=} 3 + \frac{195-15\sqrt{73}+39\sqrt{73}-3(73)}{169-73}$$

$$\frac{-176-16\sqrt{73}}{-64} \overset{?}{=} 3 + \frac{-24+24\sqrt{73}}{96}$$

$$\frac{11+\sqrt{73}}{4} \overset{?}{=} 3 + \frac{-1+\sqrt{73}}{4}$$

Finally, we add the expressions on the right-hand side:

$$\frac{11+\sqrt{73}}{4} \overset{?}{=} \frac{12}{4} + \frac{-1+\sqrt{73}}{4}$$

$$\frac{11+\sqrt{73}}{4} = \frac{11+\sqrt{73}}{4} \ .$$

You should check $\dfrac{5-\sqrt{73}}{4}$ by this method.

Clearly the method above is not efficient. This is a situation in which a calculator with a square root key, or at least a table of square roots, is very helpful. We find that $\sqrt{73} \approx 8.544$.

Then,

$$\frac{5 + \sqrt{73}}{4} \approx \frac{5 + 8.544}{4}$$

$$\approx 3.386.$$

Substituting 3.386, and calculating each side,

$$\frac{2(3.386)}{3.386 - 2} \stackrel{?}{\approx} 3 + \frac{3(3.386)}{3.386 + 2}$$

$$\frac{6.772}{1.386} \stackrel{?}{\approx} 3 + \frac{10.158}{5.386}$$

$$4.886 \stackrel{?}{\approx} 3 + 1.886$$

$$4.886 \approx 4.886.$$

For $\dfrac{5 - \sqrt{73}}{4}$ we have

$$\frac{5 - \sqrt{73}}{4} \approx \frac{5 - 8.544}{4}$$

$$\approx -.886,$$

and substituting,

$$\frac{2(-.886)}{-.886 - 2} \stackrel{?}{\approx} 3 + \frac{3(-.886)}{-.886 + 2}$$

$$\frac{-1.772}{-2.886} \stackrel{?}{\approx} 3 + \frac{-2.658}{1.114}$$

$$.614 \stackrel{?}{\approx} 3 - 2.386$$

$$.614 \approx .614.$$

Because of the approximations involved, the two sides may not always be exactly equal.

When the denominators in an equation involving rational expressions have a common factor, we find the LCD by factoring each denominator. Failing to find the least common denominator may lead to an equation we do not know how to solve.

EXAMPLE 17.4.   Solve $\dfrac{6}{x^2 - 4} = \dfrac{1}{x - 2} - \dfrac{2}{x^2 - 5x + 6}$.

Solution.   Since

$$x^2 - 4 = (x + 2)(x - 2)$$

$$x^2 - 5x + 6 = (x - 2)(x - 3),$$

the LCD is $(x + 2)(x - 2)(x - 3)$. We multiply each term by the LCD:

$$\frac{6}{x^2 - 4} = \frac{1}{x - 2} - \frac{2}{x^2 - 5x + 6}$$

$$(x + 2)(x - 2)(x - 3)\left(\frac{6}{x^2 - 4}\right) = (x + 2)(x - 2)(x - 3)\left(\frac{1}{x - 2}\right) - (x + 2)(x - 2)(x - 3)\left(\frac{2}{x^2 - 5x + 6}\right).$$

Then we divide out common factors, simplify each expression, and solve the resulting equation:

$$(x - 3)(6) = (x + 2)(x - 3)(1) - (x + 2)(2)$$

$$6x - 18 = x^2 - x - 6 - 2x - 4$$

$$0 = x^2 - 9x + 8$$

$$0 = (x - 1)(x - 8)$$

$$x - 1 = 0 \text{ and } x - 8 = 0$$

$$x = 1 \text{ and } x = 8.$$

The solutions are 1 and 8. You should check these solutions by substitution.

We must also remember to use the replacement $Q - P = (-1)(P - Q)$ when appropriate, to find the LCD.

EXAMPLE 17.5.   Solve $\dfrac{3}{x + 1} - \dfrac{1}{1 - x} = \dfrac{4}{x^2 - 1}$.

Solution.   Since $1 - x = (-1)(x - 1)$, we make the replacement

$$\frac{3}{x + 1} - \frac{1}{1 - x} = \frac{4}{x^2 - 1}$$

$$\frac{3}{x + 1} - \frac{1}{(-1)(x - 1)} = \frac{4}{x^2 - 1}$$

$$\frac{3}{x + 1} + \frac{1}{x - 1} = \frac{4}{x^2 - 1}.$$

Then, the LCD is $(x + 1)(x - 1)$:

$$(x + 1)(x - 1)\left(\frac{3}{x + 1}\right) + (x + 1)(x - 1)\left(\frac{1}{x - 1}\right) = (x + 1)(x - 1)\left(\frac{4}{x^2 - 1}\right)$$

$$(x - 1)(3) + (x + 1)(1) = 4$$

$$3x - 3 + x + 1 = 4$$

$$4x - 2 = 4$$

$$4x = 6$$

$$x = \frac{6}{4}$$

$$x = \frac{3}{2}.$$

The solution is $\frac{3}{2}$. You should check this solution.

Sometimes we will find an apparent solution which makes the LCD zero. Since we cannot multiply both sides of an equation by zero, such a number cannot actually be a solution. We call the number an **extraneous solution**. Recall that an extraneous solution of an equation is a number found by proper methods but which for some reason cannot be a solution. We do not list extraneous solutions as solutions of an equation.

EXAMPLE 17.6.   Solve $1 + \dfrac{10}{x-5} = \dfrac{2x}{x-5}$.

Solution.   The LCD is $x - 5$:

$$1 + \frac{10}{x-5} = \frac{2x}{x-5}$$

$$(x-5)(1) + (x-5)\left(\frac{10}{x-5}\right) = (x-5)\left(\frac{2x}{x-5}\right)$$

$$x - 5 + 10 = 2x$$

$$5 = x.$$

The apparent solution is 5. However, if $x = 5$, the LCD is zero. Therefore, 5 is an extraneous solution. Observe that if you try to check the extraneous solution, the denominators of the rational expressions are zero, and the rational expressions are undefined. Since the equation has no other apparent solutions, there is no solution to the equation.

Exercise 17.1

Solve for x:

1. $\dfrac{3}{x} - 2 = \dfrac{5}{x}$

2. $\dfrac{1}{x} = 2 - \dfrac{2}{x}$

3. $\dfrac{4x}{2x+3} = 1 - \dfrac{4}{2x+3}$

4. $\dfrac{6}{x-2} + 3 = \dfrac{x}{x-2}$

5. $\dfrac{1}{x-1} - 4 = \dfrac{x}{x-1}$

6. $\dfrac{1}{2x-1} = \dfrac{2x}{2x-1} + 3$

7. $\dfrac{2}{x-3} = \dfrac{12}{x} - 1$

8. $\dfrac{8}{x+3} + \dfrac{2}{x} = 2$

9. $\dfrac{1}{x} + \dfrac{3}{2} = \dfrac{8}{x+2}$

10. $\dfrac{2x}{2x-1} - \dfrac{3}{2x} - \dfrac{1}{2} = 0$

11. $\dfrac{1}{x+1} = \dfrac{1}{x+3} - 2$

12. $\dfrac{2}{x-1} + \dfrac{1}{2} = \dfrac{3}{x-2}$

13. $\dfrac{1}{x+1} + \dfrac{4}{3} = \dfrac{1}{x-1}$

14. $\dfrac{4x}{x+4} + \dfrac{x}{x+2} = 1$

15. $\dfrac{4x}{x^2+4x+4} - \dfrac{1}{x+2} = 0$

16. $\dfrac{5x}{x^2-9} - \dfrac{5}{x-3} - \dfrac{3}{x+3} = 0$

17. $\dfrac{1}{x^2-1} = \dfrac{1}{x^2-3x+2} - \dfrac{1}{x^2-x-2}$

18. $\dfrac{x}{x^2-3x+2} - \dfrac{1}{x-1} = \dfrac{3}{x^2+x-2}$

19. $\dfrac{2}{x^2-4} = \dfrac{3}{x^2+2x} + \dfrac{1}{x^2-2x}$

20. $\dfrac{3}{x^2-3x} - \dfrac{1}{2x-4} = \dfrac{1}{x^2-5x+6}$

21. $\dfrac{x}{x^2+3x+2} + \dfrac{3}{x^2+4x+3} = \dfrac{2x}{x^2+5x+6}$

22. $\dfrac{3x}{x^2-x} - \dfrac{4}{x} = \dfrac{2}{x^2-1}$

23. $\dfrac{x}{2x+4} - \dfrac{3x}{2x-4} = \dfrac{4}{x+2} - 1$

24. $\dfrac{x}{x^2-1} - \dfrac{2}{x^2-x} + \dfrac{2}{x^2+x} - \dfrac{4}{x} = 0$

<table>
<tr><td>Section
17.2</td><td></td></tr>
</table>

Applications

In this section we will consider three types of problems where the equations involve rational expressions.

Recall from Section 3.2 problems which describe a number. Using rational expressions, we can solve similar problems which describe a number and its reciprocal. The **reciprocal** of a number is a number such that the product of the number and the reciprocal is 1. The reciprocal of 2 is $\frac{1}{2}$ since $2 \cdot \frac{1}{2} = 1$. The reciprocal of $\frac{2}{3}$ is $\frac{3}{2}$ since $\frac{2}{3} \cdot \frac{3}{2} = 1$. In general, if x is any number except zero, the reciprocal of x is $\dfrac{1}{x}$. For example, if $x = 2$, $\dfrac{1}{x} = \dfrac{1}{2}$. If $x = \dfrac{2}{3}$,

$$\frac{1}{x} = \frac{1}{\frac{2}{3}} = \frac{1}{1} \cdot \frac{3}{2} = \frac{3}{2}.$$

Observe that zero has no reciprocal since $\frac{1}{0}$ is undefined.

EXAMPLE 17.7.   A number is $2\frac{1}{10}$ more than its reciprocal. Find the number.

Solution.   If the number is x, its reciprocal is $\dfrac{1}{x}$. Therefore, we have the equation

$$x = \frac{1}{x} + \frac{21}{10}.$$

This is an equation involving rational expressions. The LCD is $10x$:

$$10x(x) = 10x\left(\frac{1}{x}\right) + 10x\left(\frac{21}{10}\right)$$

$$10x^2 = 10 + 21x$$

$$10x^2 - 21x - 10 = 0$$

$$(2x - 5)(5x + 2) = 0$$

$$2x - 5 = 0 \text{ and } 5x + 2 = 0$$

$$x = \frac{5}{2} \text{ and } x = -\frac{2}{5}.$$

To check, recall that we must check the wording of the problem and not just the equation. For $x = \frac{5}{2}$, the reciprocal is $\frac{2}{5}$. Then $2\frac{1}{10}$ more than the reciprocal is:

$$\frac{2}{5} + 2\frac{1}{10} = \frac{2}{5} + \frac{21}{10}$$

$$= \frac{4}{10} + \frac{21}{10}$$

$$= \frac{25}{10}$$

$$= \frac{5}{2},$$

which is the number. If $x = -\frac{2}{5}$, the reciprocal is $-\frac{5}{2}$, and $2\frac{1}{10}$ more than the reciprocal is

$$-\frac{5}{2} + 2\frac{1}{10} = -\frac{5}{2} + \frac{21}{10}$$

$$= -\frac{25}{10} + \frac{21}{10}$$

$$= -\frac{4}{10}$$

$$= -\frac{2}{5},$$

which is the number. Since the problem did not state that the number must be positive, $-\frac{2}{5}$ is also a solution. Therefore, both $\frac{5}{2}$ and $-\frac{2}{5}$ are solutions.

We may use this kind of problem to learn to distinguish between phrases such as "twice the reciprocal of a number" and "the reciprocal of twice a number."

EXAMPLE 17.8.   Twice the reciprocal of a number is $3\frac{1}{2}$ less than the number. Find the number.

Solution.   If the number is x, the reciprocal is $\dfrac{1}{x}$. Then twice the reciprocal is $2\left(\dfrac{1}{x}\right) = \dfrac{2}{x}$. Therefore, we have the equation

$$\frac{2}{x} = x - \frac{7}{2}.$$

Solving this equation,

$$2x\left(\frac{2}{x}\right) = 2x(x) - 2x\left(\frac{7}{2}\right)$$

$$4 = 2x^2 - 7x$$

$$0 = 2x^2 - 7x - 4$$

$$0 = (x - 4)(2x + 1)$$

$$x - 4 = 0 \text{ and } 2x + 1 = 0$$

$$x = 4 \text{ and } x = -\frac{1}{2}.$$

If $x = 4$, twice the reciprocal is $2(\frac{1}{4}) = \frac{1}{2}$, and

$$4 - 3\frac{1}{2} = 4 - \frac{7}{2}$$

$$= \frac{8}{2} - \frac{7}{2}$$

$$= \frac{1}{2},$$

which is twice the reciprocal. If $x = -\frac{1}{2}$, twice the reciprocal is $2(-2) = -4$, and

$$-\frac{1}{2} - 3\frac{1}{2} = -\frac{1}{2} - \frac{7}{2}$$

$$= -\frac{8}{2}$$

$$= -4,$$

which is twice the reciprocal. Therefore, 4 and $-\frac{1}{2}$ both are solutions.

EXAMPLE 17.9.   A number is $2\frac{5}{6}$ less than the reciprocal of twice the number. Find the number.

Solution.   If the number is x, twice the number is $2x$. Then the reciprocal of twice the number is $\frac{1}{2x}$. Observe the difference between this reciprocal and the reciprocal in Example 17.8. We have the equation

$$x = \frac{1}{2x} - \frac{17}{6}.$$

Solving this equation,

$$6x(x) = 6x\left(\frac{1}{2x}\right) - 6x\left(\frac{17}{6}\right)$$

$$6x^2 = 3 - 17x$$

$$6x^2 + 17x - 3 = 0$$

$$(x + 3)(6x - 1) = 0$$

$$x + 3 = 0 \text{ and } 6x - 1 = 0$$

$$x = -3 \text{ and } x = \frac{1}{6}.$$

If $x = -3$, twice x is -6 and the reciprocal of twice x is $-\frac{1}{6}$:

$$-\frac{1}{6} - 2\frac{5}{6} = -\frac{1}{6} - \frac{17}{6}$$

$$= -\frac{18}{6}$$

$$= -3.$$

Therefore -3 is a solution. You should check to see that $\frac{1}{6}$ is also a solution.

Of course, it is possible that the number is not a rational number.

EXAMPLE 17.10.   A number is 3 more than twice its reciprocal. Find the number.

Solution.   If the number is x, the reciprocal is $\dfrac{1}{x}$, and twice the reciprocal is $\dfrac{2}{x}$:

$$x = \frac{2}{x} + 3$$

$$x(x) = x\left(\frac{2}{x}\right) + x(3)$$

$$x^2 = 2 + 3x$$

$$x^2 - 3x - 2 = 0.$$

This equation cannot be solved by factoring, so we use the quadratic formula:

$$x = \frac{-b \pm \sqrt{b^2 - 4ac}}{2a}$$

$$= \frac{-(-3) \pm \sqrt{(-3)^2 - 4(1)(-2)}}{2(1)}$$

$$= \frac{3 \pm \sqrt{9 + 8}}{2}$$

$$= \frac{3 \pm \sqrt{17}}{2}.$$

To check $\dfrac{3 + \sqrt{17}}{2}$ we approximate using $\sqrt{17} \approx 4.123$. Then

$$\frac{3 + \sqrt{17}}{2} \approx \frac{3 + 4.123}{2}$$

$$\approx 3.562.$$

The reciprocal is

$$\frac{1}{3.562} \approx .281,$$

so twice the reciprocal is approximately .562. Therefore, the number is approximately 3 more than twice the reciprocal. You should check to see that $\dfrac{3 - \sqrt{17}}{2}$ is also a solution.

Work problems offer a type of problem which leads to a similar type of equation. Work problems are based on the principle that the product of the rate at which work is done and the time spent on the work is equal to the amount of work done. For example, if a copier can copy 10 pages per minute, and works for 30 minutes, it can copy 300 pages. We will use charts similar to those used in Sections 3.3 and 3.4 to indicate the rate, time, and amount of work done.

EXAMPLE 17.11.   One copier can copy 10 pages per minute, and another can copy 15 pages per minute. If both copiers are used, how long will it take to copy a 300 page document?

Solution.   We make a chart, using x for the time:

	Rate	Time	Amount
Copier 1	10	x	$10x$
Copier 2	15	x	$15x$

Since the total amount is 300 pages,

$$10x + 15x = 300$$
$$25x = 300$$
$$x = 12 \text{ minutes.}$$

Work problems involve rational expressions when we must add the rates rather than the amounts of work.

EXAMPLE 17.12.   One copier can copy a 120 page report in 6 minutes, while another takes 10 minutes. If both copiers are used, how long will it take to copy the report?

Solution.   The first copier copies 120 pages in 6 minutes, or at a rate of $\frac{120}{6} = 20$ pages per minute. The second copier copies 120 pages in 10 minutes, or at a rate of $\frac{120}{10} = 12$ pages per minute. If both copiers take x minutes, the rate using both is $\dfrac{120}{x}$. We put a third line on the chart to represent both copiers:

	Rate	Time	Amount
Copier 1	20	6	120
Copier 2	12	10	120
Both	$\dfrac{120}{x}$	x	120

Adding the rates,

$$20 + 12 = \frac{120}{x}$$
$$32 = \frac{120}{x}$$
$$32x = 120$$
$$x = \frac{120}{32}$$
$$x = 3\frac{3}{4} \text{ minutes.}$$

To check, we find the amount each copier does in $3\frac{3}{4}$ minutes. The first copier copies

$$20\left(\frac{15}{4}\right) = 75 \text{ pages.}$$

The second copier copies

$$12\left(\frac{15}{4}\right) = 45 \text{ pages.}$$

The total amount of pages copied in $3\frac{3}{4}$ minutes is then

$$75 + 45 = 120 \text{ pages.}$$

Often a work problem is stated in terms of the amount of time to do a total job. In this case we write the amount of work for one job as 1.

EXAMPLE 17.13.   It takes 9 hours for one machine to clean the carpeting in a house, and 6 hours for another to do the same job. How long will the job take if both machines are used?

Solution.   We make a chart using 1 for the amount:

	Rate	Time	Amount
Machine 1	$\frac{1}{9}$	9	1
Machine 2	$\frac{1}{6}$	6	1
Both	$\frac{1}{x}$	x	1

Then, adding the rates,

$$\frac{1}{9} + \frac{1}{6} = \frac{1}{x}$$

$$18x\left(\frac{1}{9}\right) + 18x\left(\frac{1}{6}\right) = 18x\left(\frac{1}{x}\right)$$

$$2x + 3x = 18$$

$$5x = 18$$

$$x = \frac{18}{5}$$

$$x = 3\frac{3}{5} \text{ hours.}$$

To check, in $3\frac{3}{5}$ hours the first machine does

$$\frac{1}{9}\left(\frac{18}{5}\right) = \frac{2}{5}$$

of the job, and the second machine does

$$\frac{1}{6}\left(\frac{18}{5}\right) = \frac{3}{5}$$

of the job, so the whole job is done.

EXAMPLE 17.14.   Two tractors plow a field in $1\frac{1}{3}$ hours. If the first tractor could do the job alone in 2 hours, how long would it take the second tractor to do the job alone?

Solution.   In this example, the time for the second tractor is x:

	Rate	Time	Amount
Tractor 1	$\frac{1}{2}$	2	1
Tractor 2	$\frac{1}{x}$	x	1
Both	$\frac{3}{4}$	$\frac{4}{3}$	1

Observe that the rate for both is the reciprocal of $1\frac{1}{3}$, which is the reciprocal of $\frac{4}{3}$, or $\frac{3}{4}$. Then,

$$\frac{1}{2} + \frac{1}{x} = \frac{3}{4}$$

$$4x\left(\frac{1}{2}\right) + 4x\left(\frac{1}{x}\right) = 4x\left(\frac{3}{4}\right)$$

$$2x + 4 = 3x$$

$$4 = x$$

$$x = 4 \text{ hours.}$$

In $1\frac{1}{3}$ hours, the first tractor does

$$\frac{1}{2}\left(\frac{4}{3}\right) = \frac{2}{3}$$

of the job, and the second tractor does

$$\frac{1}{4}\left(\frac{4}{3}\right) = \frac{1}{3}$$

of the job, so the whole job is done.

EXAMPLE 17.15.   A faucet can fill a tub in 5 minutes. It takes the drain 8 minutes to empty the tub. If both the faucet and the drain are open, how long will it take to fill the tub?

Solution.   The chart is:

	Rate	Time	Amount
Faucet	$\frac{1}{5}$	5	1
Drain	$\frac{1}{8}$	8	1
Both	$\frac{1}{x}$	x	1

We subtract the rate of work of the drain since it is emptying the tub:

$$\frac{1}{5} - \frac{1}{8} = \frac{1}{x}$$

$$40x\left(\frac{1}{5}\right) - 40x\left(\frac{1}{8}\right) = 40x\left(\frac{1}{x}\right)$$

$$8x - 5x = 40$$

$$3x = 40$$

$$x = \frac{40}{3}$$

$$x = 13\frac{1}{3} \text{ minutes.}$$

In $13\frac{1}{3}$ minutes, the faucet fills

$$\frac{1}{5}\left(\frac{40}{3}\right) = \frac{8}{3} \text{ tubs}$$

and the drain empties

$$\frac{1}{8}\left(\frac{40}{3}\right) = \frac{5}{3} \text{ tubs}$$

so $\frac{8}{3} - \frac{5}{3} = 1$ tub is filled.

We must be careful not to follow the methods for solving work problems blindly. It is easy to make up problems where the methods above do not make sense. With a little common sense, we can identify such problems.

EXAMPLE 17.16. Working with an old adding machine it takes me 12 hours to do my tax returns. With my new calculator, I can do the job in 8 hours. How long will it take me using both machines?

Solution.   It will take 8 hours. I can't use both machines at once, so the fastest way is to do the whole job with the calculator.

Now, recall the rate, time, and distance problems in Section 3.4. We have an interesting variation of these problems involving rational expressions when we are given a total time rather than a total distance.

EXAMPLE 17.17. You walk to a town, mostly downhill, and return mostly uphill at an average rate of 2 miles per hour less. The total trip takes 3 hours. If the distance each way is 4 miles, what is your average rate each way?

Solution.   As in Unit 3, we make a chart for the rate, time, and distance. If the rate walking to town is r, the chart is:

	Rate	Time	Distance
To town	r	$\dfrac{4}{r}$	4
Return	$r-2$	$\dfrac{4}{r-2}$	4

Since we are given a total time of 3 hours, we add the times:

$$\frac{4}{r} + \frac{4}{r-2} = 3$$

$$r(r-2)\left(\frac{4}{r}\right) + r(r-2)\left(\frac{4}{r-2}\right) = r(r-2)(3)$$

$$(r-2)(4) + r(4) = r(r-2)(3)$$

$$4r - 8 + 4r = 3r^2 - 6r$$

$$0 = 3r^2 - 14r + 8$$

$$0 = (r-4)(3r-2)$$

$$r - 4 = 0 \text{ and } 3r - 2 = 0$$

$$r = 4 \text{ and } r = \frac{2}{3}.$$

If $r = 4$, $r - 2 = 2$. If $r = \frac{2}{3}$, $r - 2 = -\frac{4}{3}$. Since we cannot have a negative time, $r = \frac{2}{3}$ is extraneous. Your average rate going to the town is 4 miles per hour, and your average rate to return is 2 miles per hour. To check, you go 4 miles at 4 miles per hour in 1 hour. You go 4 miles at 2 miles per hour in 2 hours. Therefore, the total time is 3 hours.

EXAMPLE 17.18.   It takes a boat a total of $3\frac{1}{2}$ hours to go to an island 18 nautical miles away and to return. Because of an incoming tide, the boat returns at an average rate of 3 knots (nautical miles per hour) more than its average rate going out. What is its average rate each way?

Solution.   The chart is:

	Rate	Time	Distance
Go	r	$\dfrac{18}{r}$	18
Return	$r+3$	$\dfrac{18}{r+3}$	18

Since the total time is $3\frac{1}{2}$ hours,

$$\frac{18}{r} + \frac{18}{r+3} = \frac{7}{2}$$

$$2r(r+3)\left(\frac{18}{r}\right) + 2r(r+3)\left(\frac{18}{r+3}\right) = 2r(r+3)\left(\frac{7}{2}\right)$$

$$2(r+3)(18) + 2r(18) = r(r+3)(7)$$

$$36r + 108 + 36r = 7r^2 + 21r$$

$$0 = 7r^2 - 51r - 108$$

$$0 = (r-9)(7r+12)$$

$$r - 9 = 0 \text{ and } 7r + 12 = 0$$

$$r = 9 \text{ and } r = -\frac{12}{7}.$$

Since a rate cannot be negative, $-\frac{12}{7}$ is extraneous. The boat goes out to the island at an average rate of 9 knots and returns at an average rate of 12 knots. You should check this solution by finding that the total time is indeed $3\frac{1}{2}$ hours.

Exercise 17.2

1. A number is $\frac{5}{6}$ more than its reciprocal. Find the number.

2. The reciprocal of a number is $2\frac{2}{3}$ less than the number. Find the number.

3. Twice the reciprocal of a number is 1 less than the number. Find the number.

4. A number is $\frac{1}{2}$ less than 3 times its reciprocal. Find the number.

5. A number is $1\frac{3}{4}$ more than the reciprocal of twice the number. Find the number.

6. The reciprocal of 4 times a number is $\frac{5}{12}$ less than the number. Find the number.

7. A number is 2 more than its reciprocal. Find the number.

8. The reciprocal of twice a number is $1\frac{1}{2}$ more than the number. Find the number.

9. A copier can duplicate a 300 page document in 20 minutes, but it takes another copier 30 minutes. How long will it take if both copiers are used?

10. A new machine can cap 100 bottles in $3\frac{1}{3}$ minutes. It takes the old machine 5 minutes to cap 100 bottles. How long will it take to cap 100 bottles if both machines are used?

11. One boy scout can peel all the potatoes for the camp's dinner in 3 hours, and another takes 5 hours. If they do the job together, how long will it take?

12. A farmer has a tractor which can plow a field in 12 hours, one which takes 8 hours, and one which takes 6 hours. If he and two helpers use all three tractors, how long will it take to plow the field?

13. One washing machine can wash a load in 30 minutes, and a second machine takes 40 minutes. How long will it take the two machines to wash one load?

14. One person can walk to town in 3 hours, but his two friends take only $2\frac{1}{2}$ hours. How long will it take all three walking together?

15. Using two machines, a road can be surfaced in 4 hours. One of the machines can do the job alone in 6 hours. How long would it take the other machine to do the job alone?

16. One newspaper carrier can cover the route in 2 hours. If the route is split with a second carrier, the two together take $1\frac{1}{4}$ hours. How long would it take the second carrier to cover the route alone?

17. One pipe fills a tank in 30 minutes, and another empties it in 40 minutes. If both pipes are open, how long will it take to fill the tank?

18. A tub has two faucets and a drain. One faucet can fill the tub in $4\frac{1}{2}$ minutes, and the other can fill the tub in 6 minutes. With both faucets and the drain open, it takes 3 minutes to fill the tub. How long does it take the drain to empty the tub?

19. A bus goes to town, and returns at an average rate of 10 miles per hour less. The town is 10 miles away. If the total time for the round trip is $1\frac{1}{2}$ hours, what is the average rate of the bus each way?

20. A boat sails 3 miles across a lake, and returns against the wind at an average rate of $4\frac{1}{2}$ miles per hour less. If the total trip takes $2\frac{1}{2}$ hours, what is the average rate of the boat each way?

21. You walk for 2 miles, and then jog for 4 miles at an average rate of 4 miles per hour more. Your total time is 1 hour. What are your average rate walking and your average rate jogging?

22. A man bicycles 1 mile from his cabin to a lake, and then takes his motor boat 8 miles across the lake. The boat goes 10 miles per hour faster than he bicycles. If the total time for the trip is $\frac{2}{3}$ of an hour, what are his average rate bicycling and the average rate of the boat?

Self-test

Solve for x:

1. $\dfrac{3}{2} - \dfrac{x}{x + 6} = \dfrac{6}{x + 6}$

2. $\dfrac{3x}{2x - 4} - \dfrac{5}{2} = \dfrac{6}{x}$

3. $\dfrac{x}{x^2 - 16} = \dfrac{4}{x - 4} + \dfrac{3}{x + 4}$

4. The reciprocal of three times a number is $\frac{1}{3}$ less than the number. Find the number.

5. One mower can mow a lawn in 2 hours, while another takes $1\frac{3}{4}$ hours. If both mowers are used, how long will it take to mow the lawn?

1. ___________________

2. ___________________

3. ___________________

4. ___________________

5. ___________________

<table><tr><td>Unit
18</td><td># Equations with
Radical Expressions</td></tr></table>

INTRODUCTION

In this unit you will apply the tools of algebra you have learned in preceding units to solving equations involving radical expressions. Such equations are encountered in many situations. For example, radicals are found in applications of the distance formula. You will use the distance formula and equations involving radical expressions in an important application called conic sections, which are quadratic relations in two variables.

OBJECTIVES

When you have finished this unit you should be able to:

1. Solve equations involving radical expressions.
2. Find the equation and draw the graph of a parabola, ellipse, circle, or hyperbola centered at the origin.

<table><tr><td>Section
18.1</td><td>## Solving Equations with Radical Expressions</td></tr></table>

In Unit 14, we learned the rules for radicals. In particular, recall the rule

$$\sqrt[n]{a^n} = a,$$

since $\sqrt[n]{a^n} = (a^n)^{\frac{1}{n}} = a^1 = a$. A related rule is

$$(\sqrt[n]{a})^n = a,$$

since $(\sqrt[n]{a})^n = (a^{\frac{1}{n}})^n = a^1 = a$. In this unit we will consider only real numbers, so a must be positive when n is even.

Using the second rule, we can remove a radical expression from an equation by raising both sides of the equation to the nth power.

EXAMPLE 18.1. Solve $\sqrt{x + 3} = 4$.

Solution. Since the equation involves a square root, we square both sides:

$$\sqrt{x + 3} = 4$$

$$(\sqrt{x + 3})^2 = 4^2.$$

Then, since $(\sqrt{x + 3})^2 = x + 3$, we have

$$x + 3 = 16$$
$$x = 13.$$

To check,

$$\sqrt{x + 3} = 4$$
$$\sqrt{13 + 3} \overset{?}{=} 4$$
$$\sqrt{16} \overset{?}{=} 4$$
$$4 = 4.$$

Therefore, the solution is 13.

EXAMPLE 18.2.   Solve $\sqrt[3]{2x + 10} = 2$.

Solution.   Since the equation involves a cube root, we cube both sides:

$$\sqrt[3]{2x + 10} = 2$$
$$\left(\sqrt[3]{2x + 10}\right)^3 = 2^3.$$

Then, since $(\sqrt[3]{2x + 10})^3 = 2x + 10$, we have

$$2x + 10 = 8$$
$$2x = -2$$
$$x = -1.$$

To check,

$$\sqrt[3]{2x + 10} = 2$$
$$\sqrt[3]{2(-1) + 10} \overset{?}{=} 2$$
$$\sqrt[3]{-2 + 10} \overset{?}{=} 2$$
$$\sqrt[3]{8} \overset{?}{=} 2$$
$$2 = 2.$$

Therefore, the solution is -1.

If the radical expression is not isolated on one side of the equation, we must isolate it before raising both sides of the equation to the nth power.

EXAMPLE 18.3.   Solve $\sqrt[3]{x - 4} + 2 = 0$.

Solution.   First we isolate the radical:

$$\sqrt[3]{x - 4} + 2 = 0$$

$$\sqrt[3]{x - 4} = -2.$$

Now, we may cube both sides and solve:

$$\left(\sqrt[3]{x - 4}\right)^3 = (-2)^3$$

$$x - 4 = -8$$

$$x = -4.$$

To check, we use the original equation:

$$\sqrt[3]{x - 4} + 2 = 0$$

$$\sqrt[3]{-4 - 4} + 2 \overset{?}{=} 0$$

$$\sqrt[3]{-8} + 2 \overset{?}{=} 0$$

$$-2 + 2 = 0,$$

so the solution is -4.

It is possible for equations involving radical expressions to have extraneous solutions. This can happen when we have a radical with an even index. Since a negative number raised to an even power gives a positive number, when we raise both sides of an equation to an even power we may lose minus signs.

EXAMPLE 18.4.   Solve $\sqrt{x + 1} + 2 = 0$.

Solution.   We isolate the radical, then square both sides and solve:

$$\sqrt{x + 1} + 2 = 0$$

$$\sqrt{x + 1} = -2$$

$$\left(\sqrt{x + 1}\right)^2 = (-2)^2$$

$$x + 1 = 4$$

$$x = 3.$$

To check,

$$\sqrt{x + 1} + 2 = 0$$

$$\sqrt{3 + 1} + 2 \overset{?}{=} 0$$

$$\sqrt{4} + 2 \overset{?}{=} 0$$

$$2 + 2 \neq 0.$$

The number $x = 3$ is an extraneous solution, and the equation has no solution. In squaring both sides of the equation

$$\sqrt{x + 1} = -2$$

we lose the negative on the right-hand side. Observe that $x = 3$ is a solution to the equation

$$\sqrt{x + 1} - 2 = 0$$

or

$$\sqrt{x + 1} = 2.$$

To solve more complicated equations involving square roots, recall the square of a binomial from Section 8.1:

$$(P + Q)^2 = P^2 + 2PQ + Q^2.$$

If an equation contains a square root, we must square both sides of the equation. In doing this, we must remember to include the middle term $2PQ$.

EXAMPLE 18.5.   Solve $\sqrt{3x + 7} = x + 3$.

Solution.   Since the square root is isolated, we square both sides of the equation:

$$\sqrt{3x + 7} = x + 3$$

$$(\sqrt{3x + 7})^2 = (x + 3)^2$$

$$3x + 7 = (x + 3)^2.$$

Now, we square the binomial, remembering to include the middle term $6x$:

$$3x + 7 = x^2 + 6x + 9.$$

The resulting equation is a quadratic equation. To solve this equation, we may collect the terms and factor:

$$0 = x^2 + 3x + 2$$

$$0 = (x + 1)(x + 2)$$

$$x + 1 = 0 \text{ and } x + 2 = 0$$

$$x = -1 \text{ and } x = -2.$$

We must check both numbers since one or both could be extraneous solutions. To check $x = -1$,

$$\sqrt{3x + 7} = x + 3$$

$$\sqrt{3(-1) + 7} \overset{?}{=} -1 + 3$$

$$\sqrt{-3 + 7} \overset{?}{=} 2$$

$$\sqrt{4} \overset{?}{=} 2$$

$$2 = 2.$$

To check $x = -2$,

$$\sqrt{3(-2) + 7} \overset{?}{=} -2 + 3$$

$$\sqrt{-6 + 7} \overset{?}{=} 1$$

$$\sqrt{1} \overset{?}{=} 1$$

$$1 = 1.$$

Therefore, -1 and -2 both are solutions.

When we derive a quadratic equation from an equation involving square roots, we may not be able to solve the quadratic equation by factoring. In this case, we must use the quadratic formula.

EXAMPLE 18.6.   Solve $\sqrt{2x - 3} = x - 2$.

Solution.   Since the square root is isolated, we square both sides of the equation:

$$\sqrt{2x - 3} = x - 2$$

$$(\sqrt{2x - 3})^2 = (x - 2)^2$$

$$2x - 3 = (x - 2)^2.$$

We square the binomial, remembering to include the middle term, and collect terms:

$$2x - 3 = x^2 - 4x + 4$$

$$0 = x^2 - 6x + 7.$$

The resulting equation cannot be solved by factoring, so we use the quadratic formula:

$$x = \frac{-b \pm \sqrt{b^2 - 4ac}}{2a}$$

$$= \frac{-(-6) \pm \sqrt{(-6)^2 - 4(1)(7)}}{2(1)}$$

$$= \frac{6 \pm \sqrt{36 - 28}}{2}$$

$$= \frac{6 \pm \sqrt{8}}{2}$$

$$= \frac{6 \pm 2\sqrt{2}}{2}$$

$$= 3 \pm \sqrt{2}.$$

To check these numbers, we use a calculator or table of square roots. Since $\sqrt{2} \approx 1.414$,

$3 + \sqrt{2} \approx 4.414$. Then,

$$\sqrt{2x - 3} = x - 2$$

$$\sqrt{2(4.414) - 3} \stackrel{?}{\approx} 4.414 - 2$$

$$\sqrt{8.828 - 3} \stackrel{?}{\approx} 2.414$$

$$\sqrt{5.828} \stackrel{?}{\approx} 2.414.$$

Using a calculator, or estimating from a table of square roots,

$$\sqrt{5.828} \approx 2.414,$$

so $3 + \sqrt{2}$ is a solution. Similarly, $3 - \sqrt{2} \approx 1.586$, so we have

$$\sqrt{2x - 3} = x - 2$$

$$\sqrt{2(1.586) - 3} \stackrel{?}{\approx} 1.586 - 2.$$

Since a positive square root cannot equal a negative number, $3 - \sqrt{2}$ is an extraneous solution. The only solution is $3 + \sqrt{2}$.

When the square root is not already isolated, we must isolate it before squaring both sides of the equation.

EXAMPLE 18.7.   Solve $\sqrt{4x + 2} + 2x = 3$.

Solution.   First, we isolate the square root:

$$\sqrt{4x + 2} + 2x = 3$$

$$\sqrt{4x + 2} = 3 - 2x.$$

Now we may square both sides of the equation, square the binomial remembering to include the middle term, collect terms and solve:

$$(\sqrt{4x + 2})^2 = (3 - 2x)^2$$

$$4x + 2 = (3 - 2x)^2$$

$$4x + 2 = 9 - 12x + 4x^2$$

$$0 = 4x^2 - 16x + 7$$

$$0 = (2x - 1)(2x - 7)$$

$$2x - 1 = 0 \text{ and } 2x - 7 = 0$$

$$x = \frac{1}{2} \text{ and } x = \frac{7}{2}.$$

We check these numbers in the original equation. For $x = \frac{1}{2}$,

$$\sqrt{4x + 2} + 2x = 3$$

$$\sqrt{4\left(\frac{1}{2}\right) + 2} + 2\left(\frac{1}{2}\right) \stackrel{?}{=} 3$$

$$\sqrt{2 + 2} + 1 \stackrel{?}{=} 3$$

$$\sqrt{4} + 1 \stackrel{?}{=} 3$$

$$2 + 1 = 3,$$

so $x = \frac{1}{2}$ is a solution. For $x = \frac{7}{2}$,

$$\sqrt{4x + 2} + 2x = 3$$

$$\sqrt{4\left(\frac{7}{2}\right) + 2} + 2\left(\frac{7}{2}\right) \stackrel{?}{=} 3$$

$$\sqrt{14 + 2} + 7 \stackrel{?}{=} 3$$

$$\sqrt{16} + 7 \stackrel{?}{=} 3$$

$$4 + 7 \neq 3.$$

The number $x = \frac{7}{2}$ is an extraneous solution. The only solution is $\frac{1}{2}$.

An equation may involve two square roots. To solve such equations we must square twice because one of the square roots will appear in the middle term.

EXAMPLE 18.8.   Solve $\sqrt{x - 8} = \sqrt{x} - 4$.

Solution.   Since one of the square roots is isolated, we square both sides of the equation:

$$\sqrt{x - 8} = \sqrt{x} - 4$$

$$(\sqrt{x - 8})^2 = (\sqrt{x} - 4)^2$$

$$x - 8 = (\sqrt{x} - 4)^2.$$

Now we square the binomial, remembering that $(\sqrt{x})^2 = x$, and remembering to include the middle term:

$$x - 8 = x - 8\sqrt{x} + 16.$$

A square root appears in the middle term. We isolate this square root:

$$-8 = -8\sqrt{x} + 16$$

$$8\sqrt{x} = 24$$

$$\sqrt{x} = 3.$$

Squaring both sides again, we have

$$(\sqrt{x}\,)^2 = 3^2$$

$$x = 9.$$

Checking this number,

$$\sqrt{x - 8} = \sqrt{x} - 4$$

$$\sqrt{9 - 8} \stackrel{?}{=} \sqrt{9} - 4$$

$$\sqrt{1} \stackrel{?}{=} 3 - 4$$

$$1 \neq -1.$$

Therefore, $x = 9$ is an extraneous solution. Since there are no other possible numbers, the equation has no solution.

It is possible to determine whether a number which does not check in an equation involving square roots is an extraneous solution or simply incorrect. In the equation above, if the sign of the left-hand square root were changed, we would have

$$-\sqrt{x - 8} = \sqrt{x} - 4$$

$$-\sqrt{1} \stackrel{?}{=} 3 - 4$$

$$-1 = -1.$$

Thus 9 is the solution to the related equation

$$-\sqrt{x - 8} = \sqrt{x} - 4.$$

Once both sides have been squared, we cannot tell the related equation from the original equation. Whenever we can change one or more signs, so that our number is a solution to a related equation, we have an extraneous solution. If we cannot derive a related equation, the solution is incorrect and the procedure of solving the equation should be checked.

EXAMPLE 18.9.   Solve $\sqrt{2x + 5} + \sqrt{x + 2} = 1$.

Solution.   It is a common error simply to square each term. To do this is to forget the middle term. If the equation were squared in its stated form, there would be a middle term on the left-hand side of $2\sqrt{2x + 5}\,\sqrt{x + 2}$. We must isolate one square root, and then square both sides remembering to include the middle term:

$$\sqrt{2x + 5} + \sqrt{x + 2} = 1$$

$$\sqrt{2x + 5} = 1 - \sqrt{x + 2}$$

$$(\sqrt{2x + 5}\,)^2 = (1 - \sqrt{x + 2}\,)^2$$

$$2x + 5 = (1 - \sqrt{x + 2}\,)^2$$

$$2x + 5 = 1 - 2\sqrt{x + 2} + x + 2.$$

Now, we may isolate the remaining square root:

$$2\sqrt{x + 2} = -x - 2.$$

In this example, we will leave the coefficient 2 on the left-hand side. To square this expression, we must be sure to square the coefficient, and then multiply by each term in the square root:

$$(2\sqrt{x + 2})^2 = (-x - 2)^2$$

$$4(x + 2) = (-x - 2)^2$$

$$4x + 8 = (-x - 2)^2.$$

Now we square the binomial, collect terms, and solve:

$$4x + 8 = x^2 + 4x + 4$$

$$4 = x^2$$

$$\pm 2 = x.$$

We must check the numbers 2 and -2 in the original equation. For $x = 2$,

$$\sqrt{2x + 5} + \sqrt{x + 2} = 1$$

$$\sqrt{2(2) + 5} + \sqrt{2 + 2} \stackrel{?}{=} 1$$

$$\sqrt{9} + \sqrt{4} \stackrel{?}{=} 1$$

$$3 + 2 \neq 1.$$

We see that $x = 2$ is a solution to the related equation

$$\sqrt{2x + 5} - \sqrt{x + 2} = 1,$$

but is an extraneous solution of the original equation. For $x = -2$,

$$\sqrt{2x + 5} + \sqrt{x + 2} = 1$$

$$\sqrt{2(-2) + 5} + \sqrt{-2 + 2} \stackrel{?}{=} 1$$

$$\sqrt{1} + \sqrt{0} \stackrel{?}{=} 1$$

$$1 + 0 = 1,$$

so $x = -2$ is a solution. The only solution is -2.

Exercise 18.1

Solve for x:

1. $\sqrt{x + 2} = 3$

2. $\sqrt{2x - 3} = 2$

3. $\sqrt{4x + 3} - 3 = 0$

4. $\sqrt{4x + 6} - 2 = 0$

5. $\sqrt[3]{3x + 12} = 2$

6. $\sqrt[3]{4x + 1} + 3 = 0$

7. $\sqrt[4]{3x + 2} = 1$

8. $\sqrt[4]{4x - 4} - 2 = 0$

9. $\sqrt{x - 4} + 1 = 0$

10. $\sqrt{2x - 1} + 3 = 0$

11. $\sqrt{3x - 2} + 1 = 6$

12. $\sqrt{2x + 5} + 5 = 2$

13. $\sqrt{2x - 1} + 2 = x$

14. $\sqrt{3x + 7} + x = 7$

15. $\sqrt{2x + 5} - x = 3$

16. $\sqrt{4x - 3} + 2x = 1$

17. $\sqrt{2x + 5} - 2 = x$

18. $\sqrt{2x + 2} + x = 1$

19. $\sqrt{x + 3} - \sqrt{x} = 3$

20. $\sqrt{x - 9} + \sqrt{x} = 9$

21. $\sqrt{2x - 1} - \sqrt{x - 1} = 1$

22. $\sqrt{3x - 8} - \sqrt{x - 2} = 2$

23. $\sqrt{3x + 10} + \sqrt{x + 3} = 1$

24. $\sqrt{3x - 5} + \sqrt{x + 6} = 1$

Section 18.2   The Conic Sections

The distance formula, given in Section 6.1, is

$$d = \sqrt{(x_2 - x_1)^2 + (y_2 - y_1)^2}$$

where (x_1, y_1) and (x_2, y_2) are two points in the Cartesian coordinate system, and d is the distance between the points. Because of the square root in the distance formula, applications of it result in equations involving radical expressions. Such applications often involve two variables. Instead of solutions which are one or two distinct numbers, these applications give equations in two variables. Their solutions are ordered pairs of numbers which can be indicated by a graph. We will look at one important example of such an application in this section.

EXAMPLE 18.10.   Find the equation of all points which are equidistant from the point $(0, 2)$ and the line $y = -2$.

Solution.   The point (x, y) illustrates the points described:

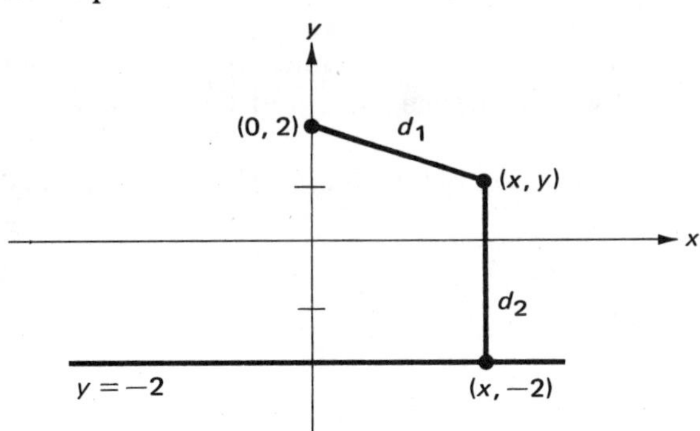

The distance between the points $(0, 2)$ and (x, y) is

$$d_1 = \sqrt{(x - 0)^2 + (y - 2)^2}$$

$$= \sqrt{x^2 + (y - 2)^2} \ .$$

The distance between the line $y = -2$ and the point (x, y) is the vertical distance

$$d_2 = y + 2.$$

Therefore, since the distances are equal,

$$\sqrt{x^2 + (y - 2)^2} = y + 2.$$

We square both sides of the equation and solve for y:

$$\left[\sqrt{x^2 + (y - 2)^2}\,\right]^2 = (y + 2)^2$$

$$x^2 + (y - 2)^2 = (y + 2)^2$$

$$x^2 + y^2 - 4y + 4 = y^2 + 4y + 4$$

$$x^2 = 8y$$

$$\frac{1}{8}x^2 = y$$

or

$$y = \frac{1}{8}x^2.$$

You should recognize this equation as an example of the quadratic function introduced in Unit 7. Its graph is a parabola:

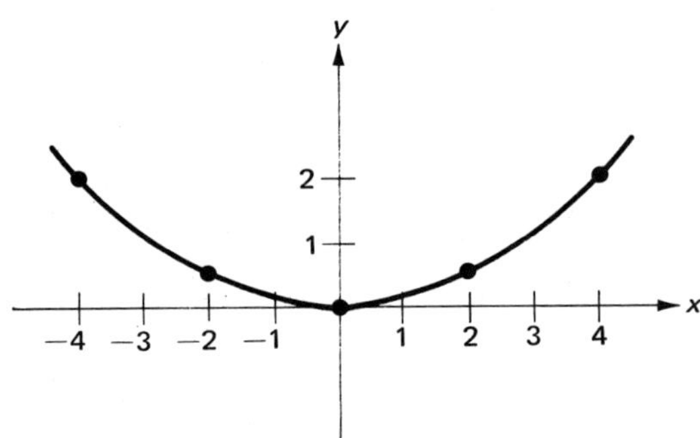

Every point of this parabola is equidistant from the point $(0, 2)$ and the line $y = -2$. The point is called the **focus** of the parabola and the line is called the **directrix** of the parabola.

EXAMPLE 18.11.   Find the equation of all points which are equidistant from the point $(-1, 0)$ and the line $x = 1$.

Solution.

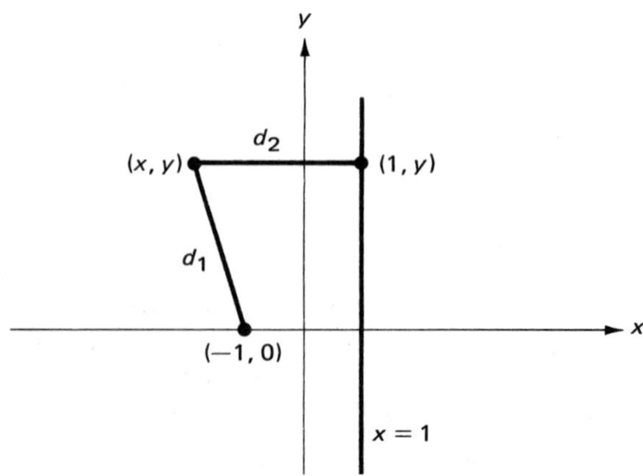

The distance between the points $(-1, 0)$ and (x, y) is

$$d_1 = \sqrt{(x + 1)^2 + (y - 0)^2}$$

$$= \sqrt{(x + 1)^2 + y^2}\ .$$

The distance between the line $x = 1$ and the point (x, y) is the horizontal distance

$$d_2 = 1 - x.$$

Since the distances are equal,

$$\sqrt{(x + 1)^2 + y^2} = 1 - x$$

$$\left[\sqrt{(x + 1)^2 + y^2}\right]^2 = (1 - x)^2$$

$$(x + 1)^2 + y^2 = (1 - x)^2$$

$$x^2 + 2x + 1 + y^2 = 1 - 2x + x^2$$

$$y^2 = -4x.$$

The graph is

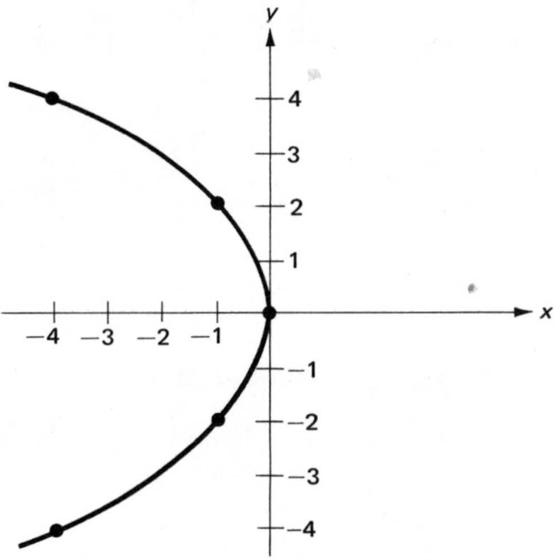

This graph is the graphs of $y = \sqrt{-4x}$ and $y = -\sqrt{-4x}$ combined. The graph is not the graph of a function because for each value of x, except $x = 0$, there are two values of y. However, the graph is a parabola. The focus of the parabola is the point $(-1, 0)$ and the directrix of the parabola is the line $x = 1$.

In general, whenever we have the equation of all points which are equidistant from a given point and a given line, the graph is a parabola. The given point is the focus of the parabola and the given line is the directrix of the parabola.

Using other combinations of distances, we derive equations which have a different kind of graph.

EXAMPLE 18.12.   Find the equation of all points such that the sum of their distances from the points (2, 0) and (−2, 0) is 6.

Solution.   The point (x, y) illustrates the points described:

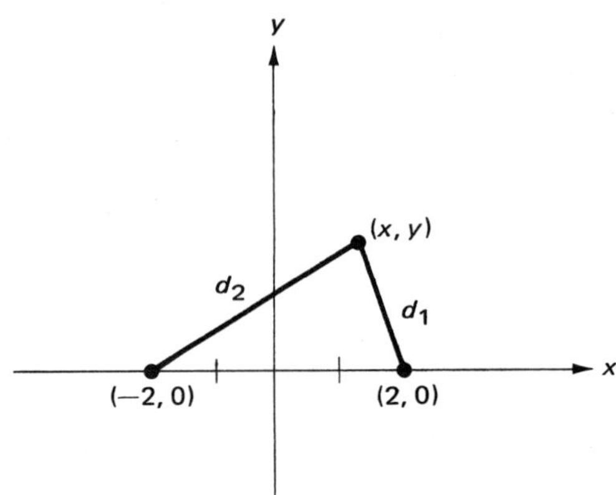

The distance between the points (2, 0) and (x, y) is

$$d_1 = \sqrt{(x - 2)^2 + (y - 0)^2}$$

$$= \sqrt{(x - 2)^2 + y^2}\ .$$

The distance between the points (−2, 0) and (x, y) is

$$d_2 = \sqrt{(x + 2)^2 + (y - 0)^2}$$

$$= \sqrt{(x + 2)^2 + y^2}\ .$$

Since the sum of the distances is 6,

$$\sqrt{(x - 2)^2 + y^2} + \sqrt{(x + 2)^2 + y^2} = 6.$$

We isolate one of the square roots and square both sides, being careful to include the middle term:

$$\sqrt{(x - 2)^2 + y^2} = 6 - \sqrt{(x + 2)^2 + y^2}$$

$$\left[\sqrt{(x - 2)^2 + y^2}\,\right]^2 = \left[6 - \sqrt{(x + 2)^2 + y^2}\,\right]^2$$

$$(x - 2)^2 + y^2 = 36 - 12\sqrt{(x + 2)^2 + y^2} + (x + 2)^2 + y^2$$

$$x^2 - 4x + 4 = 36 - 12\sqrt{(x + 2)^2 + y^2} + x^2 + 4x + 4$$

$$12\sqrt{(x + 2)^2 + y^2} = 8x + 36$$

$$3\sqrt{(x + 2)^2 + y^2} = 2x + 9.$$

Now we square both sides again:

$$\left[3\sqrt{(x + 2)^2 + y^2}\,\right]^2 = (2x + 9)^2$$

$$9\left[(x + 2)^2 + y^2\right] = (2x + 9)^2$$

$$9(x^2 + 4x + 4 + y^2) = 4x^2 + 36x + 81$$

$$9x^2 + 36x + 36 + 9y^2 = 4x^2 + 36x + 81$$

$$5x^2 + 9y^2 = 45.$$

The graph of this equation is called an **ellipse**. To draw the ellipse, we find its intercepts. If $y = 0$,

$$5x^2 = 45$$

$$x^2 = 9$$

$$x = \pm 3.$$

Therefore, the x-intercepts are $(3, 0)$ and $(-3, 0)$. If $x = 0$,

$$9y^2 = 45$$

$$y^2 = 5$$

$$y = \pm\sqrt{5}\,.$$

Therefore, the y-intercepts are $(0, \sqrt{5}\,)$ and $(0, -\sqrt{5}\,)$. The graph is

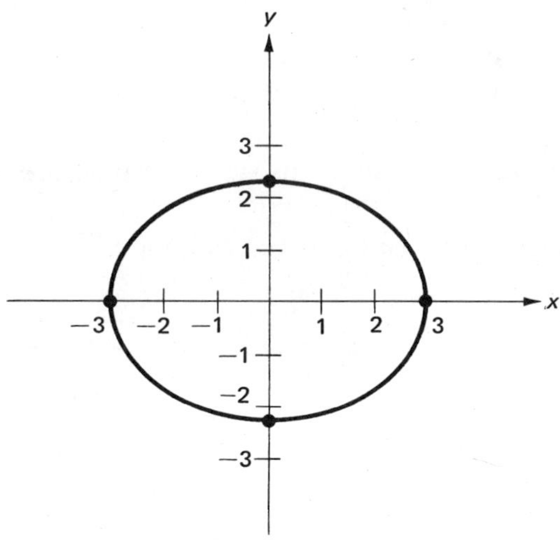

In general, the graph of all points such that the sum of their distances from two given points is a constant, is an ellipse. The two given points are called the **foci** of the ellipse. ("Foci" is the plural of "focus.") When the foci are on the x-axis, and are equidistant from the origin, the center of the ellipse is at the origin. In this case, the ellipse has an equation of the form

$$Ax^2 + By^2 = C$$

where A, B, and C all are positive, and A is less than B. If the foci of an ellipse are on the y-axis, and are equidistant from the origin, the equation is the same but A is greater than B. The graph has the form

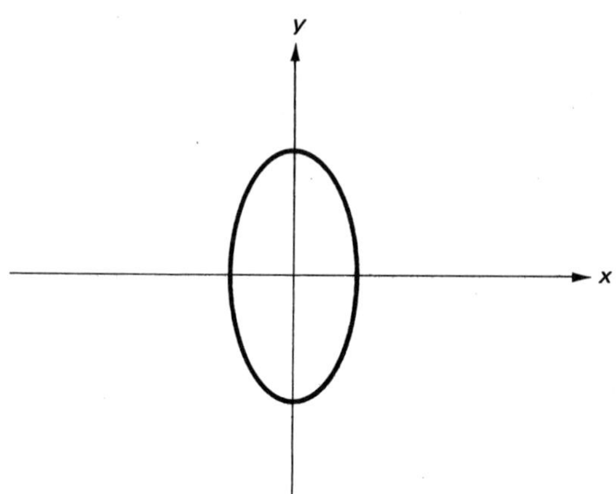

The **circle** is a special case of the ellipse. In this case, the foci are both the same point. If this point is the origin, $A = B$ and

$$Ax^2 + Ay^2 = C$$

$$x^2 + y^2 = \frac{C}{A}$$

or

$$x^2 + y^2 = r^2.$$

The center of the circle is the origin, and its radius is r.

EXAMPLE 18.13.   Find the equation of all points such that their distance from the origin is $\sqrt{5}$.

Solution.   Since the distance between $(0, 0)$ and (x, y) is $\sqrt{5}$,

$$\sqrt{5} = \sqrt{(x - 0)^2 + (y - 0)^2}$$

$$\sqrt{5} = \sqrt{x^2 + y^2}$$

$$(\sqrt{5})^2 = \left(\sqrt{x^2 + y^2}\right)^2$$

$$5 = x^2 + y^2$$

or

$$x^2 + y^2 = 5.$$

The graph is a circle with its center at the origin and radius $\sqrt{5}$:

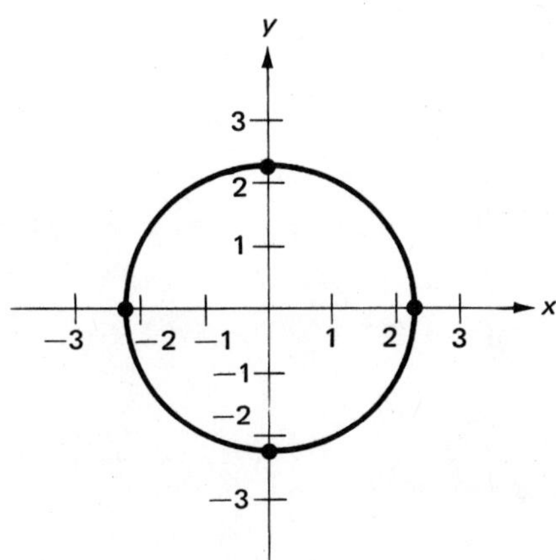

EXAMPLE 18.14.   Find the equation of all points such that the difference of their distances from $(3, 0)$ and $(-3, 0)$ is 2.

Solution.   The point (x, y) illustrates the points described:

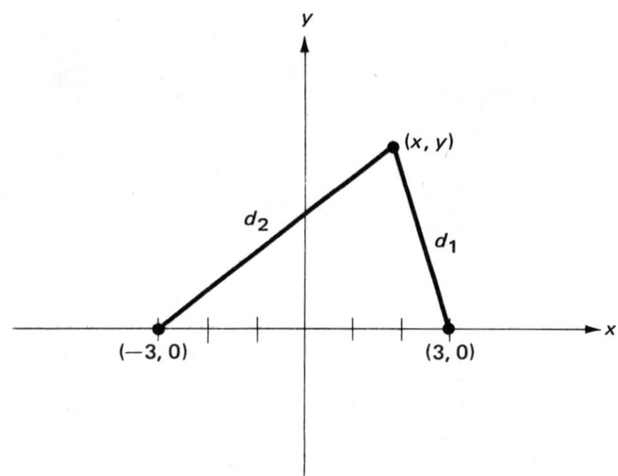

The distance between the points $(3, 0)$ and (x, y) is

$$d_1 = \sqrt{(x - 3)^2 + (y - 0)^2}$$

$$= \sqrt{(x - 3)^2 + y^2} \ .$$

The distance between the points $(-3, 0)$ and (x, y) is

$$d_2 = \sqrt{(x + 3)^2 + (y - 0)^2}$$

$$= \sqrt{(x + 3)^2 + y^2} \ .$$

Since the difference of the distances is 2,

$$\sqrt{(x+3)^2 + y^2} - \sqrt{(x-3)^2 + y^2} = 2.$$

We isolate one square root and square both sides, being careful to include the middle term:

$$\sqrt{(x+3)^2 + y^2} = 2 + \sqrt{(x-3)^2 + y^2}$$

$$\left[\sqrt{(x+3)^2 + y^2}\,\right]^2 = \left[2 + \sqrt{(x-3)^2 + y^2}\,\right]^2$$

$$(x+3)^2 + y^2 = 4 + 4\sqrt{(x-3)^2 + y^2} + (x-3)^2 + y^2$$

$$x^2 + 6x + 9 = 4 + 4\sqrt{(x-3)^2 + y^2} + x^2 - 6x + 9$$

$$12x - 4 = 4\sqrt{(x-3)^2 + y^2}$$

$$3x - 1 = \sqrt{(x-3)^2 + y^2}\ .$$

Squaring again,

$$(3x-1)^2 = \left[\sqrt{(x-3)^2 + y^2}\,\right]^2$$

$$(3x-1)^2 = (x-3)^2 + y^2$$

$$9x^2 - 6x + 1 = x^2 - 6x + 9 + y^2$$

$$8x^2 - y^2 = 8.$$

The graph of this equation is called a **hyperbola**. To draw the hyperbola, we find its intercepts. If $y = 0$,

$$8x^2 = 8$$

$$x^2 = 1$$

$$x = \pm 1.$$

The x-intercepts are $(1, 0)$ and $(-1, 0)$. If $x = 0$,

$$-y^2 = 8$$

$$y^2 = -8$$

$$y = \pm\sqrt{-8}\ .$$

Since $\sqrt{-8}$ is not a real number, there are no y-intercepts. We need more points to draw the graph. If $x = 2$ or $x = -2$,

$$8x^2 - y^2 = 8$$

$$8(4) - y^2 = 8$$

$$32 - y^2 = 8$$

$$-y^2 = -24$$

$$y^2 = 24$$

$$y = \pm\sqrt{24}$$

$$y = \pm 2\sqrt{6}\,.$$

The points $(2, 2\sqrt{6}\,)$, $(2, -2\sqrt{6}\,)$, $(-2, 2\sqrt{6}\,)$, and $(-2, -2\sqrt{6}\,)$ all are on the graph. The graph is

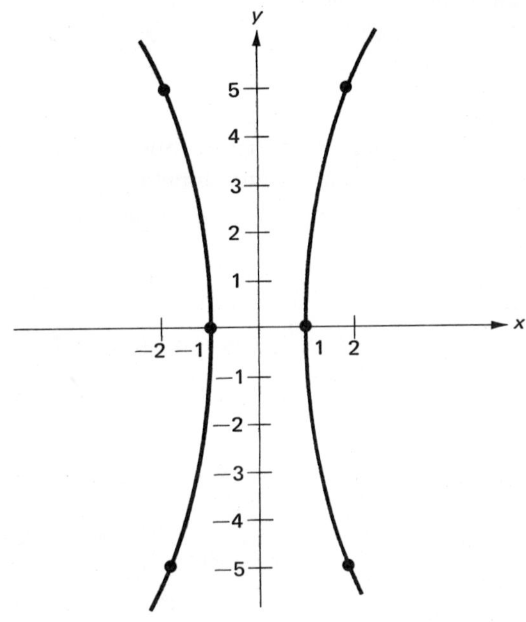

It is a common error to think of the hyperbola as two parabolas back-to-back. This is not the case. The equation of the hyperbola has a different form from that of the parabola, and the graph has different properties.

In general, the graph of all points such that the difference of their distances from two given points is a constant, is a hyperbola. The two given points are the foci of the hyperbola. When the foci are on the x-axis, and are equidistant from the origin, the center of the hyperbola is at the origin. In this case, the hyperbola has an equation of the form

$$Ax^2 - By^2 = C$$

where A, B, and C all are positive. If the foci are on the y-axis, and are equidistant from the origin, the equation has the form

$$Ay^2 - Bx^2 = C$$

where A, B, and C all are positive. The graph has the form

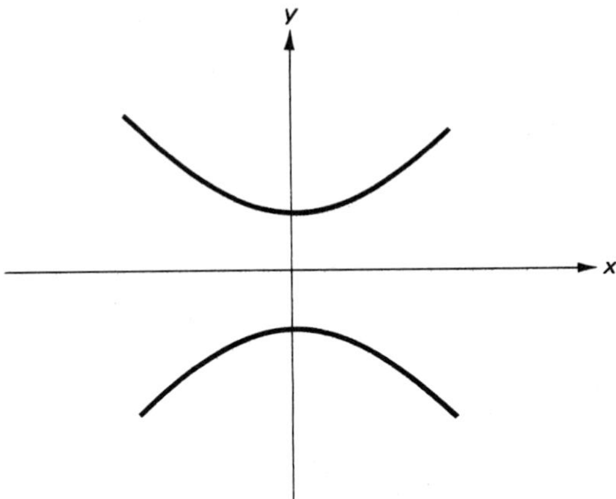

The parabola, the ellipse, the circle, and the hyperbola are called the **conic sections**. Each of the conic sections can be cut from a generalized cone consisting of two cones placed with their tips together:

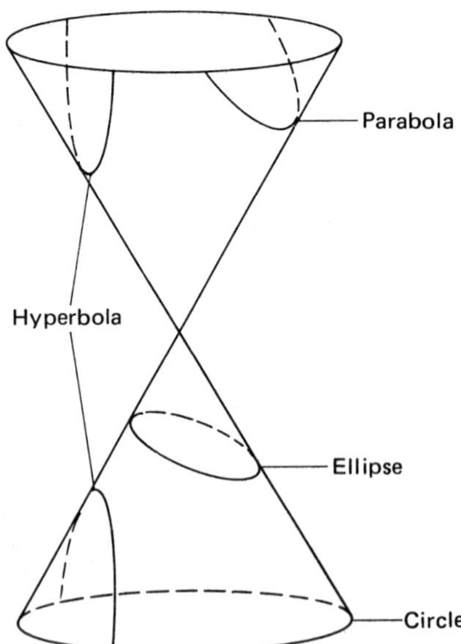

The parabola is cut parallel to one side of the cones, the ellipse is cut at a slant less than the slant of the sides, the circle is cut horizontally, and the hyperbola is cut vertically through both parts. To find out more about the conic sections, you should consult a book which includes, or has the title, *Analytic Geometry*.

Exercise 18.2

Find the equation and draw the graph of:

1. All points equidistant from the point $(0, 1)$ and the line $y = -1$.

2. All points equidistant from the point $(0, \frac{1}{4})$ and the line $y = -\frac{1}{4}$.

3. All points equidistant from the point $(2, 0)$ and the line $x = -2$.

4. All points equidistant from the point $(\frac{1}{8}, 0)$ and the line $x = -\frac{1}{8}$.

5. All points equidistant from the point $(-\frac{1}{2}, 0)$ and the line $x = \frac{1}{2}$.

6. All points equidistant from the point $(0, -1)$ and the line $y = 1$.

7. All points such that the sum of their distances from the points $(1, 0)$ and $(-1, 0)$ is 4.

8. All points such that the sum of their distances from the points $(3, 0)$ and $(-3, 0)$ is 10.

9. All points such that the sum of their distances from the points $(0, 2)$ and $(0, -2)$ is 8.

10. All points such that the sum of their distances from the points $(0, \sqrt{5})$ and $(0, -\sqrt{5})$ is 6.

11. All points such that their distance from the origin is 4.

12. All points such that their distance from the origin is $\sqrt{3}$.

13. All points such that the difference of their distances from the points $(2, 0)$ and $(-2, 0)$ is 2.

14. All points such that the difference of their distances from the points $(\sqrt{2}, 0)$ and $(-\sqrt{2}, 0)$ is 2.

15. All points such that the difference of their distances from the points $(0, 5)$ and $(0, -5)$ is 6.

16. All points such that the difference of their distances from the points $(0, \sqrt{13})$ and $(0, -\sqrt{13})$ is 4.

Self-test

Solve for x:

1. $\sqrt{3x - 5} + 5 = 1$

                  1. ________________

2. $\sqrt{2x - 2} = x - 3$

                  2. ________________

3. $\sqrt{x - 12} + \sqrt{x} = 6$

                  3. ________________

4. $\sqrt{2x - 4} + \sqrt{x - 1} = 1$

                  4. ________________

5. Find the equation and draw the graph of all points such that the sum of their distances from the points $(1, 0)$ and $(-1, 0)$ is 6.

                  5. ________________

Systems of Equations

INTRODUCTION

In previous units, you have solved various types of equations in one variable. You have also drawn graphs of equations in two variables, where the solutions are ordered pairs of real numbers and there are infinitely many solutions. In this unit you will learn to solve and graph systems of two equations in two variables. The solutions of such systems, when they exist, are ordered pairs of real numbers. You will also learn how to solve systems of three linear equations in three variables, and some applications of systems of two or three linear equations.

OBJECTIVES

When you have finished this unit you should be able to:

1. Solve systems of two linear equations in two variables by graphing.
2. Solve systems of two linear equations in two variables algebraically.
3. Solve systems of three linear equations in three variables algebraically.
4. Solve applied problems involving systems of linear equations in two or three variables.
5. Solve and graph systems of two equations in two variables involving quadratic equations.

<table>
<tr><td>Section
19.1</td><td>Solutions by Graphing</td></tr>
</table>

Suppose we have the graphs of two lines in the Cartesian coordinate system. The equations of the two lines are called a **system of two linear equations in two variables**. If the lines are neither parallel nor identical, they cross at exactly one point. We call this point the **point of intersection** of the two lines. The ordered pair of coordinates of the point of intersection is the **solution** of the system of two linear equations in two variables.

EXAMPLE 19.1.   Graph the system:

$$x + y - 5 = 0$$

$$x - 2y + 4 = 0$$

and find its solution.

Solution.   First we graph the two lines. In general, we will draw the graphs by finding the y- and x-intercepts of each line. The process of finding the y- and x-intercepts was discussed in Section 6.4. The graphs are

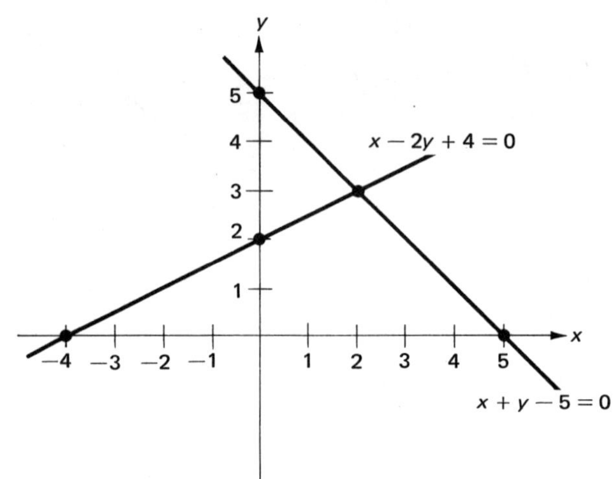

It is a common error to confuse the words "intercept" and "intersect." The *intercepts* of each line are the points where each line crosses the y-axis and the x-axis. The lines *intersect* at the point where they cross one another. We see from the graph that the lines appear to intersect at the point $(2, 3)$. To check, we substitute $(2, 3)$ into each of the equations. For $x + y - 5 = 0$,

$$2 + 3 - 5 = 0,$$

and for $x - 2y + 4 = 0$,

$$2 - 2(3) + 4 = 0.$$

Therefore, $(2, 3)$ is on both lines. Since the point is on both lines, it must be the point of the intersection. The ordered pair $(2, 3)$ is the solution of the system.

EXAMPLE 19.2.   Graph the system:
$$3y - 2x = 6$$
$$2y - 3x = 9$$

and find its solution.

Solution.   The graph is

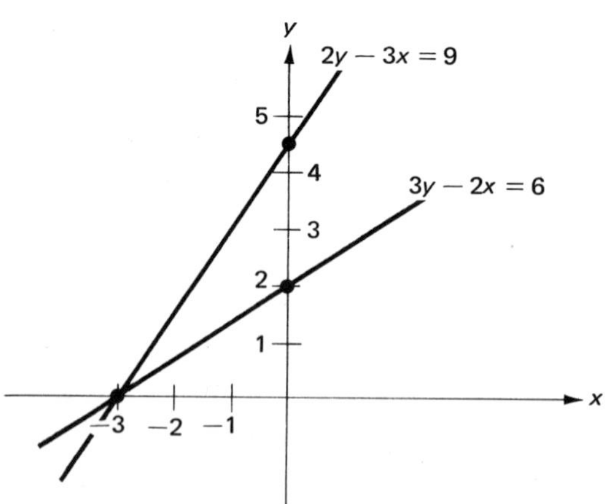

Each of the lines has x-intercept $(-3, 0)$. Clearly the point of intersection is $(-3, 0)$. Thus $(-3, 0)$ is the solution of the system. You should check $(-3, 0)$ in each equation to see that the point is on both lines.

EXAMPLE 19.3. Graph the system:

$$y = 4x + 6$$

$$y = 2$$

and find its solution.

Solution.   The graph is

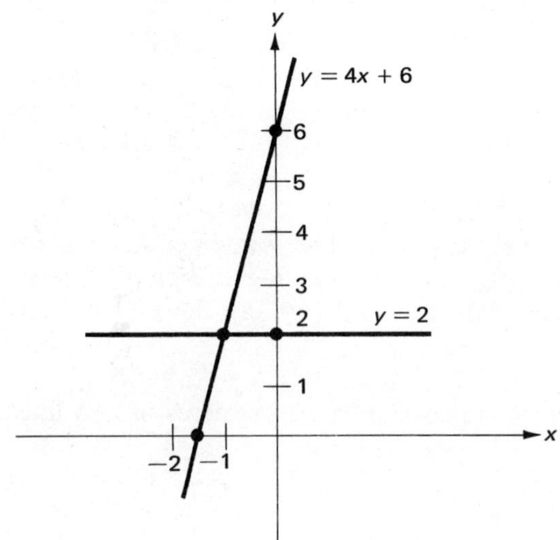

We see that the lines intersect at the point $(-1, 2)$. Thus $(-1, 2)$ is the solution of the system. Clearly $(-1, 2)$ checks for the equation $y = 2$. You should check this solution in the first equation.

There are two special cases of systems of two linear equations in two variables. In one case, the lines are parallel.

EXAMPLE 19.4. Graph the system:

$$2x + y = 3$$

$$2x + y + 5 = 0.$$

Solution.   The graph is

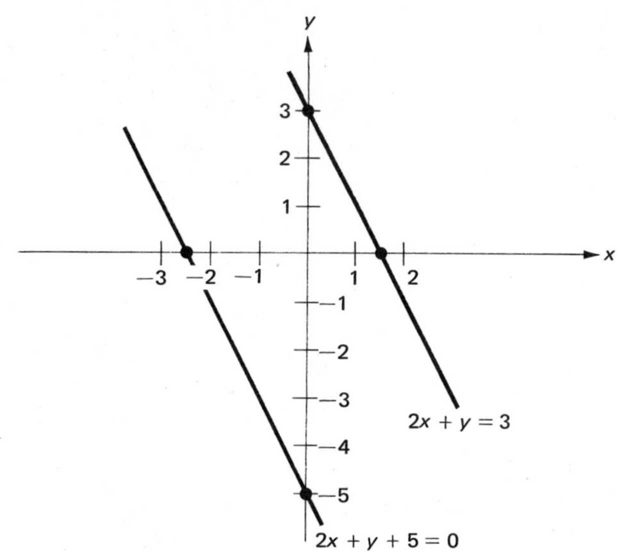

It appears that the lines are parallel. If this is the case, then the lines have no point of intersection, and the system has no solution. It is possible to show that the lines are indeed parallel. We recall the slope-intercept form of a linear function, introduced in Section 6.4:

$$y = mx + b,$$

where m is the slope of the line and $(0, b)$ is the y-intercept. If we write the equations of the given lines in this form, we have

$$y = -2x + 3$$

$$y = -2x - 5.$$

When two lines have the same slope but different y-intercepts, then the lines are parallel. Both lines have slope $m = -2$. The y-intercept of the first line is $(0, 3)$ and the y-intercept of the second line is $(0, -5)$. Thus the lines are parallel, there is no point of intersection, and the system has no solution. In this case, we say the system is **inconsistent**. An inconsistent system has no solution.

In the second special case of systems of two linear equations in two variables, the lines are identical.

EXAMPLE 19.5.   Graph the system:

$$x - \tfrac{1}{2}y = 2$$

$$y = 2x - 4.$$

Solution.   The graph is

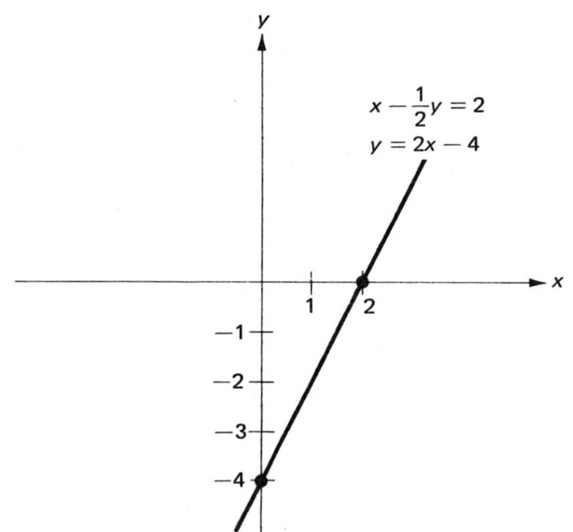

The lines are identical. Since each point of the identical lines is a point of intersection, each point of the lines is a solution, and the system has infinitely many solutions. Writing the first equation in slope-intercept form, we have

$$x - \frac{1}{2}y = 2$$

$$2x - y = 4$$

$$2x - 4 = y$$

or

$$y = 2x - 4.$$

The second equation is already in slope-intercept form, and the equations are equivalent. In this case, we say the system is **dependent**. A dependent system has infinitely many solutions.

Exercise 19.1

Graph each system, and find the solution if it exists, or state that the system is inconsistent or dependent:

1. $x+y-2=0$
   $x-y=0$

2. $2x-y=6$
   $x+2y=-2$

3. $x-3y=3$
   $3y-4x=6$

4. $2x+3y+4=0$
   $4x+y-2=0$

5. $3x+4y=12$
   $x-2y=4$

6. $5x-4y=10$
   $5x-8y=20$

7. $y=x+2$
   $y=-1$

8. $x=-1$
   $x-y+3=0$

9. $2x+y=4$
   $2x+y=2$

10. $2x-3y=6$
    $3y-2x=3$

11. $y=\frac{1}{2}x-3$
    $x-2y-6=0$

12. $y=-3x+3$
    $x+\frac{1}{3}y-1=0$

<table><tr><td>Section
19.2</td><td><h1 style="text-align:center">Algebraic Solutions</h1></td></tr></table>

In the preceding section, we have assumed that the solutions of systems were ordered pairs of integers. Clearly, if the solutions involved fractions, or irrational numbers, we would not be able to read the solutions accurately from graphs. For example, consider the system of equations $2x-y-2=0$ and $2x+3y-6=0$. The graph is

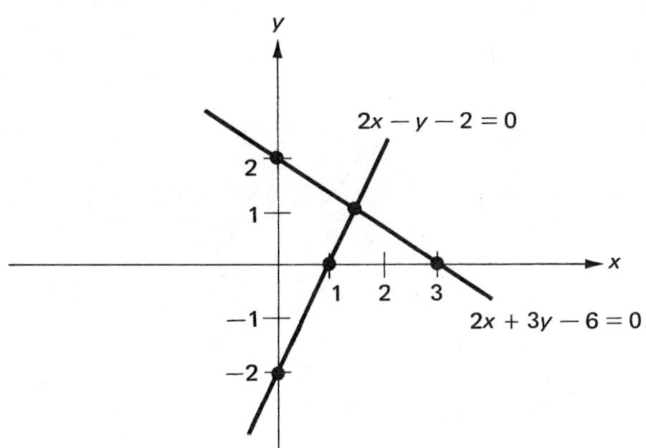

The x-coordinate of the point of intersection could be $\frac{3}{2}$, or $\frac{4}{3}$, or $\sqrt{2} \approx 1.4$, or any of many other fractions or irrational numbers.

Using algebraic methods, we can find an exact solution to a system of equations. We eliminate one variable, and then can solve an equation in the remaining one variable. One way to eliminate a variable is the **addition method**.

Addition Rule: Given a system of equations, an equivalent system can be derived by adding two equations together and replacing one of the equations by the result.

EXAMPLE 19.6. Solve the system:

$$x + y = 1$$
$$x - y = 5.$$

Solution.   We add the two equations together by adding like terms on each side:

$$\begin{array}{r} x + y = 1 \\ x - y = 5 \\ \hline 2x \quad\;\; = 6 \\ x = 3. \end{array}$$

In the equation $x = 3$, we have eliminated the variable y. Now we may replace either equation by the equation $x = 3$. If we choose to replace the second equation, we have the equivalent system

$$x + y = 1$$
$$x = 3.$$

This system is easily solved. We know $x = 3$. Replacing x by 3 in the first equation,

$$3 + y = 1$$
$$y = -2.$$

The solution is $(3, -2)$. To check, we substitute $(3, -2)$ in each equation. For $x + y = 1$,

$$3 + (-2) = 1,$$

and for $x - y = 5$,

$$3 - (-2) = 5.$$

EXAMPLE 19.7. Solve the system:

$$2x - 3y - 5 = 0$$
$$x + 3y - 7 = 0.$$

Solution.   We will generally write the equations in the form

$$2x - 3y = 5$$
$$x + 3y = 7.$$

Now, we add the equations, and replace one of them by the result:

$$\begin{array}{r} 2x - 3y = 5 \\ x + 3y = 7 \\ \hline 3x \quad\;\; = 12 \\ x = 4. \end{array}$$

An equivalent system is

$$2x - 3y = 5$$
$$x = 4.$$

Replacing x by 4 in the first equation, we have

$$8 - 3y = 5$$
$$-3y = -3$$
$$y = 1.$$

Thus the solution is (4, 1). You should check this solution by substituting in each of the original equations.

To solve the system given at the beginning of this section, we need an additional step.

EXAMPLE 19.8.   Solve the system:

$$2x - y - 2 = 0$$
$$2x + 3y - 6 = 0.$$

Solution.   We write the system in the form

$$2x - y = 2$$
$$2x + 3y = 6.$$

Observe that if we simply add the equations together, we will not eliminate a variable. We must multiply both sides of the first equation by either -1 or 3. If we choose to multiply by -1, we have

$$-1(2x - y) = -1(2)$$
$$-2x + y = -2.$$

Now, we have the system

$$-2x + y = -2$$
$$2x + 3y = 6$$
$$\overline{4y = 4}$$
$$y = 1.$$

The new equation, together with the original first equation, forms an equivalent system

$$2x - y = 2$$
$$y = 1.$$

Replacing y by 1 in the first equation,

$$2x - 1 = 2$$
$$2x = 3$$
$$x = \frac{3}{2}.$$

The solution is $(\frac{3}{2}, 1)$. To check, we substitute in the original equations. For $2x - y - 2 = 0$,

$$2\left(\frac{3}{2}\right) - 1 - 2 \overset{?}{=} 0$$

$$3 - 1 - 2 = 0.$$

For $2x + 3y - 6 = 0$,

$$2\left(\frac{3}{2}\right) + 3(1) - 6 \overset{?}{=} 0$$

$$3 + 3 - 6 = 0.$$

EXAMPLE 19.9.   Solve the system:

$$3x - 4y = -6$$

$$4x - 3y = -1.$$

Solution.   We multiply both sides of the first equation by 3, and also both sides of the second equation by -4:

$$3(3x - 4y) = 3(-6)$$

$$-4(4x - 3y) = -4(-1).$$

Now we have the system

$$9x - 12y = -18$$

$$-16x + 12y = 4$$

$$\overline{}$$

$$-7x \qquad = -14$$

$$x = 2.$$

An equivalent system is

$$3x - 4y = -6$$

$$x = 2.$$

Replacing x by 2 in the first equation,

$$3(2) - 4y = -6$$

$$6 - 4y = -6$$

$$-4y = -12$$

$$y = 3.$$

The solution is $(2, 3)$. You should check this solution by substituting in each of the original equations.

A second way to eliminate a variable is the **substitution method.** We use the substitution method when one of the equations has been solved for one variable, or can easily be solved for one variable.

Substitution Rule: Given a system of equations with one equation solved for one variable, an equivalent system can be derived by substituting for that variable in another equation and replacing one of the equations by the result.

EXAMPLE 19.10. Solve the system:

$$y = 2x - 1$$

$$3x - y = 2.$$

Solution. The first equation is solved for y. We substitute $2x - 1$ for y in the second equation, being careful to use parentheses:

$$3x - y = 2$$

$$3x - (2x - 1) = 2$$

$$3x - 2x + 1 = 2$$

$$x + 1 = 2$$

$$x = 1.$$

An equivalent system is

$$y = 2x - 1$$

$$x = 1.$$

Replacing x by 1 in the first equation,

$$y = 2(1) - 1$$

$$y = 1.$$

The solution is $(1, 1)$. To check, we substitute in each of the original equations. For $y = 2x - 1$,

$$1 = 2(1) - 1.$$

For $3x - y = 2$,

$$3(1) - 1 = 2.$$

EXAMPLE 19.11. Solve the system:

$$x = 3$$

$$2x + 4y = -3.$$

Solution. Since the first equation is solved for x, we simply replace x by 3 in the second equation:

$$2x + 4y = -3$$

$$2(3) + 4y = -3$$

$$6 + 4y = -3$$

$$4y = -9$$

$$y = -\frac{9}{4}.$$

The solution is $(3, -\frac{9}{4})$. Clearly $(3, -\frac{9}{4})$ checks for $x = 3$. You should check this solution in the second equation.

Using algebraic methods, it is easy to identify systems which are inconsistent or dependent.

EXAMPLE 19.12.   Solve the system:

$$3x - 2y = 1$$
$$6x - 4y = -1.$$

Solution.   We use the addition method, multiplying both sides of the first equation by -2:

$$3x - 2y = 1$$
$$-2(3x - 2y) = -2(1)$$
$$-6x + 4y = -2.$$

We have the system

$$-6x + 4y = -2$$
$$\underline{6x - 4y = -1}$$
$$0 = -3.$$

Since the statement $0 = -3$ is not true, there are no possible solutions. The system is inconsistent.

EXAMPLE 19.13.   Solve the system:

$$2x - 3y - 3 = 0$$
$$y = \frac{2}{3}x - 1.$$

Solution.   We use the substitution method, replacing y by $\frac{2}{3}x - 1$ in the first equation:

$$2x - 3y - 3 = 0$$
$$2x - 3\left(\frac{2}{3}x - 1\right) - 3 = 0$$
$$2x - 2x + 3 - 3 = 0$$
$$0 = 0.$$

Since the statement $0 = 0$ is true for all x and y, there are infinitely many possible solutions. The system is dependent.

Exercise 19.2

Solve the system:

1. $x + y = 8$
   $x - y = 2$

2. $x - y = -2$
   $x + y = 6$

3. $x - y = 4$
   $x + y = 5$

4. $x + y - 2 = 0$
   $x - y - 4 = 0$

5. $x + 3y = 7$
   $2x + 3y = 5$

6. $4x - 3y = 2$
   $4x + 2y = -8$

7. $4x - y = 11$
   $3x + 5y = 14$

8. $x - 5y = 11$
   $3x - 7y = 9$

9. $5x - 2y = 0$
   $-4x + 5y = 17$

10. $-3x + 4y = 0$
    $7x - 6y = 0$

11. $7x + 3y = 8$
    $3x + 5y = 9$

12. $6x - 3y + 11 = 0$
    $9x - 4y + 15 = 0$

13. $2x + y = 3$
    $4x + 2y = 5$

14. $6x - 9y - 2 = 0$
    $-4x + 6y + 3 = 0$

15. $-6x + 9y = 6$
    $10x - 15y = -10$

16. $x + \frac{1}{2}y - 2 = 0$
    $3x + \frac{3}{2}y - 6 = 0$

17. $2x - 3y = 12$
    $y = 4$

18. $x = -2$
    $3x + 4y = 4$

19. $y = 3x + 1$
    $6x - 3y + 4 = 0$

20. $x = 2y - 3$
    $3y - x = 1$

21. $y = 3x + 5$
    $y = 4x + 8$

22. $y = x + 4$
    $y = -3x + 2$

23. $y = \frac{2}{3}x - 3$
    $2x - 3y = 9$

24. $y = \frac{5}{2}x + \frac{7}{4}$
    $10x - 4y + 5 = 0$

Section 19.3

Three Linear Equations in Three Variables

A useful technique for solving a system of equations is to find an equivalent system which is in **triangular form**. A system is in triangular form if each equation has at least one of the variables of the preceding equation eliminated. The equivalent systems we have used in the preceding section are in triangular form. For example, in the system

$$2x - y = 2$$
$$y = 1$$

the variable x of the first equation has been eliminated in the second equation. A system which is in triangular form is easily solved by substitution.

EXAMPLE 19.14.   Solve the system:

$$x - 2y - 2z = 1$$
$$2x + 3y = 3$$
$$x = 3.$$

Solution.   The system is in triangular form, and we know from the last equation that $x = 3$. Replacing x by 3 in the second equation, we have

$$2x + 3y = 3$$
$$2(3) + 3y = 3$$
$$6 + 3y = 3$$
$$3y = -3$$
$$y = -1.$$

Now we know that $x = 3$ and $y = -1$. Replacing both of these variables in the first equation,

$$x - 2y - 2z = 1$$
$$3 - 2(-1) - 2z = 1$$
$$3 + 2 - 2z = 1$$
$$5 - 2z = 1$$
$$-2z = -4$$
$$z = 2.$$

Therefore, $x = 3$, $y = -1$, and $z = 2$. This result is written as the **ordered triple** $(3, -1, 2)$.

To solve a system of three linear equations in three variables, we find an equivalent system in triangular form.

EXAMPLE 19.15.   Solve the system:

$$3x + 2y - 2z = 4$$

$$2x - 2y + z = 7$$

$$4x - 3y + 3z = 11.$$

Solution.   First, we look for a variable we can eliminate from two of the equations. For example, we can eliminate y from the second equation by adding the first and second equations and replacing the second equation by the result. However, it is less easy to eliminate y from the third equation. Using a little foresight, we might plan to eliminate z from the first and third equations. It is helpful to rewrite the system with the second equation written first:

$$2x - 2y + z = 7$$

$$3x + 2y - 2z = 4$$

$$4x - 3y + 3z = 11.$$

We multiply the first equation in this system by 2 and add to the second equation:

$$4x - 4y + 2z = 14$$

$$3x + 2y - 2z = 4$$

$$\overline{}$$

$$7x - 2y = 18.$$

Also, we multiply the first equation by -3 and add to the third equation:

$$-6x + 6y - 3z = -21$$

$$4x - 3y + 3z = 11$$

$$\overline{}$$

$$-2x + 3y = -10.$$

Replacing the second and third equations by the resulting equations, we have the equivalent system

$$2x - 2y + z = 7$$

$$7x - 2y = 18$$

$$-2x + 3y = -10,$$

and we are halfway to a system in triangular form. To complete the system in triangular form, we solve the resulting system of two equations in two variables,

$$7x - 2y = 18$$

$$-2x + 3y = -10.$$

We may eliminate y by multiplying the first equation by 3 and the second equation by 2:

$$21x - 6y = 54$$

$$-4x + 6y = -20$$

$$\overline{}$$

$$17x = 34$$

$$x = 2.$$

Returning to the system of three equations, and replacing the third equation by $x = 2$, we have an equivalent system in triangular form:

$$2x - 2y + z = 7$$
$$7x - 2y = 18$$
$$x = 2.$$

Starting with $x = 2$, we replace x by 2 in the second equation:

$$7x - 2y = 18$$
$$7(2) - 2y = 18$$
$$14 - 2y = 18$$
$$-2y = 4$$
$$y = -2.$$

Then, using $x = 2$ and $y = -2$, we replace these variables in the first equation:

$$2x - 2y + z = 7$$
$$2(2) - 2(-2) + z = 7$$
$$4 + 4 + z = 7$$
$$8 + z = 7$$
$$z = -1.$$

The solution is the ordered triple $(2, -2, -1)$. To check, we substitute in each of the original equations. For the equation $3x + 2y - 2z = 4$,

$$3(2) + 2(-2) - 2(-1) \stackrel{?}{=} 4$$
$$6 - 4 + 2 = 4.$$

For $2x - 2y + z = 7$,

$$2(2) - 2(-2) + (-1) \stackrel{?}{=} 7$$
$$4 + 4 - 1 = 7.$$

For $4x - 3y + 3z = 11$,

$$4(2) - 3(-2) + 3(-1) \stackrel{?}{=} 11$$
$$8 + 6 - 3 = 11.$$

A given system of equations may be partially in triangular form.

EXAMPLE 19.16.   Solve the system:

$$x \qquad -2z = 0$$
$$2x - 3y \qquad = 5$$
$$y - z = -1.$$

Solution.   The first two equations start the triangular form. It is not necessary for the first equation to include the variable y. However, the third equation must have only the variable x or the variable y. We must eliminate z from the third equation. To do this, we need another equation in y and z. We multiply the first equation by -2 and add to the second equation:

$$-2x \qquad + 4z = 0$$
$$2x - 3y \qquad = 5$$
$$\overline{\hspace{3cm}}$$
$$-3y + 4z = 5.$$

Now, we have a system of two equations in y and z:

$$-3y + 4z = 5$$
$$y - z = -1.$$

To eliminate z, we multiply the second equation by 4 and add:

$$-3y + 4z = 5$$
$$4y - 4z = -4$$
$$\overline{\hspace{3cm}}$$
$$y = 1.$$

An equivalent system in triangular form is then

$$x - 2z = 0$$
$$2x - 3y = 5$$
$$y = 1.$$

Replacing y by 1 in the second equation,

$$2x - 3y = 5$$
$$2x - 3(1) = 5$$
$$2x - 3 = 5$$
$$2x = 8$$
$$x = 4.$$

Replacing x by 4 in the first equation,

$$x - 2z = 0$$

$$4 - 2z = 0$$

$$- 2z = -4$$

$$z = 2.$$

The solution is the ordered triple (4, 1, 2). You should check this solution in each of the original equations.

The graph of a linear equation in three variables is a plane in three-dimensional space. An ordered triple represents a point in three-dimensional space. In general, three planes in three-dimensional space have exactly one point in common:

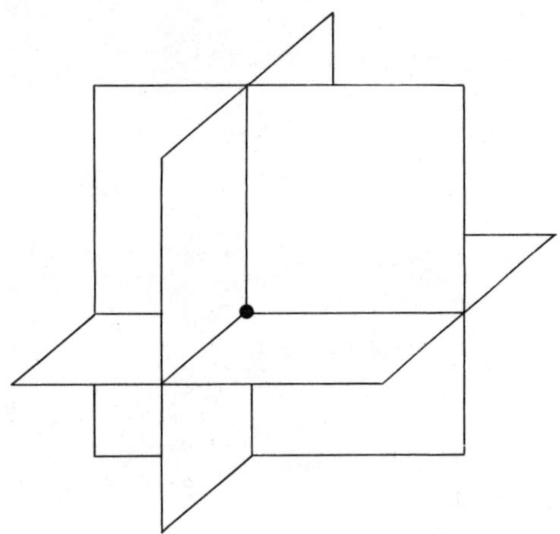

The common point is the solution of a system of three linear equations in three variables.

It is possible for three planes to have no point common to all three. These are some of the arrangements of the planes which can cause this situation:

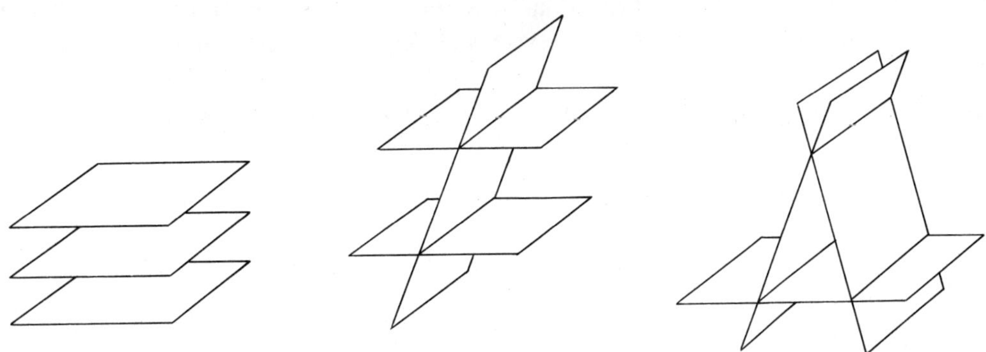

In this case, the system has no solution and is inconsistent.

It is also possible for the planes to be identical, or to have a line of points in common:

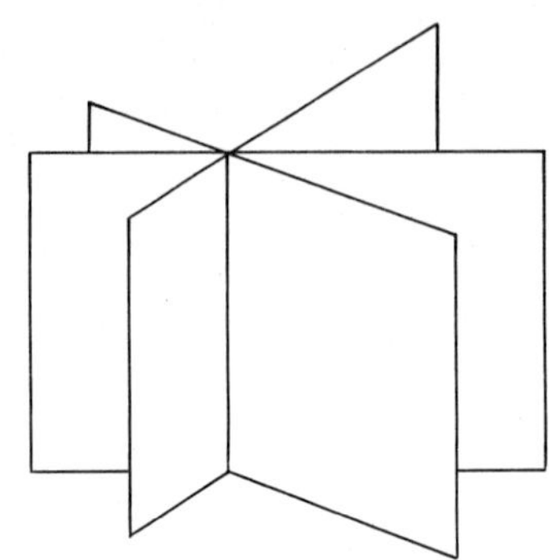

In this case, the system has infinitely many solutions and is dependent. We can recognize such inconsistent and dependent systems in the same ways as we did in the preceding section.

Exercise 19.3

Solve the system:

1. $2x - 4y - 3z = 2$
   $3x - 2y = 2$
   $x = 2$

2. $3x + 2y + 3z = 2$
   $2y - z = 4$
   $z = -1$

3. $2x - y - z = 4$
   $x - 2y + z = -1$
   $3x - 3y + 2z = 1$

4. $2x - 2y + 3z = 2$
   $x - 2y + 2z = 3$
   $-x + 3y + 2z = 4$

5. $2x + 2y - z = 2$
   $3x - y = 5$
   $3x - z = 2$

6. $y - 2z = -4$
   $x - 2z = -1$
   $2x + 2y + 3z = 1$

7. $2x - 3z = 7$
   $2x - 3y = 4$
   $2y - z = -1$

8. $y - 3z = -3$
   $x - z = 1$
   $3x + y = 3$

9. $x + 2y + 3z = -1$
   $2x + 3y + z = 2$
   $3x + 4y - z = 5$

10. $2x - 3y + z = 3$
    $3x - y - z = 0$
    $4x + y - 3z = 3$

**Section
19.4**   Applications

In many types of applications it is convenient, or necessary, to use two or more variables. For such applications, we use a system of two or more equations.

EXAMPLE 19.17.   The sum of two numbers is 45, and their difference is 33. Find the numbers.

Solution.   Let x and y represent the numbers. Then

$$x + y = 45$$

and

$$x - y = 33.$$

We must solve the system of equations

$$x + y = 45$$

$$x - y = 33$$

$$2x \quad\;\; = 78$$

$$x = 39.$$

If $x = 39$,

$$39 + y = 45$$

$$y = 6.$$

The numbers are 39 and 6. To check, we simply observe that $39 + 6 = 45$ and $39 - 6 = 33$.

It is possible to solve the preceding example using one variable. However, it is convenient to use two variables. Mixture problems are another type of problem where it is often convenient to use two variables. The following problem is solved using one variable in Section 3.3.

EXAMPLE 19.18. A 10-pound bag of potatoes contains Idahos which sell for 24¢ a pound and all-purpose Maine which sell for 16¢ a pound. The bag costs $1.84. How many pounds of each kind of potato does it contain?

Solution. We let x represent the number of pounds of Idaho potatoes and y represent the number of pounds of Maine potatoes. We may make a chart as before:

	Unit price	Number of units	Total price
Idaho	24	x	24x
Maine	16	y	16y
Mixture		10	184

The sums of the second and third columns give the system of equations

$$x + y = 10$$

$$24x + 16y = 184.$$

Multiplying the first equation by -16,

$$-16x - 16y = -160$$

$$24x + 16y = 184$$

$$8x \qquad\;\; = 24$$

$$x = 3.$$

Since

$$x + y = 10$$

$$3 + y = 10$$

$$y = 7.$$

There are 3 pounds of Idaho potatoes and 7 pounds of Maine potatoes. You should check this solution.

If a mixture has more than two components, it is usually convenient to use systems of equations. We may use as many variables as there are components in the mixture.

EXAMPLE 19.19.   A 16-ounce can of nuts contains peanuts, cashews, and almonds. There are twice as many ounces of peanuts as cashews. The peanuts are worth 6¢ an ounce, the cashews are worth 10¢ an ounce, and the almonds are worth 15¢ an ounce. If the can sells for $1.25, how many ounces of each kind of nut does it contain?

Solution.   We let x represent the number of ounces of peanuts, y the number of ounces of cashews, and z the number of ounces of almonds, and make the chart:

	Unit price	Number of units	Total price
Peanuts	6	x	$6x$
Cashews	10	y	$10y$
Almonds	15	z	$15z$
Mixture		16	125

Since there are twice as many ounces of peanuts as cashews,

$$x = 2y.$$

This equation, together with the sums of the second and third columns, give a system of three equations in three variables:

$$x = 2y$$

$$x + y + z = 16$$

$$6x + 10y + 15z = 125.$$

This system may be solved by substitution. Replacing x by $2y$ in the second equation,

$$x + y + z = 16$$

$$2y + y + z = 16$$

$$3y + z = 16,$$

and replacing x by $2y$ in the third equation,

$$6x + 10y + 15z = 125$$

$$6(2y) + 10y + 15z = 125$$

$$22y + 15z = 125.$$

The resulting system is

$$3y + z = 16$$

$$22y + 15z = 125.$$

We multiply the first equation by -15 to solve for y:

$$-45y - 15z = -240$$

$$22y + 15z = 125$$

$$\overline{}$$

$$-23y = -115$$

$$y = 5.$$

To find z,

$$3y + z = 16$$

$$3(5) + z = 16$$

$$z = 1.$$

To find x,

$$x = 2y$$

$$x = 2(5)$$

$$x = 10.$$

There are 10 ounces of peanuts, 5 ounces of cashews, and 1 ounce of almonds. You should check this solution.

We can also use systems of equations to solve a kind of rate, time, and distance problem which we cannot solve using only one variable.

EXAMPLE 19.20.   An excursion boat cruises downstream for 3 hours. The passengers take the train back, but the crew spends 9 hours taking the boat back upstream. If the boat travels 45 miles each way, what are the average rates of the boat and the current?

Solution.   We let r be the rate of the boat, and c be the rate of the current. Going downstream, the rates combine, and the actual rate is $r + c$. Going upstream, the current works against the boat, and the actual rate is $r - c$. We make the chart:

	Rate	Time	Distance
Downstream	$r+c$	3	45
Upstream	$r-c$	9	45

Since $R \cdot T = D$, we have the system of equations

$$(r + c)(3) = 45$$

$$(r - c)(9) = 45,$$

or, dividing the first equation by 3 and the second by 9,

$$r + c = 15$$

$$r - c = 5$$

$$2r = 20$$

$$r = 10.$$

If $r = 10$,

$$r + c = 15$$

$$10 + c = 15$$

$$c = 5.$$

The average rate of the boat is 10 miles per hour, and the average rate of the current is 5 miles per hour. You should check this solution by finding the actual rate each way, and seeing that the distance each way is 45.

There is also a useful application of systems of equations to linear and quadratic functions. Recall from Section 6.4 that, if a line is not vertical, its equation can be written in the slope-intercept form

$$y = mx + b.$$

Given two points on a line, we may use this form and a system of two equations in two variables to find the equation of the line.

EXAMPLE 19.21.   Use the form $y = mx + b$ to find the equation of the line determined by the points $(2, -1)$ and $(6, 1)$.

Solution.   We substitute the coordinates of each point for x and y in the form $y = mx + b$. For $(2, -1)$,

$$-1 = m(2) + b,$$

and for $(6, 1)$,

$$1 = m(6) + b.$$

These equations give the system of equations

$$2m + b = -1$$

$$6m + b = 1.$$

Multiplying the first equation by -1,

$$-2m - b = 1$$

$$6m + b = 1$$

$$4m = 2$$

$$m = \frac{1}{2}.$$

If $m = \frac{1}{2}$, we find that $b = -2$. Therefore, the equation of the line is $y = \frac{1}{2}x - 2$. You should check to see that both the given points are on this line.

Recall from Section 7.1 that the quadratic function is

$$y = ax^2 + bx + c,$$

where $a \neq 0$. Its graph is a parabola which opens either upward or downward. Three points determine such a parabola.

EXAMPLE 19.22.   Use the form $y = ax^2 + bx + c$ to find the equation of the parabola, opening either upward or downward, determined by the points $(1, -4)$, $(2, -3)$, and $(-1, 0)$.

Solution.   We substitute the coordinates of each point for x and y in the form $y = ax^2 + bx + c$. For $(1, -4)$,

$$-4 = a(1)^2 + b(1) + c$$
$$-4 = a + b + c.$$

For $(2, -3)$,

$$-3 = a(2)^2 + b(2) + c$$
$$-3 = 4a + 2b + c.$$

For $(-1, 0)$,

$$0 = a(-1)^2 + b(-1) + c$$
$$0 = a - b + c.$$

These equations give a system of three equations in three variables:

$$a + b + c = -4$$
$$4a + 2b + c = -3$$
$$a - b + c = 0.$$

Multiplying the first equation by -1 and adding to the second,

$$\begin{aligned}
-a - b - c &= 4 \\
4a + 2b + c &= -3 \\
\hline
3a + b &= 1.
\end{aligned}$$

Multiplying the first equation by -1 and adding to the third,

$$\begin{aligned}
-a - b - c &= 4 \\
a - b + c &= 0 \\
\hline
-2b &= 4 \\
b &= -2.
\end{aligned}$$

The triangular form is

$$a + b + c = -4$$
$$3a + b = 1$$
$$b = -2.$$

We find that $a = 1$ and $c = -3$, and so the equation is $y = x^2 - 2x - 3$. You should check to see that the three given points satisfy this equation.

Exercise 19.4

1. The sum of two numbers is 64, and their difference is 28. Find the numbers.

2. The sum of two numbers is 25, and their difference is 8. Find the numbers.

3. Corn and lima beans are mixed to make a pound (16 ounces) of succotash. The corn costs 3¢ an ounce and the lima beans cost 5¢ an ounce. The pound of succotash costs 59¢. How many ounces of each vegetable does it contain?

4. You are paid $2.90 for fifty pounds of recycled bottles and cans. Bottles are worth 7¢ a pound and cans are worth 3¢ a pound. How many pounds of each did you recycle?

5. One quart (32 ounces) of a salad dressing base is made from oil, wine vinegar, and cider vinegar. The amount of oil is 6 times the amount of wine vinegar. The oil costs 3¢ an ounce, the wine vinegar $2\frac{1}{2}$¢ an ounce, and the cider vinegar 2¢ an ounce. The quart of dressing base costs 90¢. How many ounces of each ingredient does it contain?

6. Herbs are planted in a mixture of potting soil, peat moss, and sand. For some herbs, there should be twice as much potting soil as sand, and equal amounts of peat moss and sand. If the potting soil costs 18¢ a quart, the peat moss costs 20¢ a quart, and the sand costs 8¢ a quart, then a bag of the mixture costs $1.60. How many quarts of each component does the mixture contain?

7. A canoeist paddles downstream 4 miles in 1 hour, but takes 2 hours to return the 4 miles upstream. What are the average rates of the canoe and the current?

8. A boat cruises downstream for $2\frac{1}{2}$ hours, but takes $12\frac{1}{2}$ hours to return upstream. The distance each way is 25 miles. What are the average rates of the boat and the current?

9. The distance between New York and Chicago is about 800 miles. A plane is scheduled to fly from New York to Chicago against the wind in 2 hours, but to return with the wind in $1\frac{2}{3}$ hours. What are the average rates of the plane and the wind expected to be?

10. The Rhine river steamer is scheduled to cruise upstream from Cologne to Mainz in $16\frac{2}{3}$ hours, and downstream from Mainz to Cologne in 10 hours. The distance between the cities is 200 kilometers. What are the average rates of the steamer and current expected to be?

11. Use the form $y = mx + b$ to find the equation of the line determined by the points $(1, 5)$ and $(-1, -1)$.

12. Use the form $y = mx + b$ to find the equation of the line determined by the points $(\frac{3}{2}, 0)$ and $(\frac{1}{2}, -\frac{1}{2})$.

13. Use the form $y = ax^2 + bx + c$ to find the equation of the parabola, opening either upward or downward, determined by the points $(-1, 2)$, $(1, 0)$, and $(2, 5)$.

14. Use the form $y = ax^2 + bx + c$ to find the equation of the parabola, opening either upward or downward, determined by the points $(2, 0)$, $(1, -1)$, and $(-2, -16)$.

15. Use the form $x = ay^2 + by + c$ to find the equation of the parabola, opening to the right or to the left, determined by the points $(-3, 1)$, $(-3 -1)$, and $(0, 2)$.

16. Use the form $x = ay^2 + by + c$ to find the equation of the parabola, opening to the right or to the left, determined by the points $(\frac{1}{4}, 1)$, $(1, 2)$, and $(4, 4)$.

<table>
<tr><td>Section
19.5</td><td><h1>Systems with Quadratic Equations</h1></td></tr>
</table>

In Section 18.2 we derived equations for parabolas, ellipses, circles, and hyperbolas centered at the origin. However, we will begin this section by reviewing the forms of these equations, and their graphs. A **parabola** with its vertex at the origin, and opening either upward or downward, has the equation

$$y = ax^2.$$

Examples of the graph of this parabola are

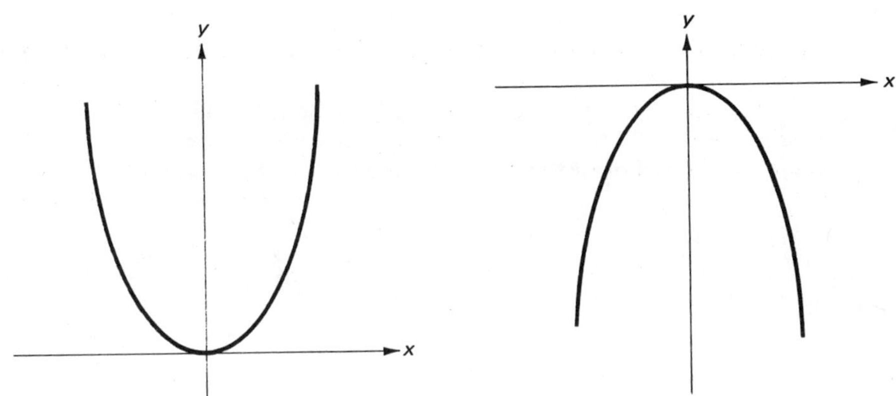

If a parabola has its vertex at the origin, and opens to the right or to the left, it has the equation

$$x = ay^2.$$

Examples of the graph of this parabola are

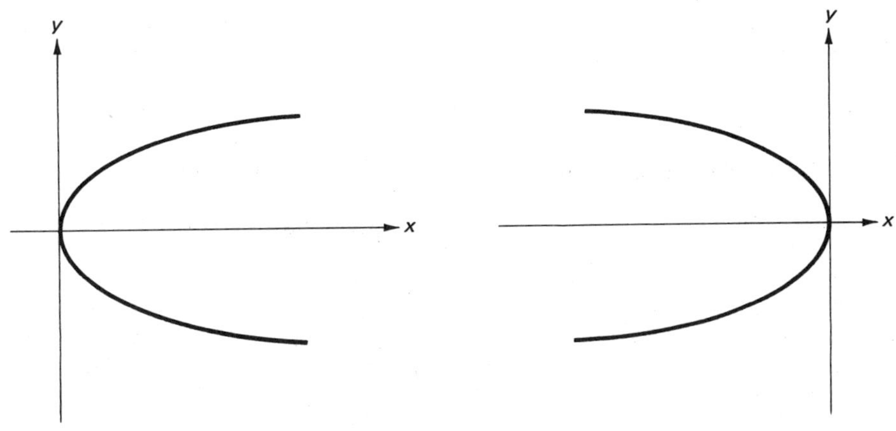

An **ellipse** with its center at the origin has the equation

$$Ax^2 + By^2 = C,$$

where A, B, and C all are positive. Examples of the graph of such ellipses are

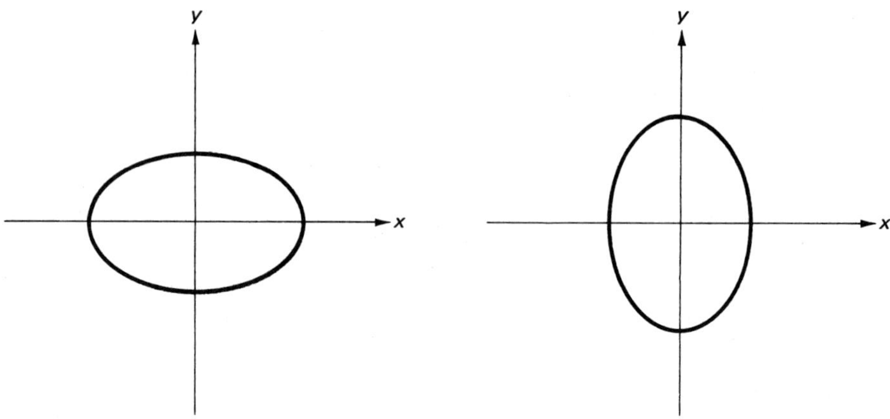

A **circle** is a special case of the ellipse where $A = B$. The equation of a circle can be written

$$x^2 + y^2 = r^2,$$

where r is the radius of the circle. An example of the graph of a circle is

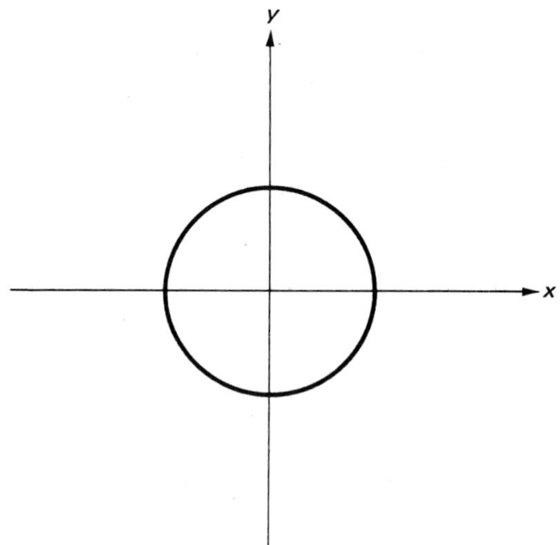

A **hyperbola** with its center at the origin has the equation

$$Ax^2 - By^2 = C,$$

or

$$Ay^2 - Bx^2 = C,$$

where A, B, and C all are positive. Examples of the graphs of such hyperbolas are

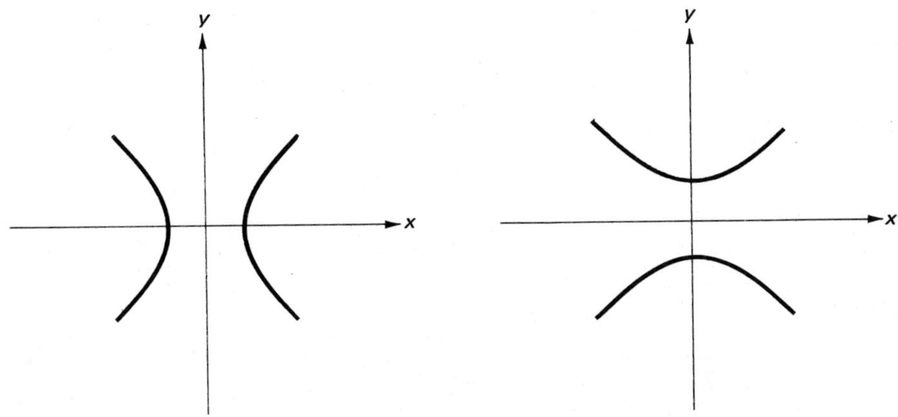

Now we consider the points of intersection of two graphs such as those above. For example, two ellipses centered at the origin can intersect four times, twice, or not at all:

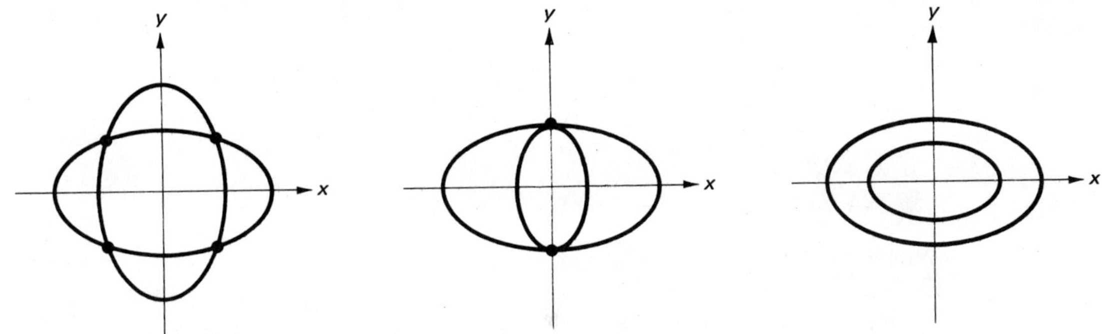

Similarly, an ellipse and a hyperbola centered at the origin can intersect four times, twice, or not at all:

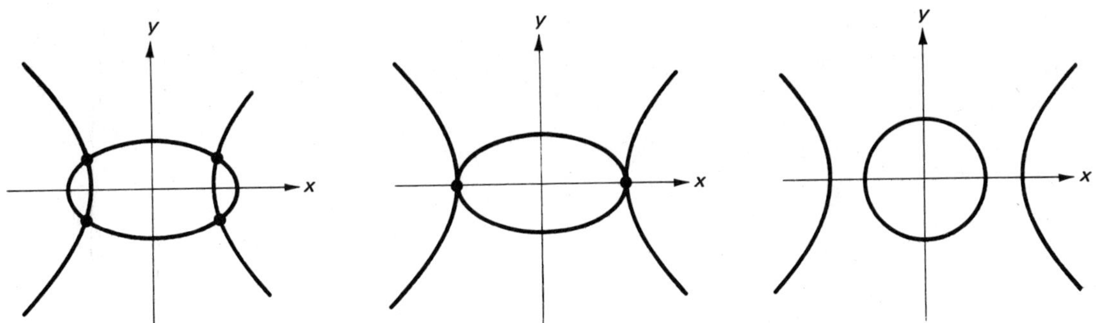

We can find the points of intersection, if any, algebraically by solving the system of equations.

EXAMPLE 19.23.   Solve and graph the system:

$$x^2 + 4y^2 = 20$$

$$2x^2 - y^2 = 4.$$

Solution.   We multiply the second equation by 4 and add to eliminate y:

$$x^2 + 4y^2 = 20$$

$$8x^2 - 4y^2 = 16$$

$$9x^2 \qquad\quad = 36.$$

Then, solving for x,

$$9x^2 = 36$$

$$x^2 = 4$$

$$x = \pm 2.$$

We may substitute 4 for x^2 in either equation to find y. Using the first equation,

$$x^2 + 4y^2 = 20$$

$$4 + 4y^2 = 20$$

$$4y^2 = 16$$

$$y^2 = 4$$

$$y = \pm 2.$$

When $x = 2$, $y = 2$ or -2, and when $x = -2$, $y = 2$ or -2. There are four solutions, $(2, 2)$, $(2, -2)$, $(-2, 2)$, and $(-2, -2)$. To check, you should substitute these solutions in each of the original equations, for this and all other examples in this section. The graph of the system is

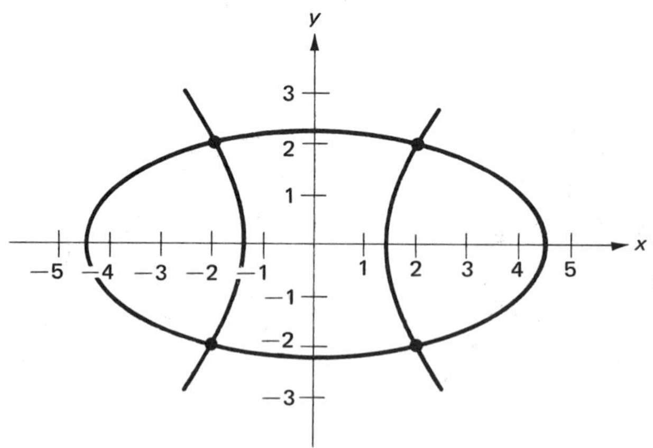

EXAMPLE 19.24.   Solve and graph the system:

$$x^2 + 2y^2 = 12$$

$$x^2 + y^2 = 3.$$

Solution.   We multiply the second equation by -1, add the equations, and solve for y:

$$x^2 + 2y^2 = 12$$
$$-x^2 - y^2 = -3$$
$$\overline{}$$
$$y^2 = 9$$
$$y = \pm 3.$$

Then, we substitute 9 for y^2 in the second equation to find x:

$$x^2 + y^2 = 3$$
$$x^2 + 9 = 3$$
$$x^2 = -6$$
$$x = \pm \sqrt{-6}\,.$$

Since x is not a real number, there are no points of intersection. The graph is

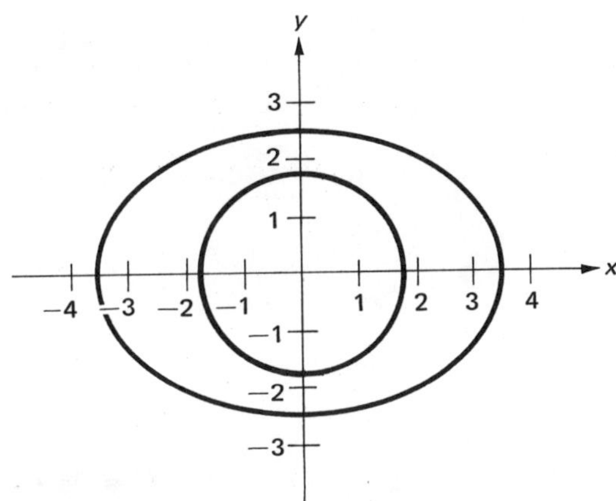

A parabola with its vertex at the origin and an ellipse centered at the origin have two points of intersection. Such a parabola and a hyperbola may have two or no points of intersection:

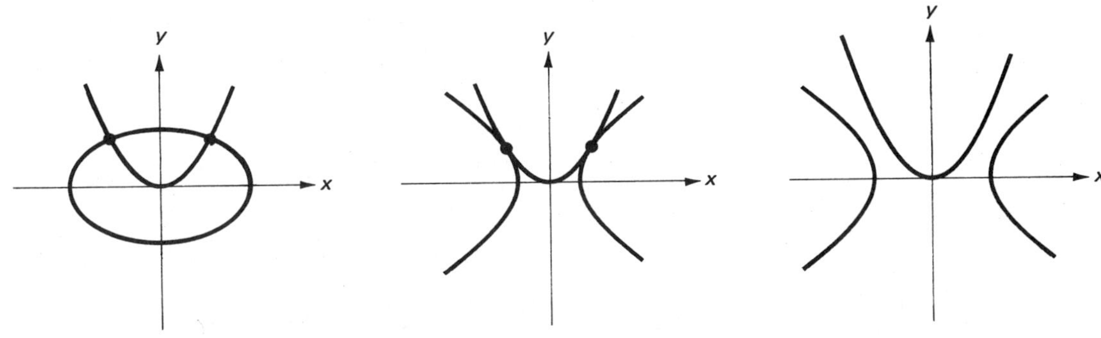

EXAMPLE 19.25.   Solve and graph the system:

$$y = \frac{1}{2}x^2$$

$$4x^2 + y^2 = 9.$$

Solution.   We solve the first equation for x^2:

$$y = \frac{1}{2}x^2$$

$$2y = x^2.$$

Then, substituting $2y$ for x^2 in the second equation, and solving for y,

$$4x^2 + y^2 = 9$$

$$4(2y) + y^2 = 9$$

$$8y + y^2 = 9$$

$$y^2 + 8y - 9 = 0$$

$$(y + 9)(y - 1) = 0$$

$$y + 9 = 0 \text{ and } y - 1 = 0$$

$$y = -9 \text{ and } y = 1.$$

If $y = -9$, there is no real number solution for x, and there are no points of intersection for this case. If $y = 1$,

$$x^2 = 2y$$

$$x^2 = 2(1)$$

$$x^2 = 2$$

$$x = \pm\sqrt{2}.$$

There are two solutions, $(\sqrt{2}, 1)$ and $(-\sqrt{2}, 1)$. Remember that you should check these solutions in both of the original equations. The graph is

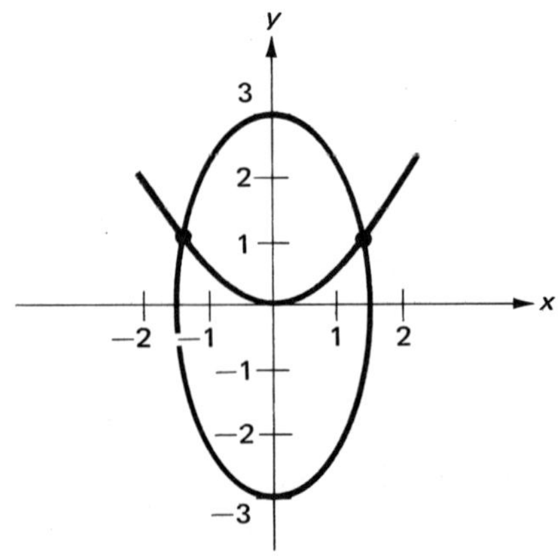

We may also have a system of one linear equation and one quadratic equation. A line may intersect a parabola, an ellipse, or a hyperbola twice, once, or not at all. For example, some possible graphs of a line and an ellipse are

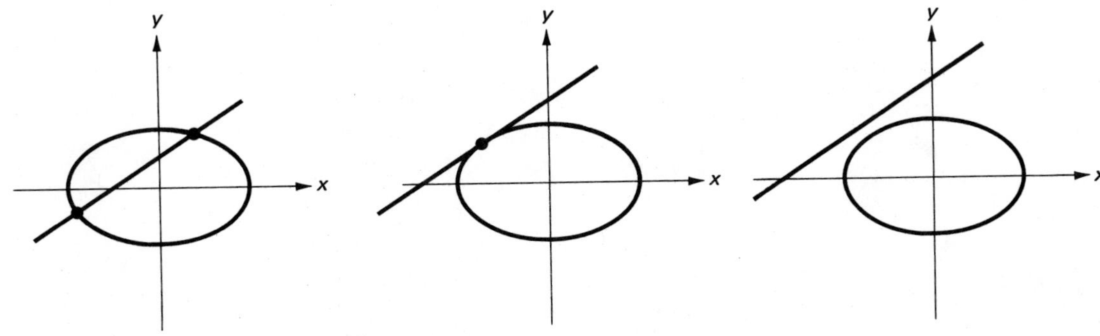

EXAMPLE 19.26.   Solve and graph the system:

$$4x^2 + y^2 = 8$$

$$2x + y - 4 = 0.$$

Solution.   We may solve the linear equation for either x or y to substitute into the quadratic equation. Solving for y,

$$2x + y - 4 = 0$$

$$y = -2x + 4.$$

Then, we substitute $-2x + 4$ for y in the quadratic equation, and solve for x:

$$4x^2 + y^2 = 8$$

$$4x^2 + (-2x + 4)^2 = 8.$$

In squaring the binomial, we must be sure to remember to include the middle term:

$$4x^2 + 4x^2 - 16x + 16 = 8$$

$$8x^2 - 16x + 8 = 0$$

$$x^2 - 2x + 1 = 0$$

$$(x - 1)(x - 1) = 0$$

$$x - 1 = 0$$

$$x = 1.$$

When $x = 1$,

$$y = -2x + 4$$

$$y = -2(1) + 4$$

$$y = 2.$$

There is just one solution, (1, 2). The graph is

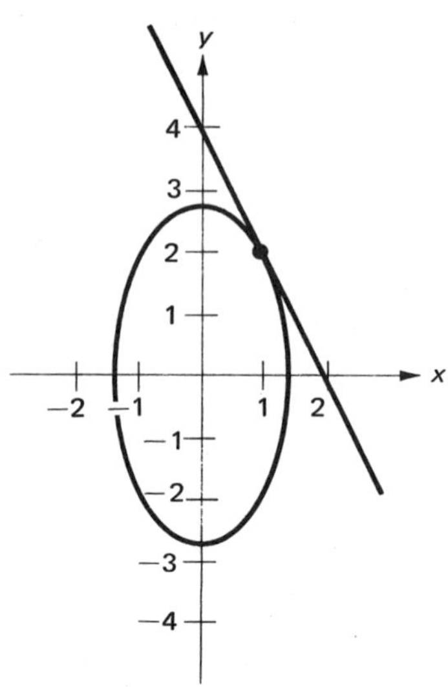

In this system, the line $2x + y - 4 = 0$ is called the **tangent line** to the ellipse $4x^2 + y^2 = 8$ at the point (1, 2).

EXAMPLE 19.27.   Solve and graph the system:

$$y^2 - x^2 = 4$$

$$x - 2y + 2 = 0.$$

Solution.   We may solve the linear equation for either x or y. Solving for x,

$$x - 2y + 2 = 0$$

$$x = 2y - 2.$$

Now, we substitute $2y - 2$ for x in the quadratic equation and solve, being careful to remember the middle term in squaring the binomial, and being careful to use parentheses:

$$y^2 - x^2 = 4$$

$$y^2 - (2y - 2)^2 = 4$$

$$y^2 - (4y^2 - 8y + 4) = 4$$

$$y^2 - 4y^2 + 8y - 4 = 4$$

$$- 3y^2 + 8y - 8 = 0$$

or

$$0 = 3y^2 - 8y + 8.$$

This equation cannot be solved by factoring. Therefore, we use the quadratic formula with y in place of x:

$$y = \frac{-b \pm \sqrt{b^2 - 4ac}}{2a}$$

$$= \frac{-(-8) \pm \sqrt{(-8)^2 - 4(3)(8)}}{2(3)}$$

$$= \frac{8 \pm \sqrt{64 - 96}}{6}$$

$$= \frac{8 \pm \sqrt{-32}}{6}.$$

Since these solutions are not real numbers, there are no points of intersection. The graph is

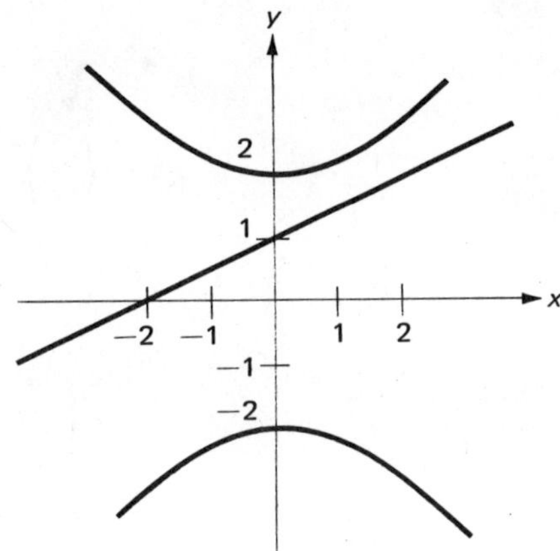

In Section 9.3 we introduced the rational function

$$y = \frac{a}{x},$$

where $a \neq 0$. Written in the form

$$xy = a,$$

this function is considered to be a quadratic function. The term xy is quadratic because it is the product of two variables, just as $x^2 = x \cdot x$ and $y^2 = y \cdot y$ are products of two variables. The graph of the function $xy = a$ is a hyperbola:

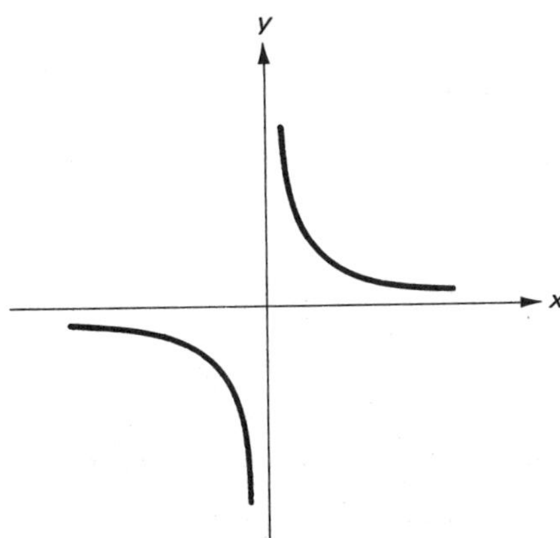

EXAMPLE 19.28.   Solve and graph the system:

$$xy = 2$$

$$x^2 - y^2 = 3.$$

Solution.   We solve the equation $xy = 2$ for either x or y. Solving for y,

$$xy = 2$$

$$y = \frac{2}{x}.$$

Then, we substitute in the second equation:

$$x^2 - y^2 = 3$$

$$x^2 - \left(\frac{2}{x}\right)^2 = 3$$

$$x^2 - \frac{4}{x^2} = 3$$

$$x^2(x^2) - x^2\left(\frac{4}{x^2}\right) = x^2(3)$$

$$x^4 - 4 = 3x^2$$

$$x^4 - 3x^2 - 4 = 0.$$

The resulting equation is called a **fourth-order** equation. In this example, it can be reduced to two quadratic equations by factoring:

$$x^4 - 3x^2 - 4 = 0$$

$$(x^2 + 1)(x^2 - 4) = 0$$

$$x^2 + 1 = 0 \text{ and } x^2 - 4 = 0$$

$$x^2 = -1 \text{ and } x^2 = 4$$

$$x = \pm\sqrt{-1} \text{ and } x = \pm 2.$$

The first two values of x are not real numbers, and so give no points of intersection. If $x = \pm 2$,

$$y = \frac{2}{2} \text{ and } y = \frac{2}{-2}$$

$$y = 1 \text{ and } y = -1.$$

There are two solutions, $(2, 1)$ and $(-2 -1)$. The graph is

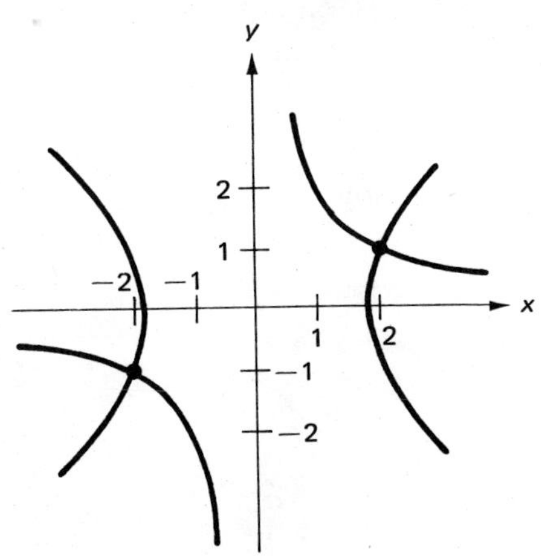

Exercise 19.5

Solve and graph the system:

1. $x^2 + y^2 = 5$
   $x^2 - y^2 = 3$

2. $4x^2 + y^2 = 4$
   $x^2 - y^2 = 1$

3. $4x^2 + y^2 = 5$
   $x^2 + y^2 = 2$

4. $4x^2 + 9y^2 = 36$
   $4x^2 + y^2 = 4$

5. $x^2 + 2y^2 = 8$
   $x^2 - y^2 = 2$

6. $x^2 + 3y^2 = 12$
   $3x^2 + y^2 = 12$

7. $4x^2 + y^2 = 4$
   $x^2 - y^2 = 16$

8. $3x^2 + 4y^2 = 12$
   $x^2 + y^2 = 1$

9. $y = x^2$
   $x^2 + y^2 = 2$

10. $y = \frac{1}{2}x^2$
    $x^2 - y^2 = 1$

11. $y = \frac{1}{2}x^2$
    $y = x + 4$

12. $x = \frac{1}{4}y^2$
    $x + y + 1 = 0$

13. $x^2 + y^2 = 13$
    $y = x + 1$

14. $x^2 + 2y^2 = 4$
    $x + y = 3$

15. $2x^2 - y^2 = 2$
    $y = x - 1$

16. $2y^2 - x^2 = 4$
    $x - 2y - 1 = 0$

17. $xy = 4$
    $x + y = 4$

18. $xy = 6$
    $x - y + 5 = 0$

19. $xy = 2$
    $x^2 + y^2 = 5$

20. $xy = 3$
    $y^2 - x^2 = 8$

Self-test

1. Graph the system and find its solution:

$$x - y = 3$$
$$3x - 2y = 12$$

1. _____________

Solve the system algebraically:

2.
$$y = 2x - 3$$
$$4x - 3y = 10$$

2. _____________

3.
$$x - y - 2z = 1$$
$$3x - y + 4z = 2$$
$$2x + y = 6$$

3. _____________

4. Solve and graph the system:

$$x^2 + 4y^2 = 8$$
$$x^2 - y^2 = 3$$

4. _____________

5. A plane flies 300 miles with the wind in 1 hour, but takes $1\frac{1}{4}$ hours to return against the wind. What are the average rates of the plane and the wind?

5. ______________________

Unit 20
Inequalities

INTRODUCTION

In previous units you have solved different kinds of equations, where two expressions are set equal. You have also drawn graphs of equations in two variables and solved systems of equations in two or more variables. In this unit you will learn how to solve inequalities, where two expressions are related so that one is less than or greater than the other. You will learn how to draw graphs of inequalities in two variables and of systems of inequalities in two variables. Finally, you will learn an important application of systems of inequalities.

OBJECTIVES

When you have finished this unit you should be able to:

1. Solve linear and quadratic inequalities in one variable.
2. Draw graphs of linear inequalities in two variables.
3. Draw graphs of systems of two linear inequalities in two variables.
4. Solve applied problems using linear programming.

Section 20.1

Solving Inequalities

You may recall the inequality symbols either from earlier use in this course or from previous courses. The **inequality symbols**, are $<$, less than, and $>$, greater than. These symbols, with equality sometimes included, compare two numbers in four different ways:

$a < b$ means a is less than b.
$a \leqslant b$ means a is less than or equal to b.
$a > b$ means a is greater than b.
$a \geqslant b$ means a is greater than or equal to b.

We often use the inequality symbols to represent sets of real numbers. For example,

$$x < 2$$

represents all real numbers less than 2, and not including 2. However,

$$x \leqslant 2$$

represents all real numbers less than 2 and also including 2. Similarly,

$$x > 2$$

represents all real numbers greater than 2, not including 2, and

$$x \geqslant 2$$

represents all real numbers greater than 2 and including 2.

An **inequality** is two algebraic expressions related by an inequality symbol. The solution to an inequality is a set of real numbers such as those illustrated above. In general, an inequality has infinitely many real number solutions.

To solve an inequality, we may add, subtract, multiply, and divide on both sides by the same number or expression. We use the same procedures as those in Section 2.1 but with one important exception. Consider the following numerical example:

$$2 < 4$$

$$(-3)(2) \overset{?}{<} (-3)(4)$$

$$-6 \not< -12.$$

In fact,

$$-6 > -12.$$

In this example, we have multiplied both sides of an inequality by a negative number. The inequality symbol reverses from less than to greater than. The inequality symbol in an inequality is called the **sense of the inequality**. When we multiply an inequality by a negative number, the sense of the inequality reverses.

Now, consider this example of division of an inequality by a negative number:

$$2 < 4$$

$$\frac{2}{-1} \overset{?}{<} \frac{4}{-1}$$

$$-2 \not< -4.$$

In fact,

$$-2 > -4.$$

When we divide an inequality by a negative number, the sense of the inequality reverses. Similar examples would indicate that, if we start with the inequality symbol greater than and multiply or divide by a negative number, the sense would reverse to the inequality symbol less than.

> **Rule for Multiplying or Dividing Inequalities:** When both sides of an inequality are multiplied or divided by a negative number, the sense of the inequality reverses.

EXAMPLE 20.1. Solve $2x - 4 < 6$.

Solution.   We may add 4 to both sides of the inequality without changing the sense of the inequality:

$$2x - 4 < 6$$

$$2x < 10.$$

Also, since 2 is positive, we may divide both sides of the inequality by 2 without changing the sense:

$$\frac{2x}{2} < \frac{10}{2}$$

$$x < 5.$$

All real numbers less than 5 are solutions. We may write the solution simply as $x < 5$. To check, we first check that 5 is the end point of the set; that is, if equality were included

$$2x - 4 = 6$$

$$2(5) - 4 \overset{?}{=} 6$$

$$10 - 4 = 6.$$

We will call 5 the **boundary**. Now, we check that the sense of the solution gives the correct side of the boundary. We may choose any number less than 5. For example, if $x = 4$,

$$2x - 4 < 6$$

$$2(4) - 4 \overset{?}{<} 6$$

$$8 - 4 \overset{?}{<} 6$$

$$4 < 6.$$

It is important to check both the boundary and the sense of the solution to an inequality.

It can be helpful to illustrate the solution of an inequality by a graph. We need only one axis on the graph since there is only one variable. We usually use a horizontal line. We indicate the boundary, in this case 5, by an open dot when it is not included in the solution. The part of the line representing the numbers less than 5 is heavier to indicate the solution $x < 5$:

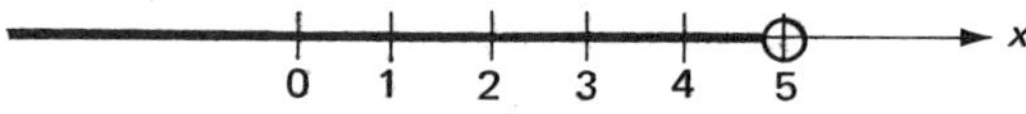

EXAMPLE 20.2.   Solve $6 - 3x < 9$.

Solution.   We may subtract 6 from both sides of the inequality without changing the sense:

$$6 - 3x < 9$$

$$-3x < 3.$$

However, if we divide by the negative number -3, we must change the sense of the inequality:

$$-3x < 3$$

$$\frac{-3x}{-3} > \frac{3}{-3}$$

$$x > -1.$$

The solution is $x > -1$. To check the boundary, if equality were included,

$$6 - 3x = 9$$

$$6 - 3(-1) \overset{?}{=} 9$$

$$6 + 3 = 9.$$

To check the sense, since $0 > -1$,

$$6 - 3x < 9$$

$$6 - 3(0) \overset{?}{<} 9$$

$$6 - 0 < 9.$$

Observe that we must check in the original inequality using the original sense. The solution $x > -1$ may be illustrated by the graph

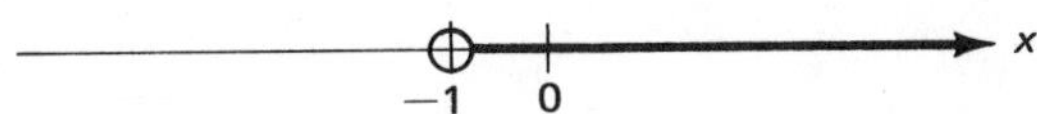

The inequality may include equality. Then, the boundary is included in the solution.

EXAMPLE 20.3. Solve $6x - 2 \geqslant 7$.

Solution. We add 2 to both sides of the inequality, and then divide both sides by 6. Neither of these operations changes the sense of the inequality:

$$6x - 2 \geqslant 7$$

$$6x \geqslant 9$$

$$x \geqslant \frac{9}{6}$$

$$x \geqslant \frac{3}{2}.$$

All real numbers greater than $\frac{3}{2}$, and also $\frac{3}{2}$, are solutions. Thus, the boundary is included in the solution. We write the solution as $x \geqslant \frac{3}{2}$. You should check the boundary and the sense of the solution in the original inequality.

We use a graph similar to those in the preceding examples to illustrate a solution which includes equality. The boundary $\frac{3}{2}$ is indicated by a solid dot when it is included in the solution. The solution $x \geqslant \frac{3}{2}$ may be illustrated by the graph

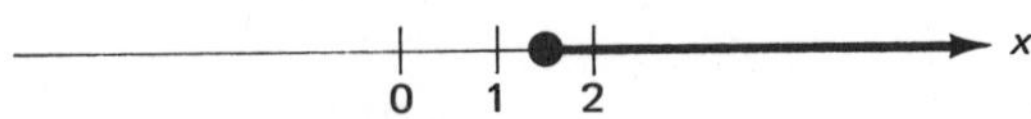

If possible, we simplify the expressions on each side of the inequality before solving an inequality.

EXAMPLE 20.4. Solve $3(x - 2) - 4x \leqslant 2$.

Solution. First, we simplify the expression on the left-hand side of the inequality:

$$3(x - 2) - 4x \leqslant 2$$

$$3x - 6 - 4x \leqslant 2$$

$$-x - 6 \leqslant 2.$$

Now, we add 6 to both sides, which does not change the sense of the inequality:

$$-x - 6 \leqslant 2$$

$$-x \leqslant 8.$$

Finally, we must multiply both sides by -1. This operation does change the sense of the inequality:

$$-x \leqslant 8$$

$$(-1)(-x) \geqslant (-1)(8)$$

$$x \geqslant -8.$$

The solution is $x \geqslant -8$. To check, we use the original inequality. The boundary is -8:

$$3(x - 2) - 4x = 2$$

$$3(-8 - 2) - 4(-8) \overset{?}{=} 2$$

$$3(-10) + 32 \overset{?}{=} 2$$

$$-30 + 32 = 2.$$

To check the sense, we may use any number greater than -8. Since $0 > -8$, and is an easy number to calculate with, we try 0 in the original inequality:

$$3(x - 2) - 4x \leqslant 2$$

$$3(0 - 2) - 4(0) \overset{?}{\leqslant} 2$$

$$3(-2) - 0 \overset{?}{\leqslant} 2$$

$$-6 \leqslant 2.$$

Often we can avoid multiplying or dividing by a negative number by our choice of how we collect the x-terms.

EXAMPLE 20.5.   Solve $x - (2x + 2) > 3(x - 2)$.

Solution.   First, we simplify the expressions on each side of the inequality:

$$x - (2x + 2) > 3(x - 2)$$

$$x - 2x - 2 > 3x - 6$$

$$-x - 2 > 3x - 6.$$

Now, we may avoid a negative coefficient in the x-term by adding x to both sides:

$$-x - 2 > 3x - 6$$

$$-2 > 4x - 6.$$

Then, adding 6 to both sides and dividing by 4 does not change the sense of the inequality:

$$-2 > 4x - 6$$

$$4 > 4x$$

$$1 > x.$$

The solution $1 > x$ is the same as $x < 1$. We will usually write such solutions in the form $x < 1$. You should check the boundary and sense of the solution using the original inequality.

If an inequality contains a quadratic expression, we find the boundaries by methods similar to those for solving quadratic equations. To find the sense, we must always consider two cases. The senses of the two cases depend on the sense of the quadratic inequality.

EXAMPLE 20.6.   Solve $x^2 - 6x + 8 > 0$.

Solution.   We factor the quadratic expression:

$$x^2 - 6x + 8 > 0$$

$$(x - 2)(x - 4) > 0.$$

We have two factors whose product is positive. The only ways this can happen are if both factors are positive or both factors are negative.

Case 1.   Suppose both factors are positive. Then,

$$x - 2 > 0 \text{ and } x - 4 > 0$$

$$x > 2 \text{ and } x > 4.$$

To have both $x > 2$ and $x > 4$, we must have $x > 4$. For example, 3 is not a solution because, although $3 > 2$, $3 \not> 4$. We may use a graph to illustrate this situation. For $x > 2$, the graph is

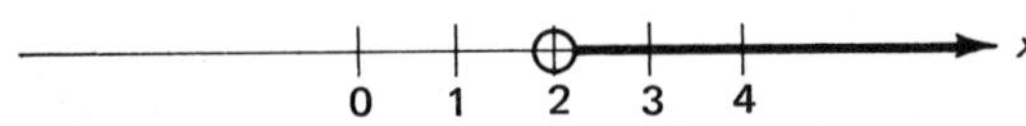

For $x > 4$, the graph is

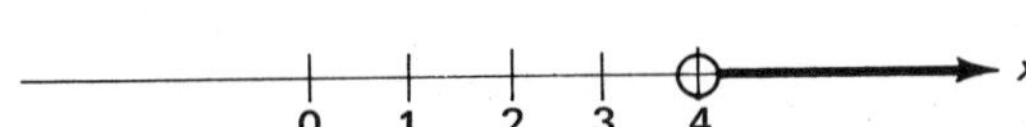

We observe that only the parts representing numbers greater than 4 are indicated on both graphs. Therefore, the solution for this case is $x > 4$.

Case 2.   Suppose both factors are negative. Then,

$$x - 2 < 0 \text{ and } x - 4 < 0$$

$$x < 2 \text{ and } x < 4.$$

To have both $x < 2$ and $x < 4$, we must have $x < 2$. For example, $3 < 4$ but $3 \not< 2$. For $x < 2$, the graph is

For $x < 4$, the graph is

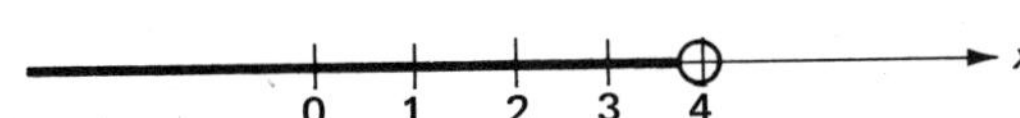

Only the parts representing numbers less than 2 are indicated on both graphs. Therefore, the solution for this case is $x < 2$.

The entire solution from both cases consists of all numbers x such that $x > 4$, from the first case, or $x < 2$, from the second case. We write the solution as

$$x > 4 \text{ or } x < 2.$$

This solution is illustrated by the graph

To check, we first check both boundaries. If $x = 4$,

$$x^2 - 6x + 8 = 0$$

$$4^2 - 6(4) + 8 \overset{?}{=} 0$$

$$16 - 24 + 8 = 0.$$

If $x = 2$,

$$2^2 - 6(2) + 8 \overset{?}{=} 0$$

$$4 - 12 + 8 = 0.$$

Now, we must check the sense of $x > 4$ and also $x < 2$. Since $5 > 4$,

$$x^2 - 6x + 8 > 0$$

$$5^2 - 6(5) + 8 \overset{?}{>} 0$$

$$25 - 30 + 8 \overset{?}{>} 0$$

$$3 > 0.$$

Since $0 < 2$,

$$0^2 - 6(0) + 8 \overset{?}{>} 0$$

$$8 > 0.$$

EXAMPLE 20.7.   Solve $2x^2 + 3x < 2$.

Solution.   First, we collect terms of the quadratic expression. Subtracting 2 from both sides does not change the sense of the inequality:

$$2x^2 + 3x < 2$$

$$2x^2 + 3x - 2 < 0.$$

Factoring the quadratic expression,

$$(x + 2)(2x - 1) < 0.$$

Since we have two factors whose product is negative, one factor must be positive and the other negative.

Case 1.   Suppose the first factor is positive and the second factor is negative. Then,

$$x + 2 > 0 \text{ and } 2x - 1 < 0$$

$$x > -2 \text{ and } x < \frac{1}{2}.$$

The numbers which are solutions for this case are greater than -2 and also less than $\frac{1}{2}$. We write this solution

$$-2 < x < \frac{1}{2}.$$

Case 2.   Suppose the first factor is negative and the second factor is positive. Then,

$$x + 2 < 0 \text{ and } 2x - 1 > 0$$

$$x < -2 \text{ and } x > \frac{1}{2}.$$

It is not possible to have a number less than -2 and also greater than $\frac{1}{2}$ at the same time. Therefore, there is no solution for this case. The entire solution from both cases consists of

$$-2 < x < \frac{1}{2}.$$

This solution is illustrated by the graph

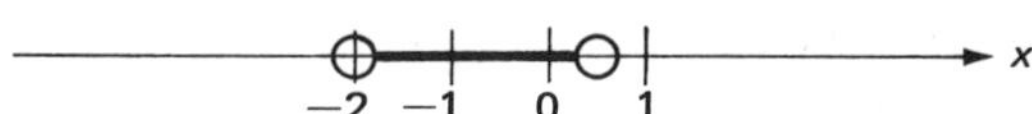

To check, you should first check the boundaries, -2 and $\frac{1}{2}$. Then, you may use any number between -2 and $\frac{1}{2}$ to check the sense. Observe that, since only one case is involved, only one number is needed to check the sense. Zero is in the set of solutions and is the easiest choice to calculate.

Further, observe how we write the solutions to the types of quadratic inequalities in Examples 20.7 and 20.6. In Example 20.7, the solution comes from just one case. A number in the solution satisfies both conditions at once, and we write

$$-2 < x < \frac{1}{2}.$$

In Example 20.6, the two parts of the solution come from different cases. A number in the solution satisfies either one condition or the other but not both at once, and we write

$$x > 4 \text{ or } x < 2.$$

If a quadratic inequality cannot be solved by factoring, we use the quadratic formula.

EXAMPLE 20.8.   Solve $1 \leqslant x^2 - 2x$.

Solution.   We collect the terms of the quadratic expression on the right-hand side:

$$1 \leqslant x^2 - 2x$$

$$0 \leqslant x^2 - 2x - 1.$$

Now we use the quadratic formula to find the boundaries:

$$x = \frac{-b \pm \sqrt{b^2 - 4ac}}{2a}$$

$$= \frac{-(-2) \pm \sqrt{(-2)^2 - 4(1)(-1)}}{2(1)}$$

$$= \frac{2 \pm \sqrt{4 + 4}}{2}$$

$$= \frac{2 \pm \sqrt{8}}{2}$$

$$= \frac{2 \pm 2\sqrt{2}}{2}$$

$$= 1 \pm \sqrt{2}.$$

The boundaries are $1 + \sqrt{2}$ and $1 - \sqrt{2}$. The solution is possibly $1 - \sqrt{2} \leqslant x \leqslant 1 + \sqrt{2}$, or possibly $x \geqslant 1 + \sqrt{2}$ or $x \leqslant 1 - \sqrt{2}$. Because the boundaries are irrational, it is difficult to determine the solution by cases of positive and negative factors. However, since we have determined the possible solutions, we may distinguish between them by means of a "check point." Suppose $x = 0$:

$$1 \leqslant x^2 - 2x$$

$$1 \stackrel{?}{\leqslant} 0^2 - 2(0)$$

$$1 \nleqslant 0.$$

Thus $x = 0$ is not a solution. But $x = 0$ is in the possible solution $1 - \sqrt{2} \leqslant x \leqslant 1 + \sqrt{2}$. Therefore, the solution must be the other possible solution,

$$x \geqslant 1 + \sqrt{2} \text{ or } x \leqslant 1 - \sqrt{2}.$$

Observe that, because the original inequality includes equality, the solution includes the boundaries. The solution is illustrated by the graph

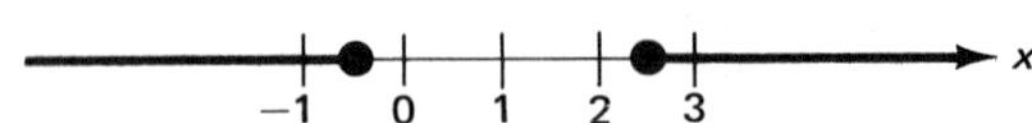

Recall that the solid dots indicate that the boundaries are included in the solution.

EXAMPLE 20.9.   Solve $x^2 > 4x - 5$.

Solution.   Collecting on the left-hand side,

$$x^2 > 4x - 5$$

$$x^2 - 4x + 5 > 0.$$

We use the quadratic formula:

$$x = \frac{-b \pm \sqrt{b^2 - 4ac}}{2a}$$

$$= \frac{-(-4) \pm \sqrt{(-4)^2 - 4(1)(5)}}{2(1)}$$

$$= \frac{4 \pm \sqrt{16 - 20}}{2}$$

$$= \frac{4 \pm \sqrt{-4}}{2}.$$

There are no real number boundaries. Therefore, either every real number is a solution to the inequality, or there are no solutions to the inequality. We may try any real number. If $x = 0$,

$$x^2 > 4x - 5$$

$$0^2 \overset{?}{>} 4(0) - 5$$

$$0 > -5.$$

Zero is a solution. Therefore, since there are no real number boundaries, every real number is a solution. The solution is the set of all real numbers.

There are many types of inequalities which are related to quadratic inequalities. We will consider one type.

EXAMPLE 20.10.   Solve $\dfrac{x - 4}{x + 2} \leqslant 3$.

Solution.   We must multiply both sides of the inequality by $x + 2$. However, we do not know whether $x + 2$ is positive or negative. Therefore, we must again consider two cases.

Case 1.   Suppose $x + 2 > 0$. Then, we may multiply by $x + 2$ without changing the sense of the inequality:

$$\frac{x - 4}{x + 2} \leqslant 3$$

$$(x + 2)\left(\frac{x - 4}{x + 2}\right) \leqslant (x + 2)(3)$$

$$x - 4 \leqslant 3x + 6.$$

Solving the resulting inequality,

$$-4 \leqslant 2x + 6$$

$$-10 \leqslant 2x$$

$$-5 \leqslant x.$$

We have $x \geqslant -5$, but we also have the condition $x + 2 > 0$, or $x > -2$. The graph for $x \geqslant -5$ is

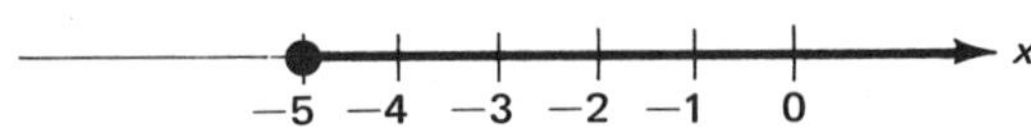

The graph for $x > -2$ is

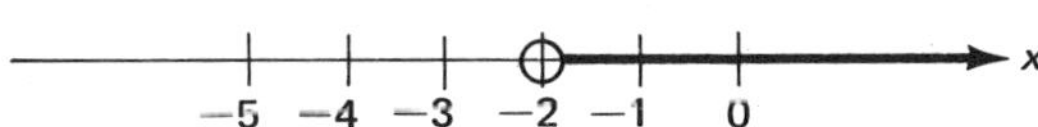

Since parts representing numbers greater than -2 are indicated on both graphs, we see that the solution for this case is $x > -2$.

Case 2.   Suppose $x + 2 < 0$. Then, when we multiply by $x + 2$, we change the sense of the inequality:

$$\frac{x - 4}{x + 2} \leqslant 3$$

$$(x + 2)\left(\frac{x - 4}{x + 2}\right) \geqslant (x + 2)(3)$$

$$x - 4 \geqslant 3x + 6$$

$$-4 \geqslant 2x + 6$$

$$-10 \geqslant 2x$$

$$-5 \geqslant x.$$

We have $x \leqslant -5$, but we also have $x + 2 < 0$, or $x < -2$. The graph for $x \leqslant -5$ is

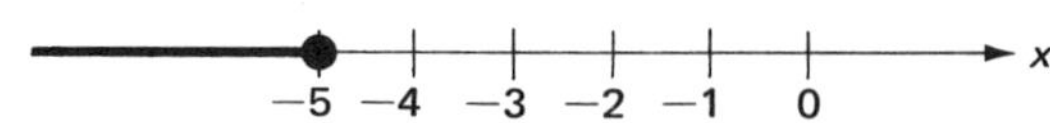

The graph for $x < -2$ is

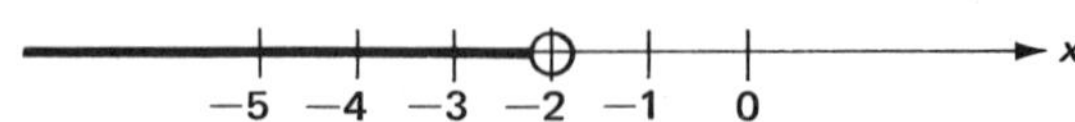

We see that the solution for this case is $x \leqslant -5$. Taking the solution from each case, the solution to the inequality is

$$x > -2 \text{ or } x \leqslant -5.$$

This solution is illustrated by the graph

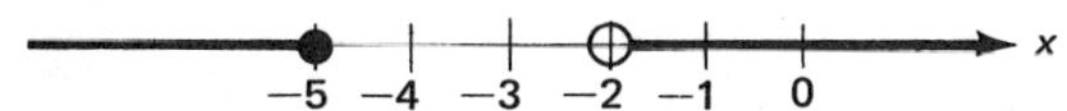

Observe that the boundary $x = -2$ results in an undefined rational expression in the original inequality. This boundary cannot be included in the solution, and also cannot be checked by substituting in the original inequality. You should check the boundary $x = -5$, and also the sense of each part of the solution.

Exercise 20.1

Solve for x:

1. $3x+2>5$

2. $4x-3<7$

3. $9-2x>3$

4. $3-4x<15$

5. $3x-3 \leqslant -4$

6. $1-9x \leqslant -5$

7. $\frac{1}{4}x+1<\frac{5}{4}$

8. $\frac{1}{6}-\frac{1}{3}x>\frac{1}{2}$

9. $2(x+1)-4x>1$

10. $2(x-3)-5x<0$

11. $2x-(x-3)<3x+2$

12. $2(x+5)<3x-(2+4x)$

13. $3(x-2)-3 \geqslant 6-4(x+2)$

14. $5(x-4)+5 \leqslant 4(2x-3)-3$

15. $x^2-6x+5>0$

16. $x^2-x \geqslant 2$

17. $x^2+3x<10$

18. $2x^2+5x+2 \leqslant 0$

19. $x^2<2x+4$

20. $x^2 \geqslant 4x-2$

21. $2x^2+2x+1 \geqslant 0$

22. $x^2-x+1 \leqslant 0$

23. $\dfrac{x-3}{x-2}<2$

24. $\dfrac{x+5}{x+1}>3$

25. $\dfrac{3x-4}{2x-1} \leqslant 1$

26. $\dfrac{3x+6}{2x+3} \leqslant 2$

Section 20.2   Linear Inequalities in Two Variables

Recall from Unit 6 that the solution of a linear equation in two variables consists of infinitely many ordered pairs. We indicate such a solution by a graph. Similarly, the solution of a linear inequality in two variables consists of infinitely many ordered pairs which we indicate by a graph.

EXAMPLE 20.11.   Draw the graph of $2x - 4y < 8$.

Solution.   First, we draw the graph of the line $2x - 4y = 8$. However, we use a dashed line because equality is not included in the inequality, so the line is not actually part of the graph:

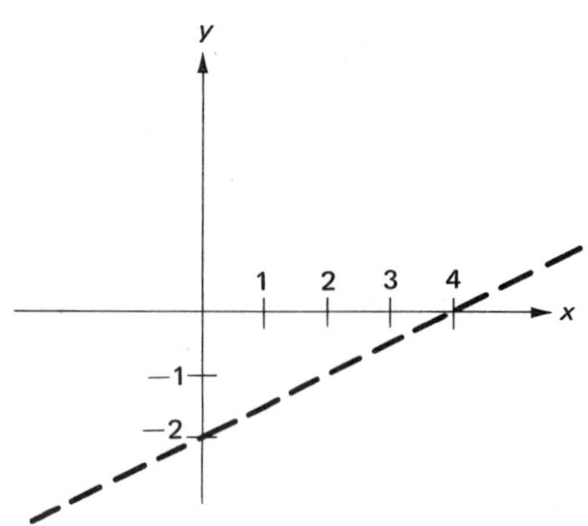

Now, we consider the inequality $2x - 4y < 8$. The points satisfying this inequality are all the points on one side of the line $2x - 4y = 8$. To determine which side of the line contains the points which satisfy the inequality, we use a sample point. The point $(0, 0)$ gives the easiest calculation:

$$2x - 4y < 8$$

$$2(0) - 4(0) \overset{?}{<} 8$$

$$0 < 8.$$

Therefore, $(0, 0)$ is a solution. Also, all points on the same side of the line as $(0, 0)$ are solutions. We indicate these points by shading the side of the line which contains $(0, 0)$. The graph is

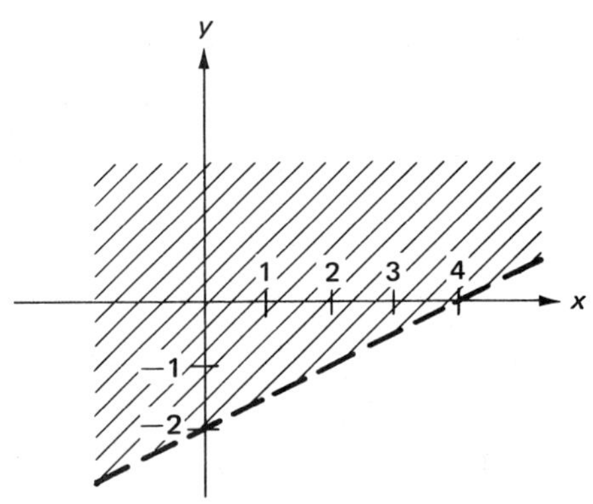

Each side of the line is called a **half-plane**. The line forms the **boundary** of each half-plane. When the inequality includes equality, the boundary is included in the graph.

EXAMPLE 20.12.   Draw the graph of $3x + 6y \leqslant 9$.

Solution.   We draw the graph of $3x + 6y = 9$, using a solid line because the boundary is included. Then, using $(0, 0)$ as a sample point,

$$3x + 6y < 9$$

$$3(0) + 6(0) \overset{?}{<} 9$$

$$0 < 9,$$

so we shade the half-plane which includes (0, 0). The graph is

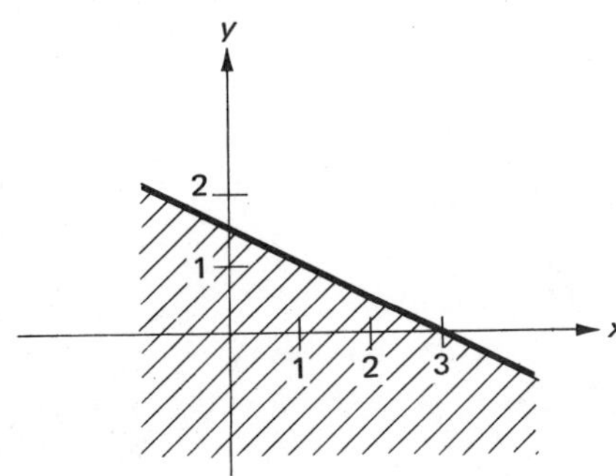

EXAMPLE 20.13.   Draw the graph of $4x - y - 4 \geqslant 0$.

Solution.   We draw the graph of $4x - y - 4 = 0$, using a solid line. Then, using (0, 0) as a sample point,

$$4x - y - 4 \geqslant 0$$

$$4(0) - 0 - 4 \overset{?}{\geqslant} 0$$

$$-4 \ngeqslant 0.$$

Therefore, (0, 0) is not in the solution. We shade the half-plane which does not include (0, 0). The graph is

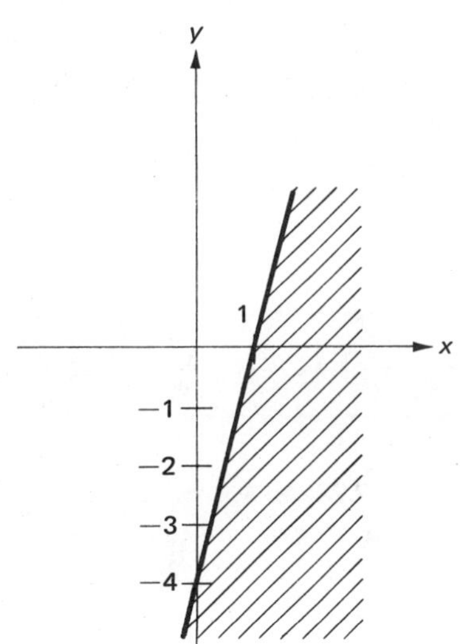

EXAMPLE 20.14.   Draw the graph of $y > x$.

Solution.   We draw the graph of $y = x$, using a dashed line. Since both intercepts are (0, 0), we must use another point such as (1, 1). Also, since (0, 0) is on the line, we cannot use it as a sample point. We may choose any other point not on the line. If we choose (1, 0),

$$y > x$$

$$0 \ngtr 1.$$

Therefore, the half-plane does not include $(1, 0)$. If we choose $(0, 1)$,

$$y > x$$

$$1 > 0.$$

Therefore, the half-plane does include $(0, 1)$. The graph is

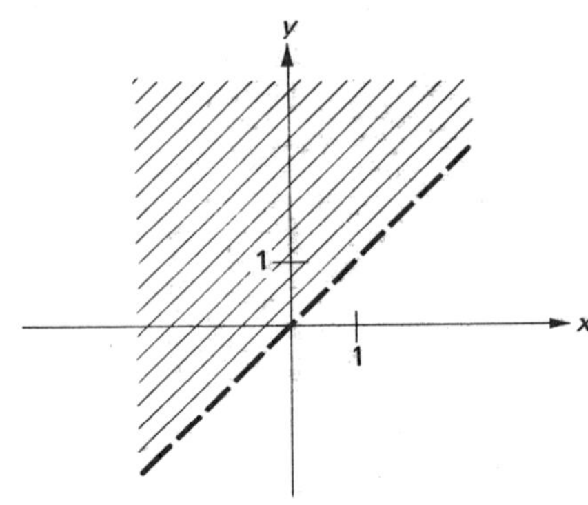

Observe that the graph of $y > x$ includes exactly those points for which the y-coordinate is greater than the x-coordinate.

Exercise 20.2

Draw the graph:

1. $2x + 3y < 6$ 2. $x - 2y < 2$ 3. $2x - 5y \leqslant 10$ 4. $3y - 4x \leqslant 12$

5. $3x - 4y > 12$ 6. $x + 2y > 4$ 7. $3x - 4y + 6 \geqslant 0$ 8. $5x - 2y - 5 \geqslant 0$

9. $y - 2 \leqslant 0$ 10. $x + 3 > 0$ 11. $x + 2y > 0$ 12. $3y - 2x \leqslant 0$

Section 20.3   # Systems of Linear Inequalities

In Sections 19.1 and 19.2 we solved systems of two linear equations in two variables. The solution, if it exists, is an ordered pair. Now we consider systems of two linear inequalities in two variables. The solutions of these systems consist of infinitely many ordered pairs. As in the preceding section, we indicate such solutions by a graph.

EXAMPLE 20.15.   Draw the graph of the system:

$$x + y < 3$$

$$x - y < 1.$$

Solution.   First, we find the point of intersection of the lines

$$x + y = 3$$

$$x - y = 1.$$

The solution of this system of equations is (2, 1). Next, we draw the graph of each line, using dashed lines since the inequalities do not include the boundaries, and indicating the point of intersection:

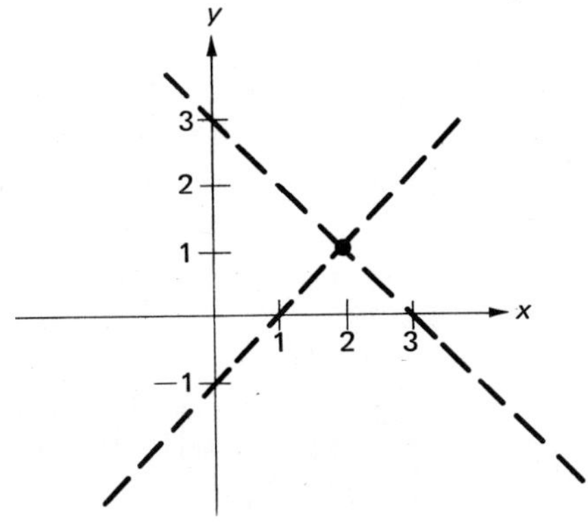

Now, we find the half-plane determined by the inequality

$$x + y < 3.$$

We have indicated this half-plane by horizontal lines:

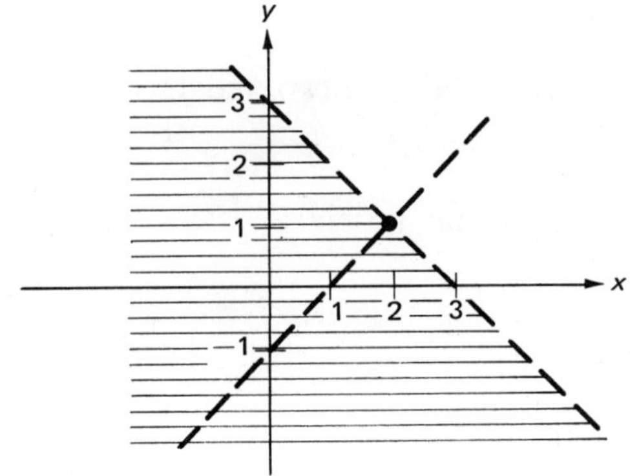

We find the half-plane determined by the inequality

$$x - y < 1.$$

We have indicated this half-plane by vertical lines:

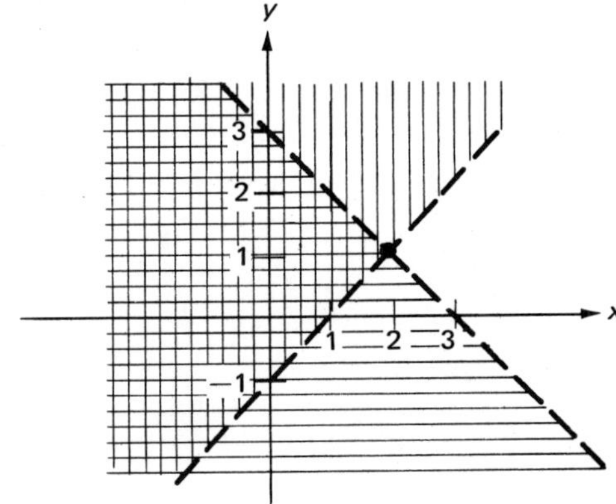

The graph of the system

$$x + y < 3$$

$$x - y < 1$$

contains all points common to the two indicated half-planes. These are the points in the part of the graph containing both horizontal and vertical lines.

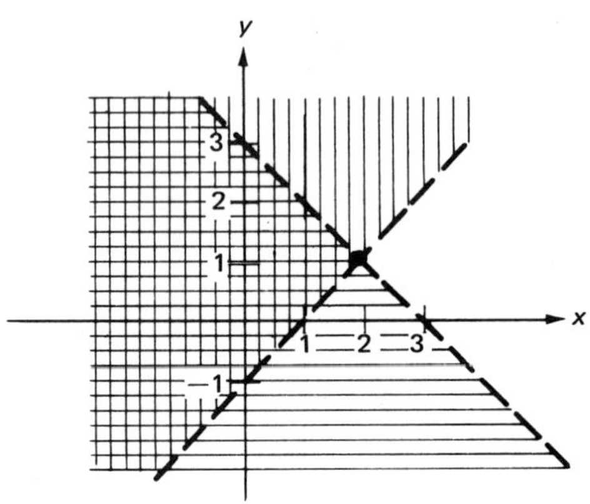

A system of inequalities in two variables may have one boundary included but the other boundary not included.

EXAMPLE 20.16.   Draw the graph of the system:

$$3x + 2y \geqslant 3$$

$$x - 3y > -10.$$

Solution.   The system of equations

$$3x + 2y = 3$$

$$x - 3y = -10$$

has the solution $(-1, 3)$. We draw the graphs of the lines, using dashed lines until we know exactly what part of the boundary is included. Then, we indicate the half-plane determined by

$$3x + 2y \geqslant 3$$

by horizontal lines, and the half-plane determined by

$$x - 3y > -10$$

by vertical lines. The part common to both half-planes is indicated by both horizontal and vertical lines. We also fill in with a solid line the part of the line $3x + 2y = 3$ which is a boundary of the common part. The graph of the system is

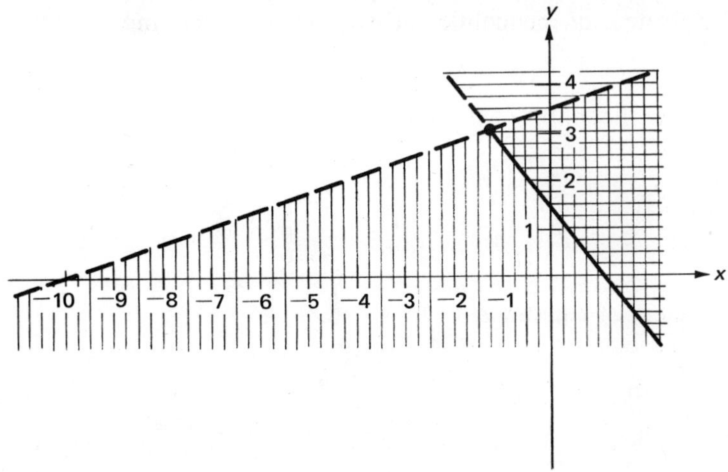

Observe that the point of intersection of the lines separates the part of the line $3x + 2y = 3$ that is a boundary from the part that is not. Each of these parts of the line is called a **half-line**. When one of the inequalities includes equality, a half-line is included in the graph of the system.

EXAMPLE 20.17.   Draw the graph of the system:

$$2x - 3y > -3$$

$$y \leqslant 5.$$

Solution.   The system of equations

$$2x - 3y = -3$$

$$y = 5$$

has the solution (6, 5). We draw the graphs of the lines using dashed lines at first, determine the half-plane for each inequality, and indicate by horizontal and vertical lines the part common to both half-planes. The common part determines the half-line of $y = 5$ which is included. The graph of the system is

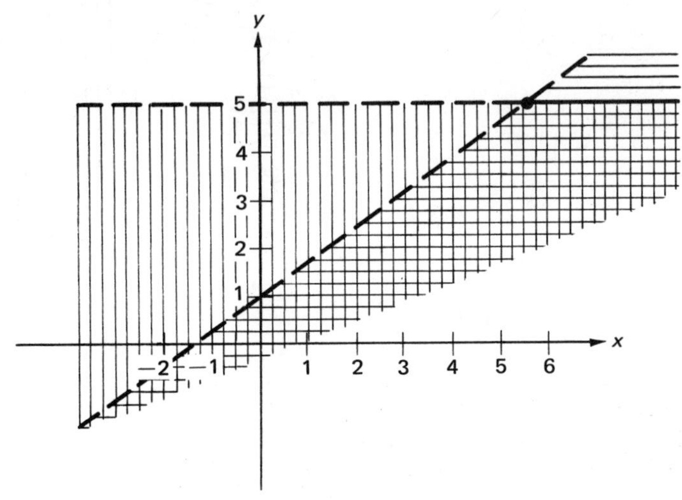

A system of inequalities in two variables may include two boundaries. Each boundary is a half-line.

EXAMPLE 20.18.   Draw the graph of the system:

$$2x - y \leqslant 7$$

$$3x + y \geqslant 3.$$

Solution.   The system of equations

$$2x - y = 7$$

$$3x + y = 3$$

has the solution $(2, -3)$. We draw the graphs of the lines, using dashed lines until we know which half-line of each is included. Then we find the half-plane determined by each inequality, and indicate by horizontal and vertical lines the part common to both half-planes. The common part determines the half-line of each boundary which is included. The graph of the system is

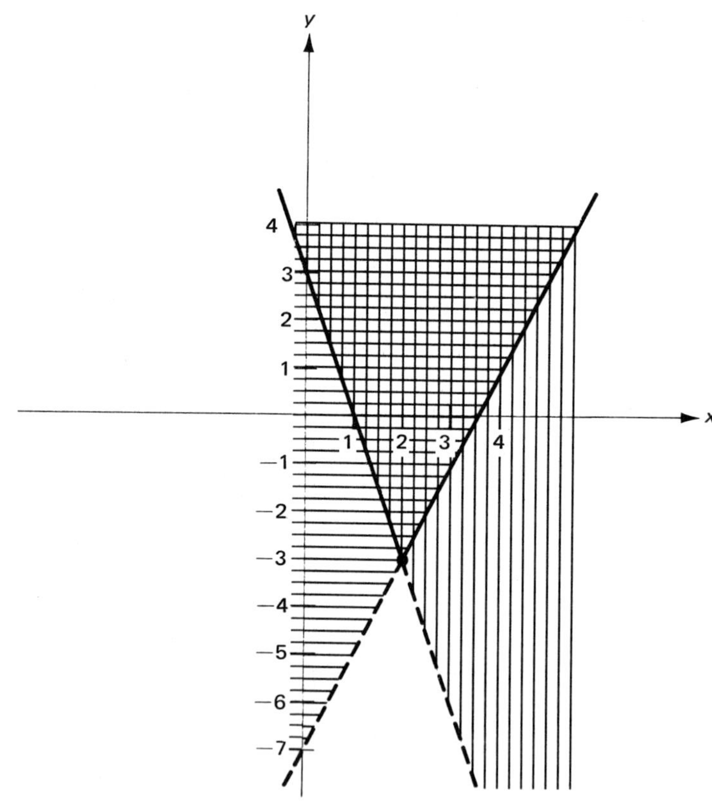

Exercise 20.3

Draw the graph of the system:

1. $x + y < 2$

    $x - y < 6$

2. $2x + 3y > 6$

    $-x + 2y < 4$

3. $x - y \leqslant -1$

    $3x + 2y < 12$

4. $2x + y \leqslant 3$
    $x - y > 3$

5. $x - y \leqslant 2$
    $x + y \geqslant 4$

6. $x + 2y \geqslant -2$
    $2x - y \geqslant -4$

7. $y < x$
    $2x + y < -3$

8. $y \geqslant 2x$
    $y \leqslant -2x$

9. $x \leqslant 1$
    $x + 2y > -1$

10. $y \geqslant -3$
    $3x + y > 3$

11. $x \geqslant -2$
    $y \leqslant 3$

12. $x \geqslant 0$
    $y \geqslant 0$

Section 20.4   Linear Programming

Systems of linear inequalities have a very important application in modern business and industry called linear programming. Linear programming is used in problems where a quantity is to be maximized or minimized. For example, we would want to maximize profits and minimize costs. For each problem we have a number of conditions, which are given by linear inequalities. In actual practice, the number of conditions is so great that linear programming problems must be solved by computers. We will consider only simple linear programming problems in which we maximize profit subject to just a few conditions. To find out more about linear programming, you should consult books entitled *Finite Mathematics*.

EXAMPLE 20.19.   A company makes two styles of jackets. Each jacket must be assembled and then finished. The inexpensive style takes 2 hours to assemble and 2 hours to finish. The better style takes 3 hours to assemble and, because it is hand-embroidered, 6 hours to finish. The company can provide at most 120 man-hours of assembling and 180 man-hours of finishing a day. If the profit on an inexpensive jacket is \$7.00 and the profit on a better jacket is \$13.00, how many jackets of each type should the company make per day to maximize profits? What is the maximum profit?

Solution.   We let x represent the number of inexpensive jackets and y represent the number of better jackets. Then the profit, in dollars, on inexpensive jackets is $7x$, and on better jackets is $13y$. The total profit is

$$P = 7x + 13y.$$

We wish to find the maximum value of P, subject to several conditions. These conditions are called **constraints**. First, we assume that the number of each type of jacket is not negative, so we have the constraints

$$x \geqslant 0$$

$$y \geqslant 0.$$

Second, we have a constraint on the number of hours available for assembling. Since it takes 2 hours to assemble each inexpensive jacket, the number of hours needed to assemble all the inexpensive jackets is $2x$. Similarly, it takes 3 hours to assemble each better jacket, so the number of hours needed to assemble all the better jackets is $3y$. The hours for assembling must total no more than 120, so the constraint is

$$2x + 3y \leqslant 120.$$

Finally, we have a constraint on the number of hours available for finishing. It takes 2 hours to finish each inexpensive jacket, or $2x$ hours for all the inexpensive jackets. It takes 6 hours to finish each better jacket, or $6y$ hours for all the better jackets. The hours for finishing must total no more than 180, so the constraint is

$$2x + 6y \leqslant 180.$$

We list all the constraints:

$$x \geqslant 0$$

$$y \geqslant 0$$

$$2x + 3y \leqslant 120$$

$$2x + 6y \leqslant 180.$$

The list of constraints forms a system of linear inequalities in two variables. We draw the graph of this system of inequalities:

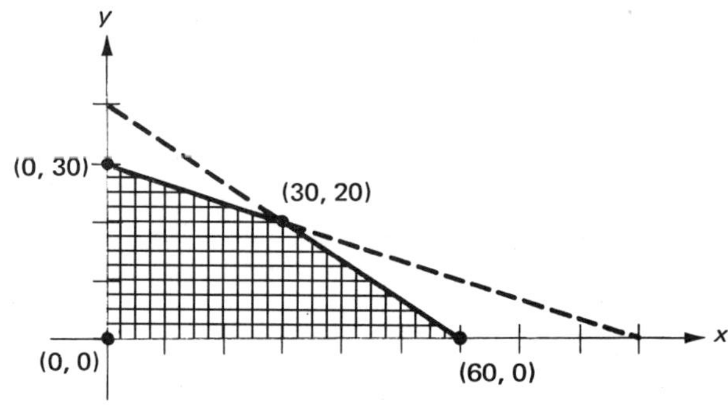

Observe that we have shown all the points of intersection of the boundaries.

The part of the graph common to all of the inequalities is indicated by horizontal and vertical lines. When this common part is bounded on all sides, it is called a **polygonal region**. The points of intersection of the boundaries are the **vertices** (each one is called a **vertex**) of the polygonal region. We now state the fundamental principle of linear programming.

> ***Fundamental Principle:*** If a linear function is given over a polygonal region, the maximum and minimum values of the function each occur at a vertex of the region.

The proof of the fundamental principle is far beyond the scope of this book. The principle, however, is very easy to apply.

Our problem is to find the maximum value of

$$P = 7x + 13y.$$

This maximum value, according to the fundamental principle, will occur at one of the vertices of our polygonal region. Therefore, we calculate P for each vertex:

At $(0, 0)$, $P = 7(0) + 13(0) = 0$.
At $(60, 0)$, $P = 7(60) + 13(0) = 420$.
At $(30, 20)$, $P = 7(30) + 13(20) = 470$.
At $(0, 30)$, $P = 7(0) + 13(30) = 390$.

Clearly, the minimum profit is at $(0, 0)$, where no jackets of either style are made and the profit is zero. The maximum profit occurs at $(30, 20)$. The company should make 30 inexpensive jackets and 20 better jackets a day. The profit is $470.

We can construct an interesting variation of the preceding example by introducing an additional constraint.

EXAMPLE 20.20.   Suppose in Example 20.19, the better jacket is so expensive that the company can sell only 10 a day. However, it also can raise the price of the inexpensive jacket slightly so the profit is $9.00. How many jackets of each type should the company make to maximize profits, and what is the maximum profit?

Solution.   Since y is the number of better jackets, we have the additional constraint

$$y \leqslant 10.$$

The graph of the system of inequalities representing the constraints becomes

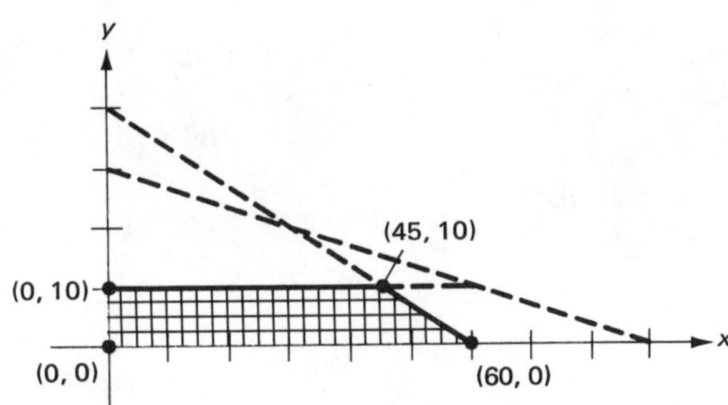

Also, because the profit on an inexpensive jacket is now $9.00, the total profit is

$$P = 9x + 13y.$$

We calculate P, using the new formula, for each vertex of the new polygonal region:

At $(0, 0)$, $P = 9(0) + 13(0) = 0.$
At $(60, 0)$, $P = 9(60) + 13(0) = 540.$
At $(45, 10)$, $P = 9(45) + 13(10) = 535.$
At $(0, 10)$, $P = 9(0) + 13(10) = 130.$

Under the new conditions, it is not profitable to make any better jackets. The company should make 60 inexpensive jackets a day, for a profit of $540.

A similar application of linear programming involves problems where we wish to maximize profit from investments.

EXAMPLE 20.21.   You plan to invest up to $12,000 in a savings account and a speculative stock. The savings account pays 6% annually and the stock pays 10% annually. Your plan calls for you to have no less than $6000 but no more than $10,000 in savings. Also, because of the speculative nature of the stock, you will invest no more than $4000 in it. How much should you invest in savings and how much in the stock to maximize your annual profit? What is the maximum profit?

Solution.   We let x represent the amount invested in savings and y the amount invested in the stock. Since the savings account pays 6% annually, the profit from it is $.06x$. Since the stock pays 10% annually, the profit from it is $.1y$. The total profit is

$$P = .06x + .1y.$$

Now, we consider the constraints. Again, we may assume that

$$x \geqslant 0$$
$$y \geqslant 0.$$

Indeed, because no less than \$6000 is to go into savings,

$$x \geqslant 6000.$$

Also, because no more than \$10,000 is to go into savings,

$$x \leqslant 10,000.$$

No more than \$4000 is to go into the stock, so

$$y \leqslant 4000.$$

At the beginning of the problem it is stated that the two investments will have a total of up to \$12.000. Therefore,

$$x + y \leqslant 12,000.$$

The list of constraints is

$$y \geqslant 0$$
$$x \geqslant 6000$$
$$x \leqslant 10,000$$
$$y \leqslant 4000$$
$$x + y \leqslant 12,000.$$

The graph of this system of inequalities is

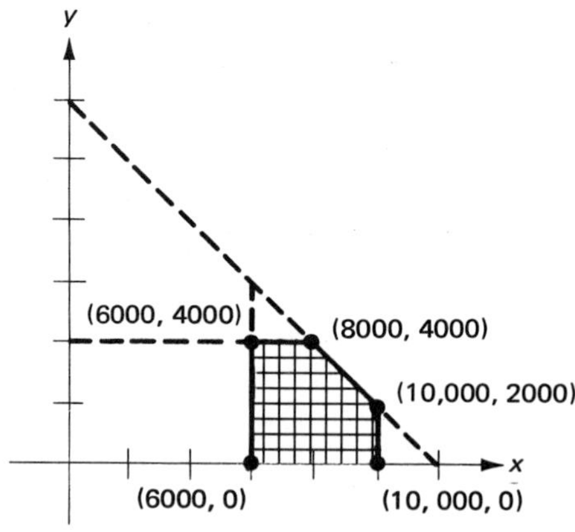

We calculate the profit for each vertex of the polygonal region:

At $(6000, 0)$, $P = .06(6000) + .1(0) = 360.$

At $(10,000, 0)$, $P = .06(10,000) + .1(0) = 600.$

At $(10,000, 2000)$, $P = .06(10,000) + .1(2000) = 800.$

At $(8000, 4000)$, $P = .06(8000) + .1(4000) = 880.$

At $(6000, 4000)$, $P = .06(6000) + .1(4000) = 760.$

For a maximum profit under the given conditions, you should invest \$8000 in savings and \$4000 in the stock. The annual profit is \$880.

Exercise 20.4

1. A factory makes portable TVs and console TVs. For each TV, a set must be assembled and a cabinet built. For the portable TV, it takes 2 hours to assemble the set and 1 hour to build the cabinet. For the console TV, it takes 2 hours to assemble the set and 4 hours to build the cabinet. The factory has up to 60 man-hours a day for assembling sets and also up to 60 man-hours a day for building cabinets. If the profit from a portable TV is $40 and the profit from a console TV is $80, how many of each kind of TV should the factory make a day to maximize profits? What is the maximum profit?

2. A company makes personal calculators and scientific calculators. To make a personal calculator takes 1 hour of labor and $1.00 worth of parts. To make a scientific calculator takes 2 hours of labor and $10.00 worth of parts. The company has up to 800 man-hours of labor available per day, and can spend up to $2000 on parts. If the profit from a personal calculator is 50¢ and the profit from a scientific calculator is $2.00, how many of each type should the company make a day to maximize profits? What is the maximum profit?

3. A silversmith makes two patterns of serving spoons. Each spoon is shaped on a machine, and then the pattern is finished by hand. A plain spoon takes $\frac{1}{2}$ hour on the machine and 1 hour of hand finishing. A decorated spoon takes 1 hour on the machine and 4 hours of hand finishing. The silversmith has staff to provide up to 30 hours of machine time and 100 hours of hand work a week. If the profit for a plain spoon is $3.00 and the profit for a decorated spoon is $8.00, how many spoons of each type should the silversmith make each week to maximize his profit? What is the maximum profit?

4. Suppose the decorated spoon in Exercise 3 contains a lot of silver, so it is very expensive. The silversmith knows he can sell no more than 10 of them a week. However, by making a slight adjustment on the plain spoon, he can raise his profit on it to $5.00 a spoon. How many spoons of each type should the silversmith make a week to maximize his profit? What is the maximum profit?

5. A farmer plans to raise potatoes and corn. He will use up to 300 acres for these crops. The potatoes cost $20 an acre to plant, and the corn costs $40 an acre to plant. The farmer can spend up to $10,000 for planting costs. If the profit from potatoes is $80 an acre and the profit from corn is $50 an acre, how many acres of each crop should the farmer plant to maximize his profit? What is the maximum profit?

6. In Exercise 5, suppose a government regulation limits the farmer to planting at most 150 acres of potatoes. All other conditions are the same. How many acres of each crop should the farmer plant to maximize his profit. What is the maximum profit?

7. You plan to invest up to $10,000 in a savings certificate and a regular savings account. The certificate pays 8% annually, and the account pays 6% annually. You plan to put no more than $8000 in the certificate and no more than $6000 in the account. Also, the certificate calls for a minimum investment of $2000. How much should you invest in the savings certificate and how much in the savings account to maximize your annual profit? What is the maximum profit?

8. You have up to $20,000 to invest. You plan to buy some conservative stock and some speculative stock. The conservative stock pays 7% annually and the speculative stock pays 10% annually. You will put at least $12,000 into the conservative stock. Also, because of the high risk nature of the speculative stock, you will put into it no more than one-third of the amount you put into the conservative stock. How much of each type of stock should you buy to maximize your annual profit? What is the maximum profit? (That the amount of the speculative stock is no more than one-third of the amount of the conservative stock gives a constraint of $y \leqslant \frac{1}{3}x$.)

Self-test

Solve for x:

1. $11 - 4x < 3$

1. _______________

2. $x^2 - 2x - 15 < 0$

2. _______________

3. Draw the graph of $3x - y > 6$.

4. Draw the graph of the system:

$$y \geqslant 1$$

$$x - y > -1$$

5. A factory makes unlined drapes and lined drapes. The unlined drapes take 1 hour to cut and 2 hours to finish. The lined drapes take 3 hours to cut and 4 hours to finish. The factory has up to 60 man-hours of cutting time and 100 man-hours of finishing time a day. If the profit for unlined drapes is $5.00 and for lined drapes is $8.00, how many of each should the factory make a day to maximize profits? What is the maximum profit?

5. __________________

Absolute Value

INTRODUCTION

In this unit you will learn the definition of absolute value. You will apply the definition, along with the other tools of algebra you have learned, to solving equations in one variable involving absolute value. Also, you will learn to graph basic equations in two variables involving absolute value.

OBJECTIVES

When you have finished this unit you should be able to:

1. State the definition of absolute value, and use the definition to evaluate or simplify expressions involving absolute value.
2. Solve equations involving absolute value.
3. Draw the graph of a function of the form $y = |x - a|$.

Section 21.1

The Definition of Absolute Value

You may have seen the symbol $|x|$ in previous courses. The symbol $|x|$ means the **absolute value** of x. You may also remember that the absolute value of a number is never negative. The absolute value of a positive number is the number itself. For example,

$$|7| = 7.$$

Also, the absolute value of zero is zero:

$$|0| = 0.$$

But the absolute value of a negative number is the corresponding positive number. For example,

$$|-7| = 7.$$

These descriptions are sufficient for basic numerical examples. However, for more sophisticated uses of absolute value we need a precise definition.

Definition:

$$|x| = \begin{cases} x, & \text{if } x \geqslant 0. \\ -x, & \text{if } x < 0. \end{cases}$$

Recall that $x \geqslant 0$ means x is positive or x is zero, and $x < 0$ means x is negative.

The definition is in two parts. If we have a positive number or zero, we use the first part. Thus, if $x = 7$, we have a positive number, so

$$|x| = x$$
$$|7| = 7.$$

Also, if $x = 0$,

$$|x| = x$$
$$|0| = 0.$$

If we have a negative number, we use the second part. It is a common error to think that the second part gives a negative number. However, when we have a negative number and use the second part of the definition, we have the negative of a negative number, which is positive. Thus, if $x = -7$,

$$|x| = -x$$

$$|-7| = -(-7) = 7.$$

The absolute value of -7 is the negative of -7, which is the positive number 7.

EXAMPLE 21.1.   Find the value of:

    a. $|-3|$ b. $\left|-\dfrac{3}{2}\right|$ c. $\left|\dfrac{3}{2}\right|$

Solutions.   a.   Since -3 is negative,

$$|-3| = -(-3) = 3.$$

    b.   Since $-\frac{3}{2}$ is negative,

$$\left|-\frac{3}{2}\right| = -\left(-\frac{3}{2}\right) = \frac{3}{2}.$$

    c.   Since $\frac{3}{2}$ is positive,

$$\left|\frac{3}{2}\right| = \frac{3}{2}.$$

A useful rule of absolute value is the **product rule**:

$$|ab| = |a|\,|b|.$$

We will not prove this rule, but will give several examples using the rule.

EXAMPLE 21.2.   Find the value of $|(2)(3)|$.

Solution.   We may write

$$|(2)(3)| = |6| = 6.$$

However, using the product rule, we also have

$$|(2)(3)| = |2|\,|3| = (2)(3) = 6.$$

EXAMPLE 21.3.   Find the value of $|(-2)(-3)|$.

Solution.   Again we may write

$$|(-2)(-3)| = |6| = 6.$$

However, using the product rule, we may also write

$$|(-2)(-3)| = |-2||-3| = (2)(3) = 6.$$

EXAMPLE 21.4.   Find the value of $|(2)(-3)|$.

Solution.   We may write

$$|(2)(-3)| = |-6| = 6,$$

and, using the product rule, we may also write

$$|(2)(-3)| = |2||-3| = (2)(3) = 6.$$

A similar rule for absolute value is the **quotient rule**:

$$\left|\frac{a}{b}\right| = \frac{|a|}{|b|}.$$

EXAMPLE 21.5.   Find the value of $\left|\dfrac{-2}{3}\right|$.

Solution.   If we think of $\dfrac{-2}{3}$ as the fraction $-\dfrac{2}{3}$, we have

$$\left|\frac{-2}{3}\right| = \left|-\frac{2}{3}\right| = \frac{2}{3}.$$

However, using the quotient rule, we also have

$$\left|\frac{-2}{3}\right| = \frac{|-2|}{|3|} = \frac{2}{3}.$$

A common error is also to try to write $|a + b|$ as $|a| + |b|$. As is often the case, what is true for factors, may not also be true for addends. There is no rule for addition of absolute values, and it is not necessarily true that $|a + b|$ is the same as $|a| + |b|$.

EXAMPLE 21.6.   Find the value of:

    a. $|2 + (-3)|$ b. $|2| + |-3|$

Solutions.   a. $|2 + (-3)| = |-1| = 1.$

    b. $|2| + |-3| = 2 + 3 = 5.$

Observe that $|2 + (-3)|$ is not the same as $|2| + |-3|$.

We may use the product and quotient rules to simplify expressions involving absolute value.

EXAMPLE 21.7.   Simplify $|5x + 10|$.

Solution.   We factor the expression inside the absolute value:

$$|5x + 10| = |5(x + 2)|.$$

Then, using the product rule,

$$|5(x + 2)| = |5|\,|x + 2|$$
$$= 5|x + 2|.$$

Since the expression remaining inside the absolute value cannot be factored further, the expression cannot be simplified further. Remember that absolute value rules may not be applied to addends.

EXAMPLE 21.8.   Simplify $\left|\dfrac{2x - 4}{8}\right|$.

Solution.   First, we reduce the expression inside the absolute value:

$$\left|\frac{2x - 4}{8}\right| = \left|\frac{x - 2}{4}\right|.$$

Then, using the quotient rule,

$$\left|\frac{x - 2}{4}\right| = \frac{|x - 2|}{|4|}$$
$$= \frac{|x - 2|}{4}.$$

Alternatively, we could use both the quotient and product rules to write

$$\left|\frac{2x - 4}{8}\right| = \frac{|2x - 4|}{|8|}$$
$$= \frac{|2(x - 2)|}{|8|}$$
$$= \frac{|2|\,|x - 2|}{|8|}$$
$$= \frac{2|x - 2|}{8}$$
$$= \frac{|x - 2|}{4}.$$

EXAMPLE 21.9.   Simplify $|-x - 4|$.

Solution.   We factor -1 from the expression $-x - 4$, and then use the product rule:

$$|-x - 4| = |(-1)(x + 4)|$$
$$= |-1|\,|x + 4|$$
$$= 1|x + 4|$$
$$= |x + 4|.$$

Exercise 21.1

Find the value of:

1. $|10|$

2. $|0|$

3. $|-10|$

4. $\left|-\dfrac{2}{5}\right|$

5. $|(-3)(5)|$

6. $|(-4)(-6)|$

7. $\left|\dfrac{-4}{3}\right|$

8. $\dfrac{|6|}{|-9|}$

9. $|7 + (-2)|$

10. $|1 + (-8)|$

11. $|15| + |-10|$

12. $\left|\dfrac{-1}{4}\right| + \left|\dfrac{-1}{2}\right|$

13. $|6| + |-6|$

14. $|6 + (-6)|$

15. $|5 - (-2)|$

16. $|5| - |-2|$

Simplify:

17. $|4x + 8|$

18. $|12x - 6|$

19. $\left|\dfrac{x - 3}{6}\right|$

20. $\left|\dfrac{5x + 10}{10}\right|$

21. $|-3x - 15|$

22. $|-x - 1|$

<table>
<tr><td>Section
21.2</td><td><h1>Equations Involving Absolute Value</h1></td></tr>
</table>

Recall from Section 21.1 that the definition of absolute value has two parts:

$$|x| = \begin{cases} x, \text{ if } x \geq 0. \\ -x, \text{ if } x < 0. \end{cases}$$

Therefore, whenever we have an equation in which there is a variable inside an absolute value, we must consider two cases. For one case, the expression inside the absolute is assumed to be positive or zero, and we use the first part of the definition. For the second case, the expression inside the absolute value is assumed to be negative, and we use the second part of the definition.

EXAMPLE 21.10.   Solve $|x| = 2$.

Solution.   We consider two cases:

Case 1.   Suppose $x \geq 0$. Then, from the definition of absolute value, $|x| = x$. Substituting x for $|x|$ in the equation,

$$|x| = 2$$

$$x = 2.$$

Case 2.   Suppose $x < 0$. Then, from the definition of absolute value, $|x| = -x$. Substituting $-x$ for $|x|$ in the equation,

$$|x| = 2$$

$$-x = 2$$

$$x = -2.$$

Therefore, there are two solutions, 2 and -2. Clearly, it is true that $|2| = 2$, and also $|-2| = 2$.

Observe that an equation such as

$$|x| = -2$$

has no solutions. For any number x, the absolute value of x is positive or zero. There is no number x such that the absolute value of x is negative. Therefore, there is no number x such that $|x| = -2$.

When there is a more complicated expression inside the absolute value, we must use the entire expression in place of x.

EXAMPLE 21.11.   Solve $|x + 1| = 4$.

Solution.   The expression inside the absolute value is $x + 1$. Therefore, we use $x + 1$ in place of x. We consider two cases, one where $x + 1$ is positive or zero, and one where $x + 1$ is negative.

Case 1.   Suppose $x + 1 \geqslant 0$. Then, from the definition of absolute value, with $x + 1$ in place of x,

$$|x + 1| = x + 1.$$

We substitute $x + 1$ for the absolute value in the equation:

$$|x + 1| = 4$$

$$x + 1 = 4$$

$$x = 3.$$

Case 2.   Suppose $x + 1 < 0$. Then, from the definition of absolute value, with $x + 1$ in place of x,

$$|x + 1| = -(x + 1)$$

$$= -x - 1.$$

We substitute $-x - 1$ for the absolute value in the equation:

$$|x + 1| = 4$$

$$-x - 1 = 4$$

$$-x = 5$$

$$x = -5.$$

The solutions are 3 and -5. To check, we substitute each solution in the original equation. If $x = 3$,

$$|x + 1| = 4$$

$$|3 + 1| \overset{?}{=} 4$$

$$|4| = 4.$$

If $x = -5$,

$$|-5 + 1| \overset{?}{=} 4$$

$$|-4| = 4.$$

EXAMPLE 21.12.   Solve $|2x - 1| = 3$.

Solution.   The expression inside the absolute value is $2x - 1$. We use $2x - 1$ in place of x, and consider two cases:

Case 1.   Suppose $2x - 1 \geqslant 0$. Then,

$$|2x - 1| = 2x - 1.$$

Substituting $2x - 1$ for the absolute value in the equation,

$$|2x - 1| = 3$$
$$2x - 1 = 3$$
$$2x = 4$$
$$x = 2.$$

Case 2.   Suppose $2x - 1 < 0$. Then,

$$|2x - 1| = -(2x - 1)$$
$$= -2x + 1.$$

Substituting $-2x + 1$ for the absolute value in the equation,

$$|2x - 1| = 3$$
$$-2x + 1 = 3$$
$$-2x = 2$$
$$x = -1.$$

The solutions are 2 and -1. You should check each of these solutions in the original equation.

The product and quotient rules may be useful in simplifying equations before considering the two cases for the absolute value.

EXAMPLE 21.13.   Solve $\left| \dfrac{3x - 1}{4} \right| = \dfrac{1}{2}$.

Solution.   First, we simplify the absolute value using the quotient rule:

$$\left| \frac{3x - 1}{4} \right| = \frac{1}{2}$$
$$\frac{|3x - 1|}{4} = \frac{1}{2}$$

Now, we may simplify the equation by multiplying both sides by 4:

$$4\left(\frac{|3x - 1|}{4} \right) = 4\left(\frac{1}{2} \right)$$
$$|3x - 1| = 2.$$

We consider two cases for the simplified equation:

Case 1.   Suppose $3x - 1 \geqslant 0$. Then,

$$|3x - 1| = 3x - 1,$$

and substituting,

$$|3x - 1| = 2$$
$$3x - 1 = 2$$
$$3x = 3$$
$$x = 1.$$

Case 2.   Suppose $3x - 1 < 0$. Then,

$$|3x - 1| = -(3x - 1)$$
$$= -3x + 1,$$

and substituting,

$$|3x - 1| = 2$$
$$-3x + 1 = 2$$
$$-3x = 1$$
$$x = -\frac{1}{3}.$$

The solutions are 1 and $-\frac{1}{3}$. To check, we substitute each solution in the original equation. For $x = 1$,

$$\left|\frac{3x - 1}{4}\right| = \frac{1}{2}$$
$$\left|\frac{3(1) - 1}{4}\right| \overset{?}{=} \frac{1}{2}$$
$$\left|\frac{3 - 1}{4}\right| \overset{?}{=} \frac{1}{2}$$
$$\left|\frac{2}{4}\right| \overset{?}{=} \frac{1}{2}$$
$$\left|\frac{1}{2}\right| = \frac{1}{2}.$$

For $x = -\frac{1}{3}$,

$$\left|\frac{3\left(-\frac{1}{3}\right) - 1}{4}\right| \overset{?}{=} \frac{1}{2}$$
$$\left|\frac{-1 - 1}{4}\right| \overset{?}{=} \frac{1}{2}$$
$$\left|\frac{-2}{4}\right| \overset{?}{=} \frac{1}{2}$$
$$\left|-\frac{1}{2}\right| = \frac{1}{2}$$

If the absolute value is not isolated on one side of the equation, we could isolate it before substituting. It is not necessary to do this, however, if we are careful in using parentheses when we substitute for the absolute value.

EXAMPLE 21.14.   Solve $x - |2x - 9| = 3$.

Solution.   We consider two cases for the absolute value:

Case 1.   Suppose $2x - 9 \geqslant 0$. Then,

$$|2x - 9| = 2x - 9.$$

We substitute for the absolute value in the equation, being careful to use parentheses:

$$x - |2x - 9| = 3$$
$$x - (2x - 9) = 3$$
$$x - 2x + 9 = 3$$
$$-x = -6$$
$$x = 6.$$

Case 2.   Suppose $2x - 9 < 0$. Then,

$$|2x - 9| = -(2x - 9)$$
$$= -2x + 9.$$

Again substituting, and being careful to use parentheses,

$$x - |2x - 9| = 3$$
$$x - (-2x + 9) = 3$$
$$x + 2x - 9 = 3$$
$$3x = 12$$
$$x = 4.$$

The solutions are 6 and 4. To check, if $x = 6$,

$$x - |2x - 9| = 3$$
$$6 - |2(6) - 9| \overset{?}{=} 3$$
$$6 - |12 - 9| \overset{?}{=} 3$$
$$6 - |3| \overset{?}{=} 3$$
$$6 - 3 = 3.$$

If $x = 4$,

$$4 - |2(4) - 9| \overset{?}{=} 3$$

$$4 - |8 - 9| \overset{?}{=} 3$$

$$4 - |-1| \overset{?}{=} 3$$

$$4 - 1 = 3.$$

We may find extraneous solutions to equations involving absolute value.

EXAMPLE 21.15.   Solve $3x - |x - 2| = 4$.

Solution.   *Case 1.*   Suppose $x - 2 \geqslant 0$. Then,

$$|x - 2| = x - 2.$$

Substituting,

$$3x - |x - 2| = 4$$

$$3x - (x - 2) = 4$$

$$3x - x + 2 = 4$$

$$2x = 2$$

$$x = 1.$$

However, if $x = 1$,

$$x - 2 = 1 - 2 = -1,$$

so $x - 2$ is negative. But $x - 2 \geqslant 0$ is a condition of this case. Since $x = 1$ does not satisfy this condition, $x = 1$ is an extraneous solution. If we try to check $x = 1$ in the original equation, we have

$$3x - |x - 2| = 4$$

$$3(1) - |1 - 2| \overset{?}{=} 4$$

$$3 - |-1| \overset{?}{=} 4$$

$$3 - 1 \neq 4.$$

Case 2.   Suppose $x - 2 < 0$. Then,

$$|x - 2| = -(x - 2)$$

$$= -x + 2.$$

Substituting,

$$3x - |x - 2| = 4$$

$$3x - (-x + 2) = 4$$

$$3x + x - 2 = 4$$

$$4x = 6$$

$$x = \frac{3}{2}.$$

When $x = \frac{3}{2}$,

$$x - 2 = \frac{3}{2} - 2 = -\frac{1}{2},$$

which is negative. The condition $x - 2 < 0$ is satisfied, so $\frac{3}{2}$ is a solution. You should check this solution in the original equation.

EXAMPLE 21.16.   Solve $x - |3x + 4| = 4$.

Solution.   *Case 1.*   Suppose $3x + 4 \geqslant 0$. Then,

$$|3x + 4| = 3x + 4,$$

and

$$x - |3x + 4| = 4$$
$$x - (3x + 4) = 4$$
$$x - 3x - 4 = 4$$
$$-2x = 8$$
$$x = -4.$$

But, if $x = -4$, $3x + 4$ is negative, so $x = -4$ is an extraneous solution.

Case 2.   Suppose $3x + 4 < 0$. Then,

$$|3x + 4| = -(3x + 4)$$
$$= -3x - 4,$$

and

$$x - |3x + 4| = 4$$
$$x - (-3x - 4) = 4$$
$$x + 3x + 4 = 4$$
$$4x = 0$$
$$x = 0.$$

But, if $x = 0$, $3x + 4$ is positive, so $x = 0$ also is an extraneous solution. The equation has no solution.

Exercise 21.2

Solve for x:

1. $|x| = 4$

2. $|x| = -4$

3. $2|x| = 6$

4. $\dfrac{|x|}{3} = 2$

5. $|x| + 3 = 4$

6. $|x| + 5 = 4$

7. $|x - 5| = 3$

8. $|x + 3| = 1$

9. $|3x + 2| = 1$

10. $|2x - 5| = 7$

11. $|x - 2| = 0$

12. $|4x + 1| = 0$

13. $|x-3|=-1$

14. $|3x+2|=-4$

15. $\left|\dfrac{x-2}{6}\right|=\dfrac{2}{3}$

16. $\left|\dfrac{2x+1}{4}\right|=2$

17. $\left|\dfrac{3x-9}{4}\right|=\dfrac{3}{2}$

18. $\left|\dfrac{2x+2}{3}\right|=\dfrac{2}{3}$

19. $x+|2x+3|=6$

20. $x+|3x-2|=10$

21. $x-|2x+1|=-5$

22. $x-|3x-4|=-2$

23. $2x-|x+5|=5$

24. $3x-|x+2|=3$

25. $x-|2x-1|=2$

26. $2x-|4x-5|=4$

<table>
<tr><td>Section
21.3</td><td></td></tr>
</table>

Absolute Value Functions

Consider the function

$$y = |x|$$

or

$$f(x) = |x|.$$

We must define this function in two parts:

$$f(x) = \begin{cases} x, & \text{if } x \geqslant 0. \\ -x, & \text{if } x < 0. \end{cases}$$

Thus, for $x \geqslant 0$, we have

$$f(0) = 0, \text{ gives } (0, 0).$$

$$f(1) = 1, \text{ gives } (1, 1).$$

$$f(2) = 2, \text{ gives } (2, 2).$$

For $x < 0$, we have

$$f(-1) = -(-1) = 1, \text{ gives } (-1, 1).$$

$$f(-2) = -(-2) = 2, \text{ gives } (-2, 2).$$

Plotting these points, we have the graph

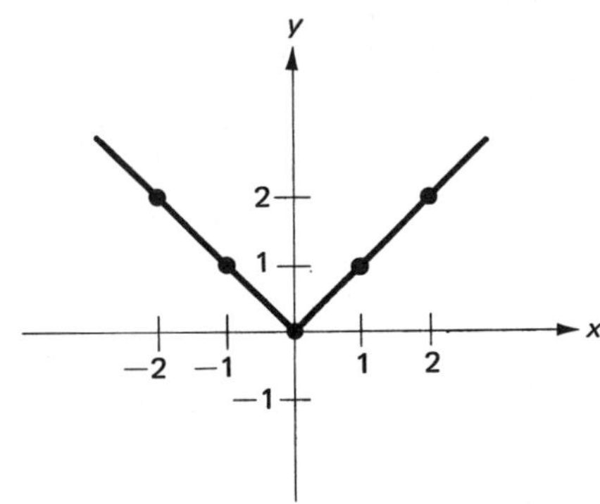

Observe that, for $x \geqslant 0$, the graph is the same as the graph of the line $y = x$. For $x < 0$, however, the graph is the same as the graph of the line $y = -x$.

A basic type of function involving absolute value has the form

$$y = |x - a|$$

or

$$f(x) = |x - a|.$$

We must use the expression $x - a$ in place of x. Thus the function is defined in two parts:

$$f(x) = \begin{cases} x - a, \text{ if } x - a \geq 0. \\ -(x - a), \text{ if } x - a < 0. \end{cases}$$

We see that the graph is the same as the graph of the line $y = x - a$ when $x - a \geq 0$, and the line $y = -x + a$ when $x - a < 0$.

EXAMPLE 21.17.   Draw the graph of $y = |x - 1|$.

Solution.   The function is defined in two parts, one for $x - 1 \geq 0$, or $x \geq 1$, and one for $x - 1 < 0$, or $x < 1$. Therefore, we will find points at $x = 1$, and on either side of $x = 1$:

$$f(1) = |1 - 1| = |0| = 0, \text{ gives } (1, 0).$$

$$f(2) = |2 - 1| = |1| = 1, \text{ gives } (2, 1).$$

$$f(3) = |3 - 1| = |2| = 2, \text{ gives } (3, 2).$$

Also,

$$f(0) = |0 - 1| = |-1| = 1, \text{ gives } (0, 1).$$

$$f(-1) = |-1 - 1| = |-2| = 2, \text{ gives } (-1, 2).$$

Plotting these points, we have the graph

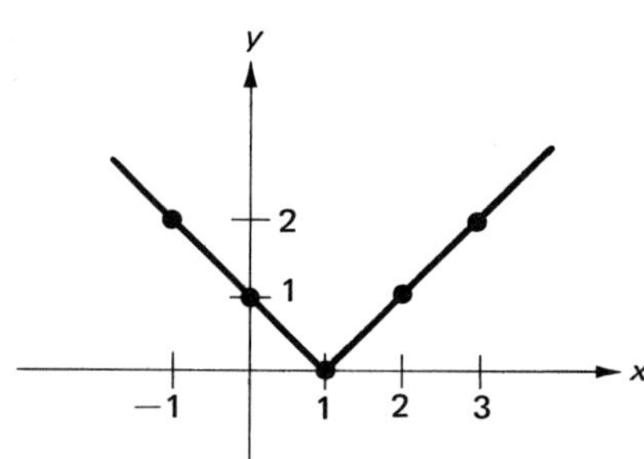

For $x \geq 1$, the graph is the same as the graph of the line $y = x - 1$. For $x < 1$, the graph is the same as the graph of the line $y = -x + 1$. Observe that the domain of the function is all real numbers. However, the y-values all are positive or zero. Therefore, the range of the function is $y \geq 0$.

Exercise 21.3

Draw the graph:

1. $y = |x - 2|$ 2. $y = |x - 3|$ 3. $y = |x + 1|$ 4. $y = |x + 2|$

5. $y = |x + 3|$ 6. $y = |x - 5|$ 7. $y = |x - 4|$ 8. $y = |x + 4|$

Self-test

1. Find the value of:

   a. $|4 + (-9)|$

   b. $|4| + |-9|$

1a. ______________

1b. ______________

Solve for x:

2. $|3x| + 4 = 3$

2. ______________

3. $|4x - 3| = 5$

3. ______________

4. $3x - |x + 4| = 8$

4. ______________

5. Draw the graph of $y = |x + 5|$.

Cumulative Review

INTRODUCTION

In this unit you should make certain you remember all the material in all of the preceding units.

OBJECTIVE

When you have finished this unit you should be able to demonstrate that you can fulfill every objective of each preceding unit.

To prepare for this unit you should review the Self-Tests for Units 1 through 21. Do each problem in each Self-Test over again. If you cannot do a problem, or even if you have the slightest difficulty, you should:

1. Find out from the answer section the Objective for the unit to which the problem relates.
2. Review all the material in the section which has the same number as the objective, and redo all the Exercises for the section.
3. Try the Self-Test for the unit again.

Repeat these steps until you can do each problem in each Self-Test for each of Units 1 through 21 easily and accurately.

Self-test

1. Find the value of:

    a. $(64)^{-\frac{2}{3}}$

    b. $(-16)^{\frac{1}{4}}$

    1a. ________________

    1b. ________________

2. Add: $\dfrac{3}{x-1} + \dfrac{2x}{1-x^2}$

    2. ________________

3. Rationalize the denominator of $\dfrac{6+\sqrt{6}}{2+\sqrt{6}}$.

    3. ________________

Solve for x:

4. $x^2 + 4x = 3$

    4. ________________

5. $\sqrt{3x - 5} - x = -5$

5. _________________

6. $2x - |x - 5| = 13$

6. _________________

7. $3x - (5x + 2) < 10$

7. _________________

8. You bicycle to a picnic ground, and return at an average rate of 2 miles per hour less. The distance each way is 10 miles. If your total time is $2\frac{1}{4}$ hours, what is your average rate going to the picnic ground and your average rate returning?

8. _________________

9. Solve and graph the system:

$$x + 2y = 4$$
$$4x + 5y = 20$$

9. _________________

10. Match each function with the graph which represents it:

$$y = \left(\tfrac{1}{4}\right)^x$$

$$y = \frac{-4}{x}$$

$$y = |x - 4|$$

$$y = x^2 - 4$$

$$y = \sqrt{x + 4}$$

None of these

a.

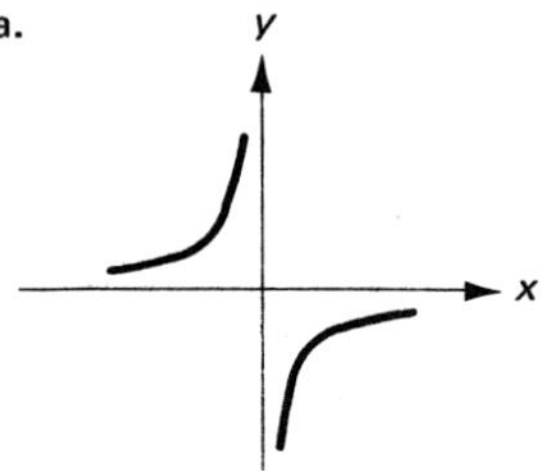

b.

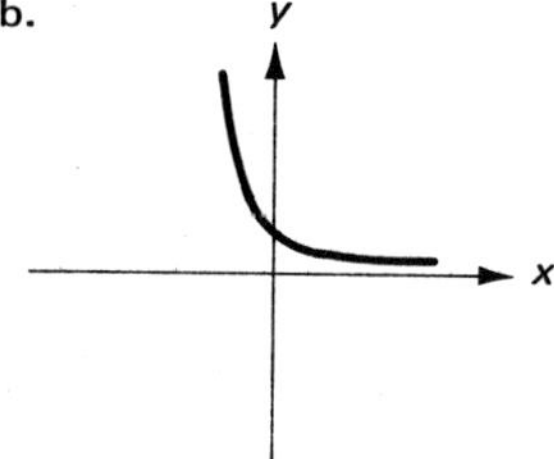

c.

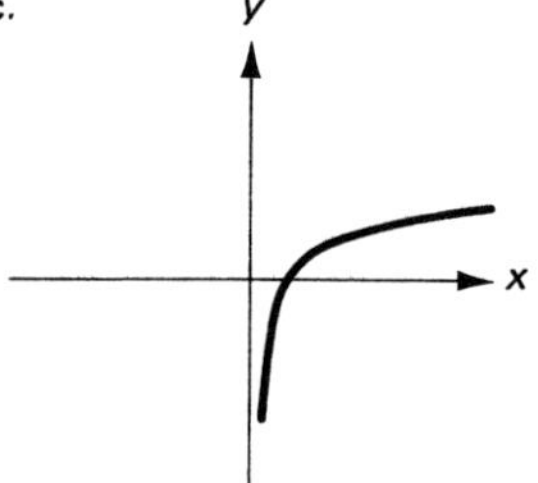

d.

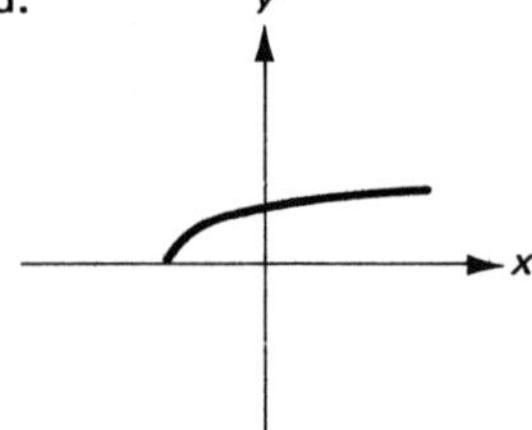

10a. _______________

10b. _______________

10c. _______________

10d. _______________

e.

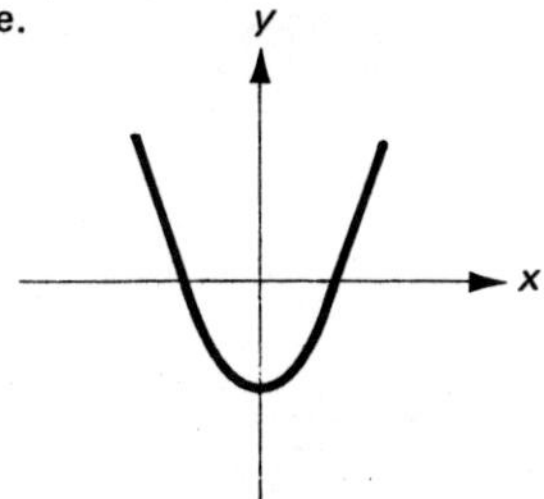

10e. _______________

f.

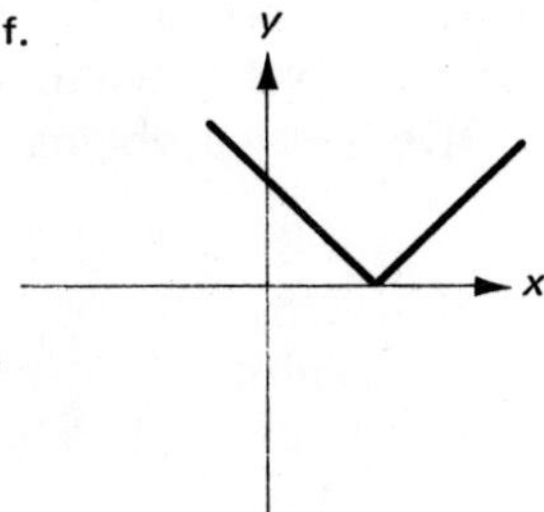

10f. _______________

Unit 23

Logarithms

INTRODUCTION

In this unit you will learn about a different kind of function called the logarithmic function. The logarithmic function has many applications. Logarithms were an important computational tool from the time of their invention by John Napier in the early 1600s until the very recent general use of handheld calculators. Now, the logarithmic function is used to solve many problems which cannot be solved using ordinary algebraic operations.

OBJECTIVES

When you have finished this unit you should be able to:

1. Use the definition of a logarithm to write an exponential statement in logarithmic form, or a logarithmic statement in exponential form.
2. Find the value of a logarithmic expression, where the value can be found by inspection from the exponential form.
3. Draw the graph of a logarithmic function.
4. Use properties of logarithms to write a logarithmic expression as sums, differences, and multiples of logarithms.
5. Use a table of common logarithms to find the common logarithm of a number, or the common antilogarithm of a number.

Section 23.1

Definition of the Logarithm

We begin this unit by introducing the definition of the inverse of a function, and a few of its properties.

> **Definition:** The **inverse** of a function is the relation in two variables obtained by reversing the first and second coordinates in the ordered pairs of the function.

For a function given by an equation in x and y, the inverse can be found by interchanging x and y in the equation of the function. The domain of the function is the range of the inverse, and the range of the function is the domain of the inverse. For example, recall from Section 7.4 the

350

equations $y = x^2$ and $y^2 = x$, and their graphs:

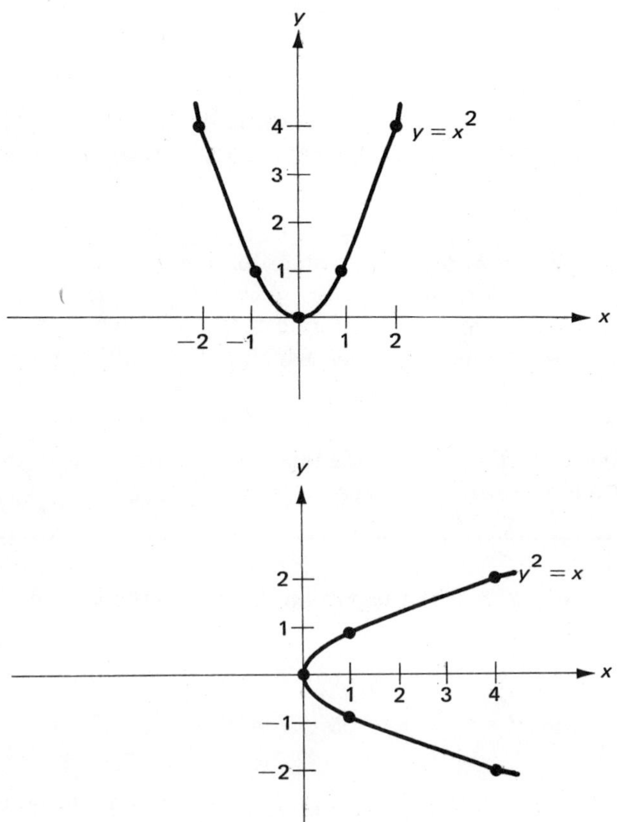

The inverse of the function $y = x^2$ is the relation $x = y^2$, or $y^2 = x$, found by interchanging x and y. The domain of $y = x^2$ is all real numbers and the range is $y \geqslant 0$. The domain of $y^2 = x$ is $x \geqslant 0$ and the range is all real numbers. Observe that $y^2 = x$, the inverse of $y = x^2$, is not a function.

    If the inverse of a function is also a function, then it is called the **inverse function**. For example, consider $y = x^2$ with just the domain $x \geqslant 0$. Its inverse is $y^2 = x$ with just the range $y \geqslant 0$, which we may write as $y = \sqrt{x}$. We have drawn $y = x^2$, $x \geqslant 0$, and $y = \sqrt{x}$, on one set of axes:

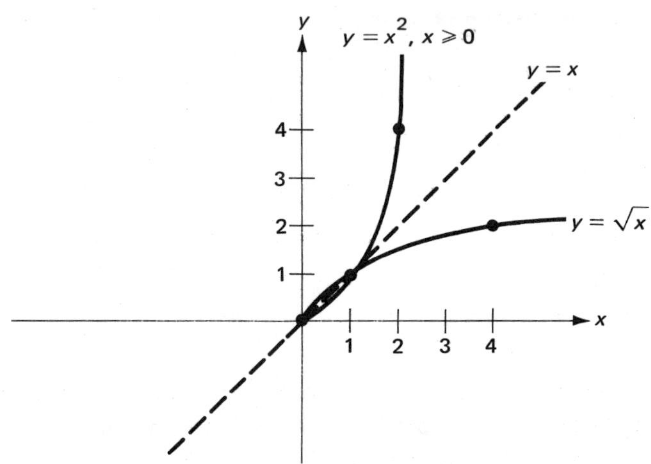

Observe that the graph of $y = \sqrt{x}$ is the reflection of the graph of $y = x^2$, $x \geqslant 0$, in the line $y = x$. We say the graphs are **symmetric** with respect to the line $y = x$.

Now, recall the exponential function

$$y = b^x,$$

where $b > 0$ and $b \neq 1$, from Section 12.4. Its domain is all real numbers, and its range is $y > 0$. To construct the inverse of the exponential function, we write

$$x = b^y,$$

where $b > 0$ and $b \neq 1$. The domain of $x = b^y$ is $x > 0$ and its range is all real numbers. It has just one y-value for each x-value, so it is an inverse function.

However, there is no algebraic method by which the equation $x = b^y$ can be solved for y. Therefore, we define a function which is the solution for y of the equation $x = b^y$.

Definition: The **logarithmic function** $y = \log_b x$, where $b > 0$ and $b \neq 1$, is the solution for y of the exponential function $x = b^y$.

We say $y = \log_b x$ is the **logarithm (or log) to the base b of x**.

EXAMPLE 23.1.   Solve for y:

      a. $x = 2^y$ b. $x = 3^y$ c. $x = 10^y$,

Solutions.   We use the definition of a logarithm: $y = \log_b x$ is the solution of $x = b^y$.

    a. Since

$$x = 2^y,$$

    we have $b = 2$, and

$$y = \log_2 x.$$

    b. Since

$$x = 3^y,$$

    we have $b = 3$, and

$$y = \log_3 x.$$

    c. Since

$$x = 10^y,$$

    we have $b = 10$, and

$$y = \log_{10} x.$$

The form

$$y = \log_b x$$

is called the **logarithmic form** of $x = b^y$. The form

$$x = b^y$$

is called the **exponential form** of $y = \log_b x$.

EXAMPLE 23.2. Write in logarithmic form:

  a. $10^2 = 100$ b. $2^{-3} = \dfrac{1}{8}$ c. $9^{\frac{1}{2}} = 3$

Solutions.   We use the definition $y = \log_b x$ is the logarithmic form of $x = b^y$.

  a. For

$$10^2 = 100,$$

we have

$$b^y = x,$$

where $b = 10$, $y = 2$, and $x = 100$. Therefore, the logarithmic form

$$y = \log_b x$$

is

$$2 = \log_{10} 100$$

or

$$\log_{10} 100 = 2.$$

  b. For

$$2^{-3} = \frac{1}{8},$$

we have

$$b^y = x,$$

where $b = 2$, $y = -3$, and $x = \frac{1}{8}$. Therefore, the logarithmic form

$$y = \log_b x$$

is

$$-3 = \log_2 \frac{1}{8}$$

or

$$\log_2 \frac{1}{8} = -3.$$

  c. For

$$9^{\frac{1}{2}} = 3,$$

we have

$$b^y = x,$$

where $b = 9$, $y = \frac{1}{2}$, and $x = 3$. Therefore, the logarithmic form

$$y = \log_b x$$

is

$$\frac{1}{2} = \log_9 3$$

or

$$\log_9 3 = \frac{1}{2}.$$

EXAMPLE 23.3.   Write in exponential form:

    a. $\log_3 81 = 4$ b. $\log_{64} 4 = \frac{1}{3}$ c. $\log_{10} .001 = -3$

Solutions.   We use the definition, $x = b^y$ is the exponential form of $y = \log_b x$.

    a.   For

$$\log_3 81 = 4,$$

we have

$$\log_b x = y,$$

where $b = 3$, $x = 81$, and $y = 4$. Therefore, the exponential form

$$x = b^y$$

is

$$81 = 3^4$$

or

$$3^4 = 81.$$

    b.   For

$$\log_{64} 4 = \frac{1}{3},$$

we have

$$\log_b x = y,$$

where $b = 64$, $x = 4$, and $y = \frac{1}{3}$. Therefore, the exponential form

$$x = b^y$$

is

$$4 = 64^{\frac{1}{3}}$$

or

$$64^{\frac{1}{3}} = 4.$$

c.   For

$$\log_{10} .001 = -3,$$

we have

$$\log_b x = y,$$

where $b = 10$, $x = .001$, and $y = -3$. Therefore, the exponential form

$$x = b^y$$

is

$$.001 = 10^{-3}$$

or

$$10^{-3} = .001.$$

Exercise 23.1

Write in logarithmic form:

1. $4^3 = 64$ 2. $10^{-2} = .01$ 3. $64^{\frac{1}{2}} = 8$ 4. $64^{\frac{1}{3}} = 4$

5. $4^{-3} = \frac{1}{64}$ 6. $4^{\frac{1}{2}} = 2$ 7. $10^4 = 10,000$ 8. $10^{-4} = .0001$

Write in exponential form:

9. $\log_2 64 = 6$ 10. $\log_{27} 3 = \frac{1}{3}$ 11. $\log_{10} 100 = 2$ 12. $\log_{10} 1 = 0$

13. $\log_{36} 6 = \frac{1}{2}$ 14. $\log_2 \frac{1}{64} = -6$ 15. $\log_{10} .1 = -1$ 16. $\log_{10} .0001 = -4.$

**Section
23.2**   Values of Logarithms

Values of certain logarithmic expressions can be found using the definition of the logarithm. These expressions have exponential forms in which the exponent can be found by inspection.

EXAMPLE 23.4.   Find the value of $\log_2 16$.

Solution.   We write

$$\log_2 16 = y,$$

and then write the exponential form

$$2^y = 16.$$

By inspection, it is clear that $y = 4$, and therefore,

$$\log_2 16 = 4.$$

In some cases, we must do a little more work with the exponential form to find the value of a logarithmic expression.

EXAMPLE 23.5.   Find the value of $\log_3 \frac{1}{9}$.

Solution.   We write

$$\log_3 \frac{1}{9} = y,$$

and so the exponential form is

$$3^y = \frac{1}{9}$$

$$= \frac{1}{3^2}$$

$$= 3^{-2}.$$

Therefore, $y = -2$, and

$$\log_3 \frac{1}{9} = -2.$$

EXAMPLE 23.6.   Find the value of $\log_9 3$.

Solution.   We write

$$\log_9 3 = y,$$

and so the exponential form is

$$9^y = 3$$

$$= \sqrt{9}$$

$$= 9^{\frac{1}{2}}.$$

Therefore, $y = \frac{1}{2}$, and

$$\log_9 3 = \frac{1}{2}.$$

Observe that, for each example, the goal is to produce a power on the right-hand side for which the base is the same as the base of the exponential form. This base is also the base of the logarithm.

Logarithms to the base 10 are of particular importance, and are called **common logarithms**. We can use the definition of the logarithm to find the value of the common logarithm of any number which can be written as an integral power of 10.

EXAMPLE 23.7.   Find the value of $\log_{10} 100$.

Solution.   If

$$\log_{10} 100 = y,$$

the exponential form is

$$10^y = 100,$$

so $y = 2$. Therefore,

$$\log_{10} 100 = 2.$$

EXAMPLE 23.8.   Find the value of $\log_{10} .01$.

Solution.   If

$$\log_{10} .01 = y,$$

the exponential form is

$$10^y = .01$$

$$= \frac{1}{100}$$

$$= \frac{1}{10^2}$$

$$= 10^{-2}.$$

Thus $y = -2$, and

$$\log_{10} .01 = -2.$$

Exercise 23.2

Find the value of:

1. $\log_2 8$ 2. $\log_5 125$ 3. $\log_4 16$ 4. $\log_4 16$

5. $\log_{16} 4$ 6. $\log_{16} 2$ 7. $\log_3 27$ 8. $\log_{27} 3$

9. $\log_3 \frac{1}{27}$ 10. $\log_2 \frac{1}{16}$ 11. $\log_5 \frac{1}{125}$ 12. $\log_{125} 5$

13. $\log_{10} 1000$ 14. $\log_{10} 10,000$ 15. $\log_{10} 100,000$ 16. $\log_{10} .001$

17. $\log_{10} .0001$ 18. $\log_{10} .00001$ 19. $\log_{10} 10$ 20. $\log_{10} 1$

<table>
<tr><td>Section
23.3</td><td><h1>Graphs of Logarithmic Functions</h1></td></tr>
</table>

To draw the graph of a logarithmic function, we use the exponential form. We may compare the exponential form with the exponential function for which it is the inverse function.

EXAMPLE 23.9.   Draw the graph of $y = \log_2 x$.

Solution.   The exponential form of $y = \log_2 x$ is $x = 2^y$. Therefore, it is the inverse function of $y = 2^x$. Recall from Section 12.4 that some points on the graph of $y = 2^x$ are:

$$f(0) = 2^0 = 1, \text{ gives } (0, 1).$$

$$f(1) = 2^1 = 2, \text{ gives } (1, 2).$$

$$f(2) = 2^2 = 4, \text{ gives } (2, 4).$$

$$f(3) = 2^3 = 8, \text{ gives } (3, 8).$$

$$f(-1) = 2^{-1} = \frac{1}{2}, \text{ gives } \left(-1, \frac{1}{2}\right).$$

$$f(-2) = 2^{-2} = \frac{1}{4}, \text{ gives } \left(-2, \frac{1}{4}\right).$$

$$f(-3) = 2^{-3} = \frac{1}{8}, \text{ gives } \left(-3, \frac{1}{8}\right).$$

The graph is

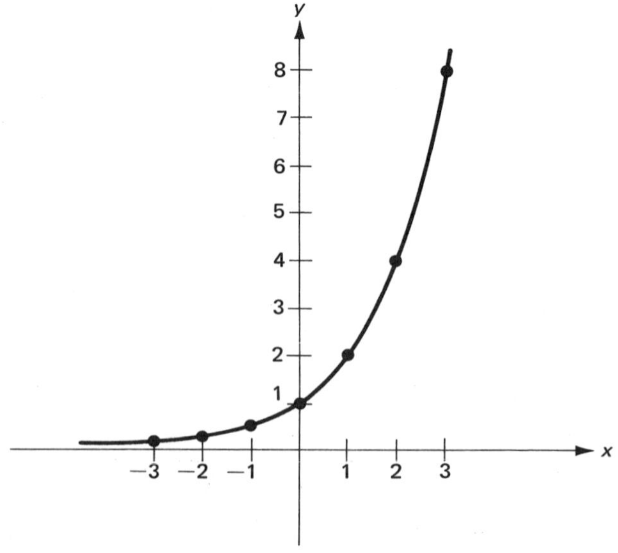

Observe that the domain is all real numbers, but the range is $y > 0$. Also, observe that the graph approaches the negative x-axis asymptotically. We now reverse the coordinates of the ordered pairs of $y = 2^x$ to draw the graph of its inverse $x = 2^y$, or $y = \log_2 x$:

$$(1, 0) \text{ gives } \log_2 1 = 0.$$

$$(2, 1) \text{ gives } \log_2 2 = 1.$$

$$(4, 2) \text{ gives } \log_2 4 = 2.$$

$$(8, 3) \text{ gives } \log_2 8 = 3.$$

$$\left(\frac{1}{2}, -1\right) \text{ gives } \log_2 \frac{1}{2} = -1.$$

$$\left(\frac{1}{4}, -2\right) \text{ gives } \log_2 \frac{1}{4} = -2.$$

$$\left(\frac{1}{8}, -3\right) \text{ gives } \log_2 \frac{1}{8} = -3.$$

We show the graphs of $y = 2^x$ and $y = \log_2 x$ on one set of axes:

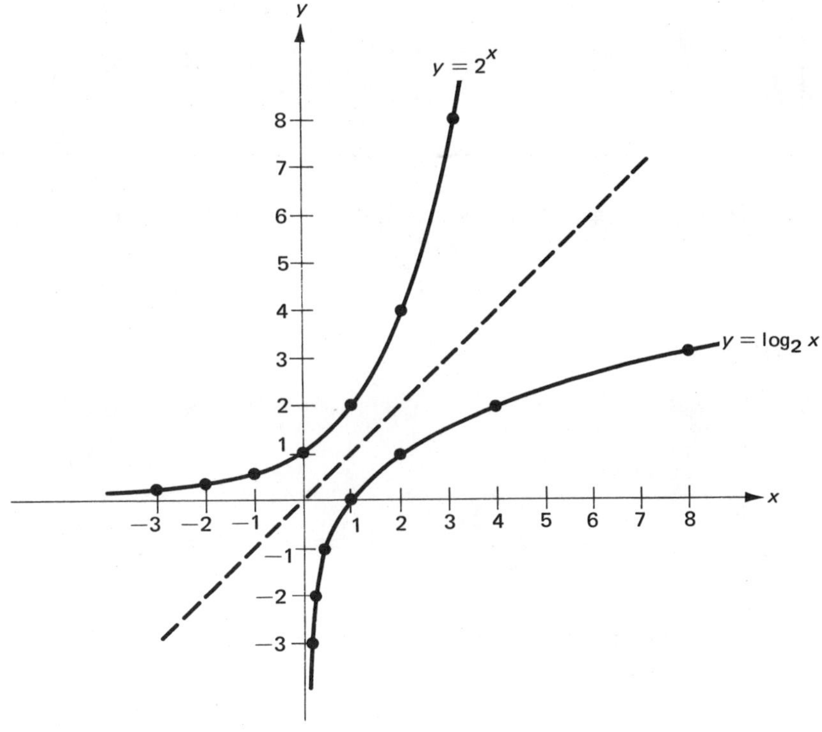

The graphs of $y = 2^x$ and its inverse $y = \log_2 x$ are symmetric with respect to the line $y = x$. Also, since x and y reverse, the graph of $y = \log_2 x$ approaches the negative y-axis asymptotically. The domain of $y = \log_2 x$ is $x > 0$ and its range is all real numbers. It is important to remember that, since the domain of a logarithmic function is $x > 0$, $\log_b A$ is defined only if $A > 0$.

Exercise 23.3

Draw the graph:

1. $y = \log_3 x$

2. $y = \log_4 x$

3. $y = \log_5 x$

4. $y = \log_{10} x$ (Use a scale of 100, 200, 300, . . . , on the x-axis.)

<table><tr><td>Section
23.4</td><td align="center"># Properties of Logarithms</td></tr></table>

The definition of the logarithm can be used to prove three basic properties of logarithms. The proofs of these properties are given at the end of this unit.

> **Properties of Logarithms:** For $A > 0$, $B > 0$, $b > 0$ and $b \neq 1$, and any real number n:
> 1. $\log_b AB = \log_b A + \log_b B$
> 2. $\log_b \dfrac{A}{B} = \log_b A - \log_b B$
> 3. $\log_b A^n = n \log_b A$

We use Properties 1, 2, and 3 to simplify logarithmic expressions.

EXAMPLE 23.10.   Write $\log_b (xy^2)$ as sums and multiples of logarithmic expressions.

Solution.   Using Property 1,

$$\log_b (xy^2) = \log_b x + \log_b y^2.$$

Then, using Property 3,

$$\log_b x + \log_b y^2 = \log_b x + 2 \log_b y.$$

EXAMPLE 23.11.   Write $\log_b \left(\dfrac{m^3}{n^2} \right)$ as differences and multiples of logarithmic expressions.

Solution.   Using Property 2,

$$\log_b \left(\frac{m^3}{n^2} \right) = \log_b m^3 - \log_b n^2.$$

Then, using Property 3,

$$\log_b m^3 - \log_b n^2 = 3 \log_b m - 2 \log_b n.$$

EXAMPLE 23.12. Write $\log_b\left(\dfrac{xy}{z}\right)^3$ as sums, differences, and multiples of logarithmic expressions.

Solution. Since the exponent applies to the entire expression inside the logarithm, we use Property 3 first:

$$\log_b\left(\frac{xy}{z}\right)^3 = 3\log_b\left(\frac{xy}{z}\right).$$

Then, using Property 2,

$$3\log_b\left(\frac{xy}{z}\right) = 3(\log_b xy - \log_b z).$$

Finally, using Property 1,

$$3(\log_b xy - \log_b z) = 3(\log_b x + \log_b y - \log_b z)$$
$$= 3\log_b x + 3\log_b y - 3\log_b z.$$

If there is a radical expression inside a logarithm, we write the radical as a rational exponent. Then, we apply Property 3 to the rational exponent.

EXAMPLE 23.13. Write $\log_b\sqrt[3]{\dfrac{p}{qr}}$ as sums, differences, and multiples of logarithmic expressions.

Solution. We write

$$\log_b\sqrt[3]{\frac{p}{qr}}$$

as

$$\log_b\left(\frac{p}{qr}\right)^{\frac{1}{3}}.$$

Then,

$$\log_b\left(\frac{p}{qr}\right)^{\frac{1}{3}} = \frac{1}{3}\log_b\left(\frac{p}{qr}\right)$$
$$= \frac{1}{3}(\log_b p - \log_b qr)$$
$$= \frac{1}{3}\left[\log_b p - (\log_b q + \log_b r)\right]$$
$$= \frac{1}{3}(\log_b p - \log_b q - \log_b r)$$
$$= \frac{1}{3}\log_b p - \frac{1}{3}\log_b q - \frac{1}{3}\log_b r.$$

EXAMPLE 23.14. Write $\log_b\left(\dfrac{p}{\sqrt[3]{qr}}\right)$ as sums, differences, and multiples of logarithmic expressions.

Solution. We write

$$\log_b\left(\frac{p}{\sqrt[3]{qr}}\right).$$

as

$$\log_b\left(\frac{p}{(qr)^{\frac{1}{3}}}\right).$$

Then,

$$\log_b\left(\frac{p}{(qr)^{\frac{1}{3}}}\right) = \log_b p - \log_b (qr)^{\frac{1}{3}}$$

$$= \log_b p - \frac{1}{3}\log_b (qr)$$

$$= \log_b p - \frac{1}{3}(\log_b q + \log_b r)$$

$$= \log_b p - \frac{1}{3}\log_b q - \frac{1}{3}\log_b r.$$

Observe that the coefficient $\frac{1}{3}$ does not apply to $\log_b p$ because p was not included in the original radical expression.

Exercise 23.4

Write as sums, differences, and multiples of logarithmic expressions:

1. $\log_b (xyz)$ 2. $\log_b (x^2y^2)$ 3. $\log_b\left(\frac{xy}{z}\right)$ 4. $\log_b\left(\frac{x}{yz}\right)$

5. $\log_b (k^2mn^3)$ 6. $\log_b\left(\frac{k^2}{mn^3}\right)$ 7. $\log_b\left(\frac{\sqrt{s}}{t^3}\right)$ 8. $\log_b\left(\frac{u^2}{\sqrt{vw}}\right)$

9. $\log_b \sqrt[3]{\frac{pq^2}{r}}$ 10. $\log_b \sqrt[4]{\frac{p^2}{qr^3}}$

Section 23.5   Common Logarithms

We mentioned in Section 23.2 that logarithms to the base 10, called common logarithms, are of particular importance. It is the common logarithm which was devised for use as a computational aid. With the widespread use of handheld calculators, common logarithms are decreasing in importance as a calculating tool. However, they still have importance in solving applied problems.

It is conventional to write a common logarithm as log x, leaving out the base 10. Thus, when you see log x, you may assume the base is 10. If you are using a scientific calculator, the function name "log" finds the common logarithm of a number.

We know from the definition of a logarithm how to find common logarithms of numbers which are integral powers of 10. Recall from the examples and exercises in Section 23.2:

$$\log 100{,}000 = \log 10^5 = 5$$
$$\log 10{,}000 = \log 10^4 = 4$$
$$\log 1000 = \log 10^3 = 3$$
$$\log 100 = \log 10^2 = 2$$
$$\log 10 = \log 10^1 = 1$$
$$\log 1 = \log 10^0 = 0$$
$$\log .1 = \log 10^{-1} = -1$$
$$\log .01 = \log 10^{-2} = -2$$
$$\log .001 = \log 10^{-3} = -3$$
$$\log .0001 = \log 10^{-4} = -4$$
$$\log .00001 = \log 10^{-5} = -5$$

It is clear that the value of the common logarithm of an integral power of 10 is the integral exponent. In general,

$$\log 10^n = n.$$

To find common logarithms of numbers other than integral powers of 10, we must use tables of common logarithms. The values given in these tables are found by sophisticated methods beyond the scope of this book, as are the methods used by calculators to find common logarithms.

A part of a table of logarithms is reproduced here. The full table is given in Table II at the back of this book.

N	0	1	2	3	4	5	6	7	8	9
0.0	.0000	.0043	.0086	.0128	.0170					
3.5	.5441	.5453	.5465	.5478	.5490	.5502	.5514	.5527	.5539	.5551
3.6	.5563	.5575	.5587	.5599	.5611	.5623	.5635	.5647	.5658	.5670
3.7	.5682	.5694	.5705	.5717	.5729	.5740	.5752	.5763	.5775	.5786
3.8	.5798	.5809	.5821	.5832	.5843	.5855	.5866	.5877	.5888	.5899
3.9	.5911	.5922	.5933	.5944	.5955	.5966	.5977	.5988	.5999	.6010
4.0	.6021	.6031	.6042	.6053	.6064	.6075	.6085	.6096	.6107	.6117
4.1	.6128	.6138	.6149	.6160	.6170	.6180	.6191	.6201	.6212	.6222
4.2	.6232	.6243	.6253	.6263	.6274	.6284	.6294	.6304	.6314	.6325
4.3	.6335	.6345	.6355	.6365	.6375	.6385	.6395	.6405	.6415	.6425
4.4	.6435	.6444	.6454	.6464	.6474	.6484	.6493	.6503	.6513	.6522
4.5	.6532	.6542	.6551	.6561	.6571	.6580	.6590	.6599	.6609	.6618

Notice that all of the numbers in the column marked N are between 1 and 10. Thus we can find the common logarithm of any three-digit number between 1 and 10 directly from the table.

EXAMPLE 23.15.   Find log 4.37.

Solution. We look under N until we find 4.3. Then we go over to the right to the number in the column under 7. This number is the common logarithm of 4.37. Thus log 4.37 = .6405.

It should be mentioned that, except for the common logarithms of rational powers of 10, common logarithms are nonterminating decimals. The values given in Table II are four-place approximations of those decimals. The values given by calculators are approximations to a greater number of places, depending on the calculator. However, having observed that we are dealing with approximations, we will use the ordinary equal sign throughout the rest of this unit and the next unit. In dealing with logarithms, we do not usually say "approximately" every time a logarithm is used, or to use the "approximately equal to" symbol.

To find common logarithms of numbers which are not between 1 and 10, we use Property 1 of Section 23.4.

EXAMPLE 23.16. Find log 3820.

Solution. From the table, we can find log 3.82 = .5821. To find log 3820, we write

$$\log 3820 = \log 3.82(10^3)$$
$$= \log 3.82 + \log 10^3$$
$$= .5821 + 3$$
$$= 3.5821.$$

Thus, log 3820 = 3.5821. The number 3 is called the **characteristic**, and is the exponent of an integral power of 10. The number .5821 is called the **mantissa**, and is found from the table.

EXAMPLE 23.17. Find log .0414.

Solution. From the table, we can find log 4.14 = .6170. To find log .0414, we write

$$\log .0414 = \log 4.14(10^{-2})$$
$$= \log 4.14 + \log 10^{-2}$$
$$= .6170 + (-2)$$
$$= .6170 - 2.$$

You must *never* write a negative characteristic in front of a positive mantissa as, for example, -2.6170. The number -2.6170 is $-2 - .6170$. However, the correct number is $-2 + .6170$. When the characteristic is negative, it is best to write the logarithm in the form

$$\log .0414 = .6170 - 2.$$

We will call this form the **characteristic-mantissa form.** If you are using a scientific calculator, you will find it gives the answer in the form -1.3830. This number is the ordinary subtraction of $.6170 - 2$:

$$0.6170 - 2.0000 = -1.3830.$$

In this section, we will expect logarithms to be written in characteristic-mantissa form. However, when we come to applications in the next unit, the subtracted form will be useful.

Finally, we need to know how to find a number given its logarithm. This process is known formally as finding the **antilogarithm**, written antilog x, but you might refer to it informally as "unlogging" a number.

EXAMPLE 23.18.   Find:

   a.   antilog .5944   b.   antilog 1.5786   c.   antilog $(.5911-4)$

Solutions.

   a.   We find .5944 to the right of 3.9 and under 3, thus antilog .5944 = 3.93.

   b.   The mantissa is .5786. We find .5786 to the right of 3.7 and under 9, thus antilog .5786 = 3.79. Since the characteristic is 1,

$$\text{antilog } 1.5786 = 3.79(10^1)$$

$$= 37.9.$$

   c.   The mantissa is .5911. We find .5911 to the right of 3.9 and under 0, thus antilog .5911 = 3.90. Since the characteristic is -4,

$$\text{antilog } (.5911 - 4) = 3.90(10^{-4})$$

$$= .00039.$$

Exercise 23.5

Use Table II to find:

1. $\log 2.19$	2. $\log 8.30$	3. $\log 84.6$	4. $\log .323$
5. $\log .0055$	6. $\log 46,700$	7. $\log 9300$	8. $\log 4,230,000$
9. $\log .00645$	10. $\log .0000399$	11. antilog .5366	12. antilog .2041
13. antilog $(.8312-1)$	14. antilog 2.7364	15. antilog 4.3729	16. antilog 5.7709
17. antilog $(.7152-3)$	18. antilog $(.9494-4)$	19. antilog $(.9175-5)$	20. antilog 9.9832

Proofs of the Properties of Logarithms

PROPERTY 1. $\log_b AB = \log_b A + \log_b B$

Proof.   Let $\log_b A = x$ and $\log_b B = y$. Then, by the definition of a logarithm,

$$A = b^x \text{ and } B = b^y$$

$$AB = b^x b^y$$

$$AB = b^{x+y}.$$

Again, by the definition of a logarithm,

$$\log_b AB \cdot = x + y.$$

Substituting back for x and y,

$$\log_b AB = \log_b A + \log_b B.$$

PROPERTY 2. $\log_b \dfrac{A}{B} = \log_b A - \log_b B$

Proof.   Proceeding as in Property 1,

$$\frac{A}{B} = \frac{b^x}{b^y}$$

$$\frac{A}{B} = b^{x-y}$$

$$\log_b \frac{A}{B} = x - y$$

$$\log_b \frac{A}{b} = \log_b A - \log_b B.$$

PROPERTY 3. $\log_b A^n = n \log_b A$

Proof.   Let $\log_b A = x$. Then, by definition of a logarithm,

$$A = b^x$$

$$A^n = (b^x)^n$$

$$A^n = b^{nx}.$$

Again, by the definition of a logarithm,

$$\log_b A^n = nx.$$

Substituting back for x,

$$\log_b A^n = n \log_b A.$$

Self-test

1. Write $\log_b \left(\dfrac{x^2 \sqrt{y}}{z} \right)$ as sums, differences, and multiples of logarithmic expressions.

1. _______________________

2. Find the value of :

    a. $\log_3 \frac{1}{81}$

    b. $\log_{81} 3$

2a. _______________________

2b. _______________________

3. Write $3^2 = 9$ in logarithmic form.

3. _______________________

4. Use Table II to find:

    a. $\log .634$

    b. antilog 2.7033

4a. _______________________

4b. _______________________

5. Draw the graph of $y = \log_6 x$.

Unit 24 — Exponential Equations

INTRODUCTION

An exponential equation is an equation with a variable in the exponent. In this unit you will learn how to solve exponential equations, and how to use exponential equations to solve some of the many types of problems they apply to. Also, you will have a brief introduction to the second important type of logarithm, the natural logarithm, and one of its applications.

OBJECTIVES

When you have finished this unit you should be able to:

1. Use common logarithms to solve exponential equations.
2. Given an appropriate formula, use common logarithms to solve applied problems involving exponential equations.
3. Given an appropriate formula, use natural logarithms to solve applied problems involving exponential equations with the base e.

Section 24.1 — Solving Exponential Equations

An **exponential equation** is an equation with the variable in an exponent. Exponential equations cannot be solved by ordinary algebraic methods. The problem is to get the variable out of the exponent. This can be accomplished by using the properties of logarithms, in particular, Property 3. It must be assumed that we can take the logarithm of each side of an equation, just as we can add to, or multiply, or take the square root of each side.

EXAMPLE 24.1. Solve $5^x = 16$.

Solution. We take the common logarithm of each side of the equation:

$$\log 5^x = \log 16.$$

Applying Property 3 removes the variable from the exponent:

$$x \log 5 = \log 16.$$

Thus,

$$x = \frac{\log 16}{\log 5}$$

$$x = \frac{1.2041}{.6990}$$

$$x = 1.72.$$

You should keep in mind that this solution is only a decimal approximation of x. Indeed, it is not an awfully good approximation since there are two logarithms involved, themselves both approximations, and a round-off on the division. The only way to write the exact answer, however, is to write $x = \dfrac{\log 16}{\log 5}$, which is not very useful for practical applications. Therefore, we will write solutions as decimal approximations, usually to three significant digits.

EXAMPLE 24.2.   Solve $3^{x-1} = 8$.

Solution.   Following the method of the preceding example, we take the common logarithm of each side, apply Property 3, and then approximate x:

$$\log 3^{x-1} = \log 8$$

$$(x - 1)\log 3 = \log 8$$

$$x - 1 = \frac{\log 8}{\log 3}$$

$$x = \frac{\log 8}{\log 3} + 1$$

$$x = \frac{.9031}{.4771} + 1$$

$$x = 2.89.$$

Exercise 24.1

Solve:

1. $7^x = 24$ 2. $15^x = 200$ 3. $4^{x-1} = 11$ 4. $6^{x+1} = 21$

5. $3^{2x} = 15$ 6. $4^{-3x} = 7$ 7. $2 = 100^{2x}$ 8. $.6 = 2^{-.4x}$

<table>
<tr><td>Section
24.2</td><td><h1 style="text-align:center">Applications</h1></td></tr>
</table>

One application of common logarithms is to investment of an amount of money with interest compounded over a period of time. The original amount of money invested is called the **principal**. We say the interest is **compounded annually** if it is calculated once a year, **semiannually** if twice a year, **quarterly** if four times a year. How often the interest is compounded is important because interest is calculated on the total amount present at the time, the principal plus interest previously paid, so we get "interest on the interest."

The formula for calculating the total amount present at any time is

$$A = P\left(1 + \frac{r}{n}\right)^{nt},$$

where A is the total amount, P is the principal, r is the interest rate written in decimal form, n is the number of times a year the interest is calculated, and t is the number of years during which the interest accumulates.

EXAMPLE 24.3. If \$2000 is invested at a rate of 6% compounded semiannually, how long will it take to accumulate to \$2500?

Solution. $P = 2000$, $r = .06$, $n = 2$, and $A = 2500$. We must find t. Using the formula $A = P(1 + \frac{r}{n})^{nt}$,

$$2500 = 2000\left(1 + \frac{.06}{2}\right)^{2t}$$

$$2500 = 2000(1.03)^{2t}$$

$$\frac{2500}{2000} = 1.03^{2t}$$

$$1.25 = 1.03^{2t}.$$

Thus, we have an exponential equation. Taking the common logarithm of each side,

$$\log 1.25 = \log 1.03^{2t}$$

$$\log 1.25 = 2t \log 1.03$$

$$t = \frac{\log 1.25}{2 \log 1.03}$$

$$t = 3.77, \text{ or about 4 years.}$$

We have rounded the answer to 4 years because, since the interest is compounded semiannually, it will be calculated at $3\frac{1}{2}$ years, when the accumulated amount will be somewhat less than \$2500, and at 4 years, when the accumulated amount will be somewhat more than \$2500.

EXAMPLE 24.4. If you invest \$1000 at $7\frac{1}{2}$% compounded quarterly, how much will you have after 10 years?

Solution. $P = 1000$, $r = .075$, $n = 4$, and $t = 10$. We must find A:

$$A = 1000\left(1 + \frac{.075}{4}\right)^{4(10)}$$

$$A = 1000(1.01875)^{40}.$$

This equation is not an exponential equation, since the variable A is not in an exponent. However, common logarithms provide an efficient way to find the value of the right-hand side.

Taking the common logarithm of each side, and carefully applying the properties of logarithms,

$$\log A = \log 1000(1.01875)^{40}$$

$$\log A = \log 1000 + \log 1.01875^{40}$$

$$\log A = \log 1000 + 40 \log 1.01875.$$

To find log 1.01875, you may use a scientific calculator or approximate by finding log 1.02 from Table II. We used a calculator to find

$$\log A = 3.3227.$$

Observe that this number is clearly not the answer, since we should have significantly more than the principal of $1000. Indeed, what we have is the logarithm of the answer. To find A, we must "unlog"; that is, find the antilogarithm.

$$A = \text{antilog } 3.3227$$

$$A = \$2100.$$

(If you used the approximation 1.02, you will get $A = \$2210$. By using the approximation, you would have actually used an interest rate of 8% rather than $7\frac{1}{2}\%$, and so the answer will be significantly higher.)

Most banks now compound interest continuously, as if it were being calculated not once, or twice, or four times a year, but every instant. The formula for this calculation is

$$A = Pe^{rt},$$

where A, P, r, and t are as before, and e is a special constant. Like π, e is a number that is found many places in mathematics and science. Also like π, e is an irrational number, which may be approximated to as many places as we wish by a decimal. A common approximation for e is 2.718. If you are using Table II, you may use 2.72, but your answers might be slightly different from ours.

EXAMPLE 24.5.   Suppose you put $500 in a savings account on which the interest is $5\frac{1}{4}\%$ compounded continuously. How much will you have at the end of a year?

Solution. $P = 500$, $r = .0525$, and $t = 1$. We must find A:

$$A = 500(2.718)^{.0525(1)}$$

$$A = 500(2.718)^{.0525}$$

$$\log A = \log 500(2.718)^{.0525}$$

$$\log A = \log 500 + \log 2.718^{.0525}$$

$$\log A = \log 500 + .0525 \log 2.718$$

$$\log A = 2.7218$$

$$A = \text{antilog } 2.7218$$

$$A = \$527.$$

Another application in which the constant e appears is growth and decay problems. Actually, the preceding example is a growth problem, specifically, growth of money. Often growth problems involve the growth of a population such as a bacteria culture. Decay problems often involve disintegration of radioactive materials. For both, the formula is

$$N = N_0 e^{kt},$$

where N is the amount after time t, N_0 is the initial amount, and k is a given constant. For growth problems k is positive, and for decay problems k is negative.

EXAMPLE 24.6.   A radioactive material decays according to the formula $N = N_0 e^{-.2t}$, where t is in seconds. If there are 600 grams of the material initially, how long will it take for the material to be reduced to 100 grams?

Solution. $N = 100$, $N_0 = 600$, and we must find t. Since $N = N_0 e^{-.2t}$,

$$100 = 600(2.718)^{-.2t}$$

$$\frac{100}{600} = 2.718^{-.2t}$$

$$\frac{1}{6} = 2.718^{-.2t}$$

To avoid unnecessary rounding approximations, we will leave $\frac{1}{6}$ in fraction form. Then, solving the exponential equation,

$$\log \frac{1}{6} = \log 2.718^{-.2t}$$

$$\log 1 - \log 6 = -.2t \log 2.718$$

$$t = \frac{\log 1 - \log 6}{-.2 \log 2.718}$$

$$t = \frac{0 - .7782}{-.2(.4342)}$$

$$t = \frac{.7782}{.2(.4342)}$$

$$t = 8.96 \text{ seconds.}$$

(If you used Table II and $e = 2.72$, you will get $t = 8.95$ seconds.) Observe that, since t is in an exponent, we do not have $\log t$ and we do not find an antilog.

Exercise 24.2

Use $A = P(1 + \frac{r}{n})^{nt}$:
   (The interest rates are chosen so that if you are using Table II you should not get a significantly different answer.)

1. If you invest $3000 at 9% compounded annually, how much will you have after 5 years?

2. If you invest $600 at 4% compounded quarterly, how much will you have after 2 years?

3.   How long will it take for $5000 invested at 6% compounded semiannually to grow to $10,000?

4.   If $1500 is invested at 8% compounded quarterly, how long will it take to reach $2000?

Use $A = Pe^{rt}$ for problems 5 and 6:

5.   How long will it take for $1000 invested at 7% compounded continuously to reach $1500?

6.   If you invest $10,000 at $7\frac{1}{2}$% compounded continuously, how much will you have after 10 years?

7.   A bacteria culture grows according to the formula $N = N_0 e^{.5t}$, where t is in days. If there are 100 bacteria initially, how many will there be after 7 days?

8.   A bacteria culture grows according to the formula $N = N_0 e^{.02t}$, where t is in hours. If the initial amount is 3000 bacteria, how long will it take for the population to double?

9.   A radioactive material decays according to the formula $N = N_0 e^{-.2t}$, where t is in seconds. After 10 seconds, how much is left of an initial amount of 50 grams?

10.   A radioactive material decays according to the formula $N = N_0 e^{-.004t}$, where t is in hours. How long will it take an initial amount of 1000 grams to reduce to 100 grams?

11.* Dating of ancient remains can be done using a radioactive isotope of carbon called carbon 14. If carbon 14 decays according to the formula $N = N_0 e^{-.0001245t}$, where t is in years, what is its half-life; that is, how long will it take an initial amount of N_0 to reduce to $\frac{1}{2}N_0$?

12.* Suppose it is known that the initial amount of carbon 14 in a piece of wood is N_0, and now only $\frac{1}{5}N_0$ remains. How old is the piece of wood?

<table>
<tr><td>Section
24.3</td><td><h1 style="text-align:center">Natural Logarithms</h1></td></tr>
</table>

In the examples and exercises in Section 24.2, we saw several exponential equations involving the constant e. This constant occurs so often in such equations that it is defined as the base of another type of logarithm. A **natural logarithm** is a logarithm to the base e. We use $\ln x$ to indicate the natural logarithm of x. When you see $\ln x$, you may assume $\log_e x$ is meant. If you are using a scientific calculator, the function name "ln" finds the natural logarithm of a number.

First, we observe that $\ln e = 1$. Since $\ln e = \log_e e$, if $y = \log_e e$, then by the definition of a logarithm, $e^y = e$, and so $y = 1$.

To find other values of natural logarithms we must use tables. Table III at the back of this book is a table of natural logarithms for three-digit numbers between 1 and 10.

EXAMPLE 24.7.   Find $\ln 3.92$.

Solution.   Referring to Table III, we look to the right of 3.9 and under 2. We find the number 3661. However, the first number in this group has a whole number 1 in front. Thus $\ln 3.92 = 1.3661$.

If you look at the numbers next to 2.7, and also the last number next to 7.3, you will see asterisks in front of them. The asterisks mean we should use the whole numbers, 1 or 2, for the

---

*Problems 11 and 12 are adapted from William E. Boyce and Richard C. DiPrima, *Elementary Differential Equations*, 2nd Edition, John Wiley & Sons, Inc., 1969, p. 58.

group which follows with these numbers. If there is no indication of a whole number with the first number of a group, the number is written as a decimal less than 1.

EXAMPLE 24.8. Find:

a. ln 1.73   b. ln 2.77

Solutions.

a. Since the number to the right of 1.7 and under 3 is 5481, with no whole number in front of the group, ln 1.73 = .5481.

b. Since the number to the right of 2.7 and under 7 is 0188, and there is an asterisk in front of the number, ln 2.77 = 1.0188.

We cannot use characteristics which are powers of 10 to find natural logarithms of numbers not between 1 and 10. Powers of 10 go with common logarithms, which are to the base 10. However, we can use a power of 10 times ln 10, given that ln 10 = 2.3026.

EXAMPLE 24.9. Find ln 578.

Solution.

$$\ln 578 = \ln 5.78(10^2)$$

$$= \ln 5.78 + \ln 10^2$$

$$= \ln 5.78 + 2 \ln 10$$

$$= 1.7544 + 2(2.3026)$$

$$= 6.3596.$$

If you have a calculator with the function "ln," you do not have to go through the process in the preceding example. Even without a scientific calculator, exponential equations involving e are often easier to solve using natural logarithms than using common logarithms.

EXAMPLE 24.10. A radioactive material decays according to the formula $N = N_0 e^{-.5t}$, where t is in seconds. If there are 800 grams of the material initially, how long will it take for the material to reduce to 200 grams?

Solution. $N = 200$, $N_0 = 800$, and we must find t:

$$200 = 800e^{-.5t}$$

$$\frac{200}{800} = e^{-.5t}$$

$$.25 = e^{-.5t}.$$

We use natural logarithms to solve the exponential equation, recalling that $\ln e = 1$:

$$\ln .25 = \ln e^{-.5t}$$

$$\ln .25 = -.5t \ln e$$

$$\ln .25 = -.5t$$

$$t = \frac{\ln .25}{-.5}$$

$$t = 2.77 \text{ seconds.}$$

If you are using Table III, you can find ln .25 by calculating

$$\ln .25 = \ln 2.5(10^{-1})$$
$$= \ln 2.5 - \ln 10$$
$$= .9163 - 2.3026$$
$$= -1.3863.$$

To use Table III to find a number with a natural logarithm greater than $\ln 10 = 2.3026$, we subtract ln 10 as many times as necessary.

EXAMPLE 24.11.   Find N if $\ln N = 6.5823$.

Solution.   We subtract 2.3026 twice:

$$6.5823 - 2(2.3026) = 1.9771.$$

Then, from Table III,

$$\ln 7.22 = 1.9771.$$

Therefore, we have

$$\ln N = 6.5823$$
$$= 1.9771 + 2(2.3026)$$
$$= \ln 7.22 + 2 \ln 10$$
$$= \ln 7.22(10)^2$$
$$= \ln 722,$$

and so $N = 722$.

EXAMPLE 24.12.   Suppose you invest \$5000 in a savings account at 6% compounded continuously. How much will you have in the account at the end of 3 years?

Solution.   The formula, from Section 24.2, is $A = Pe^{rt}$. We have $P = 5000$, $r = .06$, $t = 3$, and we must find A:

$$A = 5000e^{.06(3)}.$$

Using natural logarithms,

$$\ln A = \ln 5000e^{.18}$$
$$= \ln 5000 + .18 \ln e$$
$$= \ln 5000 + .18.$$

To find ln 5000,

$$\ln 5000 = \ln 5(10)^3$$
$$= \ln 5 + 3 \ln 10$$
$$= 1.6094 + 3(2.3026)$$
$$= 8.5172.$$

Therefore,

$$\ln A = 8.5172 + .18$$

$$= 8.6972.$$

Now, to find the natural antilogarithm of 8.6972, we use

$$8.6972 - 3(2.3026) = 1.7894,$$

and so

$$\ln A = 8.6972$$

$$= 1.7894 + 3(2.3026)$$

$$= \ln 5.99 + 3 \ln 10$$

$$= \ln 5.99(10)^3$$

$$= \ln 5990.$$

The amount is $A = \$5990$. Clearly, this problem is much easier to solve using a calculator. Indeed, it can be solved on a calculator without using logarithms, because the variable A is not in an exponent. If you have an "e^x" key, you may use it to find

$$e^{.18} = 1.1972.$$

Then,

$$5000e^{.18} = 5000(1.1972)$$

$$= 5986,$$

which rounds off to $5990, to three significant digits.

Exercise 24.3

Solve using natural logarithms:

1. A bacteria culture grows according to the formula $N = N_0 e^{.69t}$, where t is in hours. If there are 500 bacteria initially, how long will it take for the culture to grow to 1000 bacteria?

2. A radioactive material decays according to the formula $N = N_0 e^{-.092t}$, where t is in seconds. If 10 grams are present initially, how long will it take to reduce the material to 4 grams?

3. Suppose you invest $2000 at $5\frac{1}{2}\%$ compounded continuously. Use $A = Pe^{rt}$ to find how much you will have at the end of 1 year.

4. Suppose you invest $1000 at $6\frac{1}{2}\%$ compounded continuously. Use $A = Pe^{rt}$ to find how long it will take for your investment to double.

5. Radium decays according to the formula $N = N_0 e^{-.000411t}$, where t is in years. If there are 6 milligrams of radium in a substance initially, how much will there be 100 years later?

6.* The formula $P = P_0 e^{-kt}$ represents the amount of pollutant P in a lake after t years, where P_0 is the initial amount of pollutant, and then only clear water enters the lake. The constant $k = \dfrac{r}{V}$, where r is the rate at which the clear water enters the lake and V is the volume of the lake. Find how long it will take to reduce the amount of pollution to 10% of its current amount, that is, to reduce P_0 to $\frac{1}{10} P_0$, for each of these Great Lakes:

Lake	V in km³	r in km³/yr
Superior	12,200	65.2
Michigan	4900	158
Erie	460	175
Ontario	1600	209

---

*Adapted from Boyce and DiPrima, *Elementary Differential Equations*, 2nd Edition, pp. 60-61.

Solve using common logarithms:

1. $5^{x+1} = 20$

1. _______________

2. $500 is invested at 8% compounded quarterly. Use $A = P(1 + \dfrac{r}{n})^{nt}$ to find how much will accumulate in 5 years.

2. _______________

Solve using common or natural logarithms:

3. $500 is invested at $5\frac{1}{2}\%$ compounded continuously. Use $A = Pe^{rt}$ to find how long it will take to grow to $550.

3. _______________

4. A bacteria culture grows according to the formula $N = N_0 e^{.04t}$, where t is in hours. If there are 300 bacteria initially, how many will there be after 12 hours?

4. _______________

5. A radioactive material decays according to the formula $N = N_0 e^{-.5t}$, where t is in seconds. How long will it take an initial amount of 20 grams to reduce to 2 grams?

5. _______________

Answers

Unit 1

Exercise 1.1

1. $4x - 4y$
2. $-3x + 3y$
3. $-x + y$
4. $2x^2 + 6x$
5. $-4x^2 + 8x + 4$
6. $10x^2 + 20x$
7. $4x^3 + 6x^2 + 4x$
8. $-x^2 + 2x + 3$
9. $-2x^2 + 20x$
10. $-x^2 - 6x$
11. $3(x + y)$
12. $2(x - 2)$
13. $3(2x + 1)$
14. $2(x^2 - 3x + 1)$
15. $x(x - 2)$
16. $2x(x + 2)$
17. $5(2x^2 - 2x - 1)$
18. $4(3x^2 + 2x - 4)$
19. $x(x^2 + 2x + 2)$
20. $3x(x^2 - 2x - 1)$

Exercise 1.2

1. $x - 1$
2. $-x + 9$
3. $10x$
4. $7x$
5. $3x - 1$
6. $x + 7$
7. 1
8. $5x - 13$
9. $-3x - 12$
10. $-2x$
11. $x^2 + x - 6$
12. $-x^2$
13. $-4x^2 + x - 3$
14. $2x^2 - 8x + 1$
15. $8x^2 - 14x$
16. $-7x^3 + 2x^2 + 11x$

Exercise 1.3

1. $\dfrac{x + 2}{2}$
2. $1 - 2x$
3. $\dfrac{3 + 4x}{6}$
4. $\dfrac{3x - 4}{2}$
5. $x - 4 - z$

6. $2 + x + y$ 7. $y - z + 2$ 8. $\dfrac{x^2 - x + 2}{2x}$ 9. $\dfrac{1 - \sqrt{2}}{2}$ 10. $\dfrac{-4 + 6\sqrt{5}}{5}$

Exercise 1.4

1. $x^2 + 3x + 2$

2. $x^2 + x - 6$

3. $10x^2 - 21x + 8$

4. $18x^2 + 15x - 12$

5. $3x^2 + 10xy + 3y^2$

6. $10x^2 - 3xy - y^2$

7. $9x^2 + 30x + 25$

8. $4x^2 - 4x + 1$

9. $x^2 - 10x + 25$

10. $16x^2 + 24x + 9$

11. $16x^2 - 4$

12. $36x^2 - 9y^2$

Exercise 1.5

1. $(x + 1)(x + 4)$

2. $(x + 3)(x + 6)$

3. $(x - 3)(x - 4)$

4. $(x - 2)(x - 8)$

5. $(x + 3)(x - 5)$

6. $(x - 6)(x + 8)$

7. $(x + 2)(2x - 3)$

8. $(x - 5)(3x - 1)$

9. $(x + 6)(x + 6)$

10. $(2x - 3)(2x - 3)$

11. $(2x + 1)(2x - 1)$

12. $(3x + 5)(3x - 5)$

13. $(2x + 3)(3x + 8)$

14. $(x - 10)(8x + 5)$

15. $3(x + 1)(x + 3)$

16. $5(x - 4)(2x + 3)$

Self-test

1. $1 - 5x$ (Objective 3)

2. $(x - 6)(2x + 5)$ (Objective 5)

3. $2x(x - 4)$ (Objective 1)

4. $-x - 12$ (Objective 2)

5. $12x^2 - 17x - 5$ (Objective 4)

Unit 2

Exercise 2.1

1. 8

2. -6

3. 2

4. -3

5. 8

6. 5

7. -4

8. -9

9. $\frac{8}{3}$ 10. $\frac{9}{10}$ 11. $\frac{3}{2}$ 12. 4

13. $\frac{11}{5}$ 14. 4   15. -8 16. 13

Exercise 2.2

1. 15   2. -12 3. $\frac{15}{2}$ 4. -9 5. $-\frac{28}{3}$

6. $\frac{20}{7}$ 7. $\frac{7}{5}$ 8. -5 9. $\frac{10}{3}$ 10. $-\frac{1}{2}$

11. 6   12. $-\frac{3}{4}$ 13. $\frac{1}{2}$ 14. $\frac{13}{6}$

Exercise 2.3

1. 2   2. -2 3. 6   4. $\frac{3}{2}$ 5. $-\frac{2}{3}$

6. $-\frac{3}{4}$ 7. $-\frac{1}{2}$ 8. 3   9. -2 10. 3

11. -1 12. $-\frac{1}{3}$ 13. $\frac{4}{3}$ 14. $\frac{3}{2}$

Exercise 2.4

1. $0, 5$ 2. $0, \frac{1}{2}$ 3. $0, 3$ 4. $0, -3$

5. $0, -\frac{3}{2}$ 6. $0, \frac{1}{3}$ 7. $0, -\frac{5}{3}$ 8. $0, -4$

Exercise 2.5

1. $1, -1$ 2. $5, -5$ 3. $\frac{1}{2}, -\frac{1}{2}$ 4. $\frac{2}{3}, -\frac{2}{3}$

5. $0, \frac{1}{2}$ 6. $0, -3$ 7. $2, 3$ 8. $6, -10$

9. $1, -\frac{2}{3}$ 10. $-2, \frac{3}{4}$ 11. -3 12. 1

13. $1, -2$ 14. $5, \frac{1}{5}$ 15. $4, \frac{2}{3}$ 16. $-1, \frac{1}{2}$

Self-test

1. $0, \frac{2}{3}$ (Objective 4)   2. $\frac{1}{2}$ (Objective 3)   3. $-\frac{2}{7}$ (Objective 2)

4. 2   (Objective 1)   5. $2, \frac{1}{2}$ (Objective 5)

Unit 3

Exercise 3.1

1. $x + 14$ 2. $3 + x$ 3. $x - 5$ 4. $30 - x$

5. $x - 7$ 6. $9 - x$ 7. $\frac{1}{3}x$ 8. $\frac{5}{2}x$

9. $\frac{2}{5}(x + 8)$ 10. $\frac{1}{5}x - 10$ 11. $\frac{2}{3}x + 18$ 12. $\frac{3}{4}(x - 12)$

Exercise 3.2

1. 2 2. $\frac{5}{4}$ 3. $\frac{5}{9}$ 4. 32 5. 84

6. -6 7. 29 8. 53 9. -10 10. $\frac{17}{8}$

11. $\frac{21}{2}$ 12. $-\frac{13}{2}$ 13. -2 14. $\frac{25}{2}$ 15. 5

16. $-\frac{9}{4}$

Exercise 3.3

1. 2 pounds Bosc, 3 pounds Bartlett

2. 10 pounds fine sand, 40 pounds coarse sand

3. 20 ball-point pens, 10 felt-tip pens

4. 15 cans light tuna, 9 cans white tuna

5. 4 pounds kidney beans, 6 pounds green beans

6. 60 ounces beef flavor, 20 ounces liver flavor

7. $7\frac{1}{2}$ pecks peat moss, $17\frac{1}{2}$ pecks potting soil

8. $22\frac{1}{2}$ quarts olive oil, $27\frac{1}{2}$ quarts vegetable oil

9. $1\frac{1}{2}$ quarts 8% salt, $7\frac{1}{2}$ quarts 2% salt

10. 14 gallons 20% detergent, 6 gallons 5% detergent

11. 3 quarts   12. $7\frac{1}{2}$ gallons

13. 6 gallons   14. $\frac{2}{3}$ cup

15. 8 ounces   16. $24\frac{3}{4}$ gallons

Exercise 3.4

1. 2 hours to go, 5 hours to return

2. 1 hour to go, $\frac{1}{2}$ hour to return

3. 8 miles per hour jogging, 4 miles per hour walking

4. 125 miles per hour going, 175 miles per hour returning

5. 2 hours   6. $\frac{1}{3}$ hour   7. $1\frac{2}{3}$ hours

8. 6 minutes   9. 2 hours   10. 140 miles per hour, 110 miles per hour

Exercise 3.5

1. 7 centimeters, 10 centimeters   2. $2\frac{1}{2}$ inches, $3\frac{1}{2}$ inches

3. $1\frac{2}{3}$ meters, $1\frac{1}{3}$ meters   4. 2 feet, $\frac{1}{2}$ foot

5. 7 inches, 3 inches   6. $\frac{1}{2}$ meter, $1\frac{1}{2}$ meters

7. 3 yards, 7 yards   8. 1 foot, 4 feet

9. 3 meters, 6 meters   10. $1\frac{1}{2}$ feet, $3\frac{1}{2}$ feet

Self-test

1. $\frac{2}{3}(x - 5)$ (Objective 1)

2. -5 (Objective 2)

3. $5\frac{1}{2}$ quarts   (Objective 3)

4. 38 miles per hour, 42 miles per hour   (Objective 4)

5. $1\frac{1}{2}$ inches, $3\frac{1}{2}$ inches   (Objective 5)

Unit 4

Exercise 4.1

1. $x^2 + 2x + 1 = 0,\ a = 1,\ b = 2,\ c = 1$

2. $-3x^2 - x + 3 = 0,\ a = -3,\ b = -1,\ c = 3$

3. $-2x^2 + 3x = 0,\ a = -2,\ b = 3,\ c = 0$

4. $4x^2 - x = 0,\ a = 4,\ b = -1,\ c = 0$

5. $x^2 - 1 = 0,\ a = 1,\ b = 0,\ c = -1$

6. $-2x^2 + 5 = 0,\ a = -2,\ b = 0,\ c = 5$

7. $-x^2 + x - 1 = 0,\ a = -1,\ b = 1,\ c = -1$

8. $-x^2 + 5x - 5 = 0,\ a = -1,\ b = 5,\ c = -5$

9. $2x^2 - 5x - 2 = 0,\ a = 2,\ b = -5,\ c = -2$

10. $-3x^2 + 2x - 4 = 0,\ a = -3,\ b = 2,\ c = -4$

Exercise 4.2

1. ± 4
2. ± 7
3. $\pm \dfrac{\sqrt{5}}{2}$
4. $\pm \frac{5}{8}$

5. No real number solution
6. No real number solution

7. $\pm 5\sqrt{3}$
8. $\pm \dfrac{4\sqrt{2}}{3}$
9. ± 5
10. $\pm 2\sqrt{2}$

Exercise 4.3

1. $\frac{1}{2}, 2$
2. $\frac{1}{2}, -\frac{3}{2}$
3. $\frac{2}{3}, -\frac{4}{3}$
4. $\frac{1}{4}, 4$

5. $2 \pm \sqrt{7}$
6. $-1 \pm \sqrt{6}$
7. $-2 \pm 2\sqrt{2}$
8. $3 \pm \sqrt{7}$

9. $\dfrac{1 \pm \sqrt{3}}{2}$
10. $\dfrac{-2 \pm \sqrt{31}}{3}$
11. $\dfrac{-3 \pm \sqrt{65}}{4}$
12. $\dfrac{5 \pm \sqrt{61}}{6}$

13. No real number solution
14. No real number solution

15. $3 \pm 2\sqrt{3}$
16. No real number solution

Exercise 4.4

1. 2.87 feet, 4.87 feet
2. .854 inches, 5.85 inches

3. 1.89 meters, 2.39 meters
4. 5.52 feet, 6.52 feet

5. 1.19 meters, 2.19 meters
6. 1.70 inches, 3.70 inches

7. .732 feet, 2.73 feet
8. 2.89 feet, 5.89 feet

9. 3 inches
10. 3.34 meters (3.32 meters using Table I)

Self-test

1. $2x^2 - x - 4 = 0, a = 2, b = -1, c = -4$ (Objective 1)
2. $\pm \dfrac{\sqrt{5}}{3}$ (Objective 2)

3. $2, -\frac{3}{4}$ (Objective 3)
4. $-4 \pm \sqrt{10}$ (Objective 3)

5. 2.32 inches, 4.32 inches (Objective 4)

Unit 5

Self-test

1. $(3x - 2)(x + 4)$ (Unit 1)

2. $2 - 5x$ (Unit 1)

3. 2 (Unit 2)

4. $0, \frac{3}{2}$ (Unit 2)

5. $\pm \frac{1}{2}$ (Unit 4)

6. $\dfrac{2 \pm \sqrt{14}}{2}$ (Unit 4)

7. $\frac{19}{2}$ (Unit 3)

8. 6 grams marjoram, 12 grams thyme   (Unit 3)

9. $2\frac{1}{2}$ hours   (Unit 3)

10. 2.07 feet, 5.07 feet   (Unit 4)

Unit 6

Exercise 6.1

1. 5   2. 7   3. 8   4. 4   5. 5

6. $2\sqrt{5}$ 7. $\sqrt{26}$ 8. $5\sqrt{2}$ 9. 13   10. $2\sqrt{13}$

11. $\sqrt{10}$ 12. $\sqrt{29}$ 13. $4\sqrt{2}$ 14. $5\sqrt{2}$ 15. $2\sqrt{17}$

16. 5

Exercise 6.2

1. $\frac{5}{2}$ 2. $\frac{3}{5}$ 3. -2 4. $-\frac{3}{2}$ 5. $-\frac{7}{5}$

6. $-\frac{5}{4}$ 7. -1 8. 2   9. $-\frac{1}{2}$ 10. $\frac{1}{5}$

11. 3   12. $-\frac{3}{2}$ 13. undefined   14. 0   15. 0   16. undefined

Exercise 6.3

1. $4x - y - 1 = 0$ 2. $2x + y - 10 = 0$ 3. $x + y + 5 = 0$

4. $2x - y - 2 = 0$ 5. $3x - 4y - 11 = 0$ 6. $5x - 2y + 26 = 0$

7. $2x+3y+21=0$

8. $3x+2y-7=0$

9. $4x-y-7=0$

10. $2x+y-10=0$

11. $3x-2y-13=0$

12. $3x+2y-7=0$

13. $x+2y+7=0$

14. $x-4y-16=0$

15. $3x-2y-12=0$

16. $2x+3y-6=0$

17. $y=2$ or $y-2=0$

18. $y=-1$ or $y+1=0$

19. $x=-2$ or $x+2=0$

20. $x=6$ or $x-6=0$

Exercise 6.4

1. (0, 2) , (−4, 0)

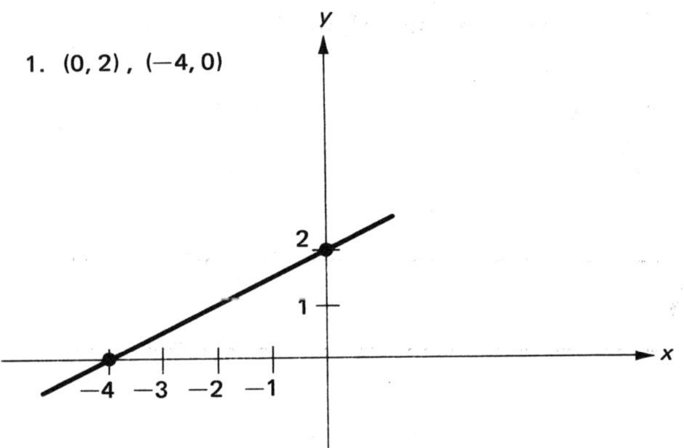

2. (0, 6) , (2, 0)

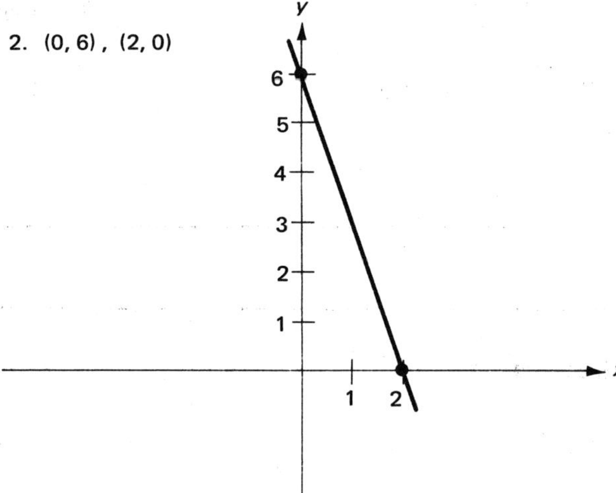

3. (0, −2) , (5, 0)

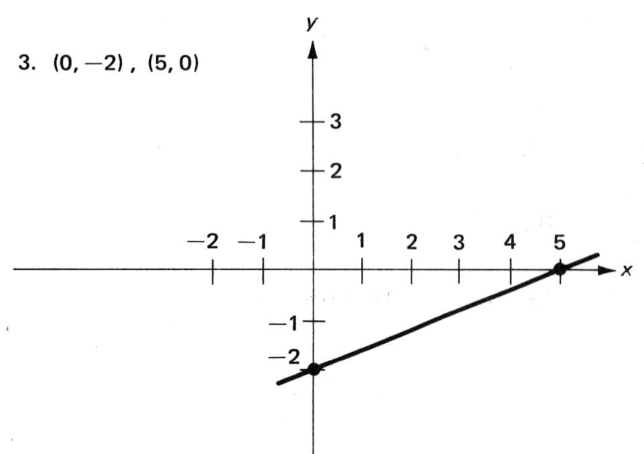

4. (0, 7) , (−2, 0)

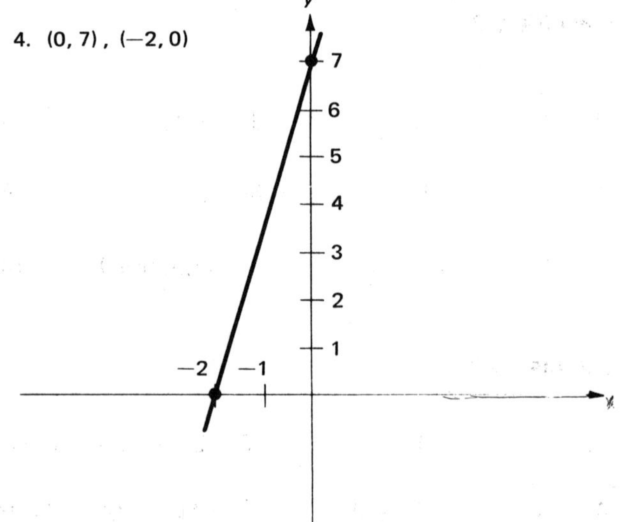

5. $(0, -2)$, $(-\frac{12}{5}, 0)$

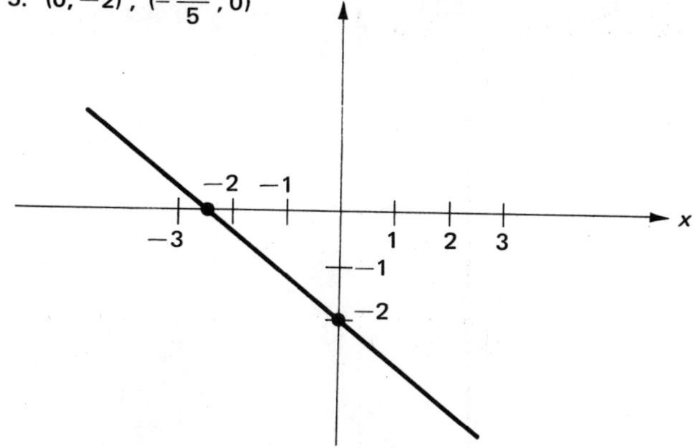

6. $(0, -\frac{9}{5})$, $(3, 0)$

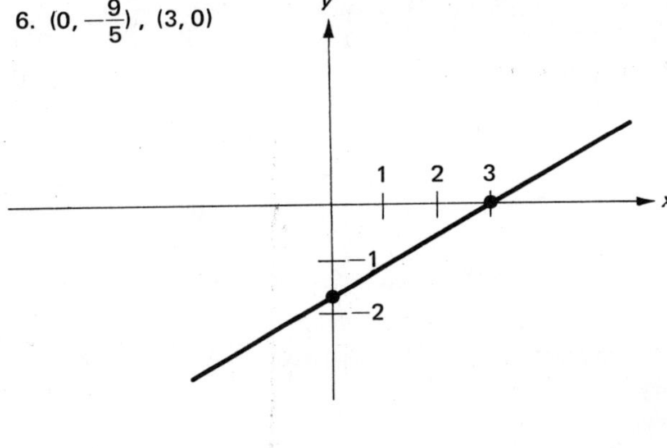

7. $(0, 0)$, $(0, 0)$

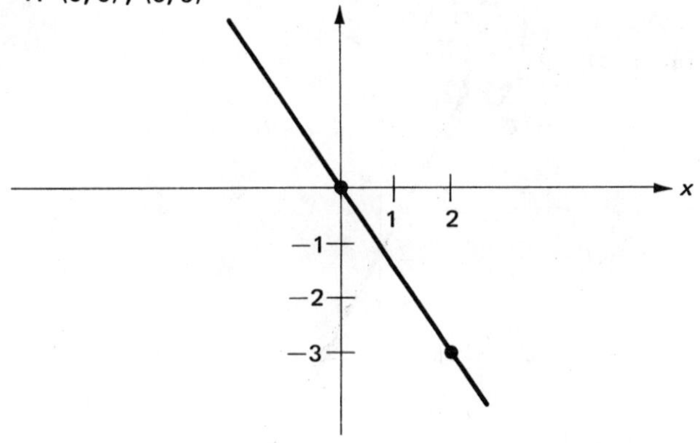

8. $(0, 0)$, $(0, 0)$

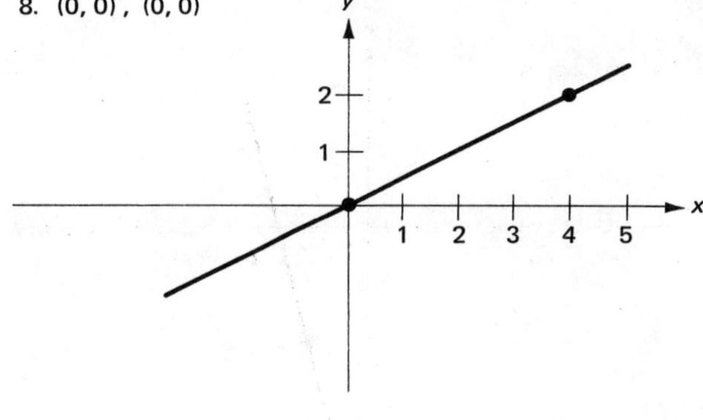

9. $(0, 3)$, no x-intercept

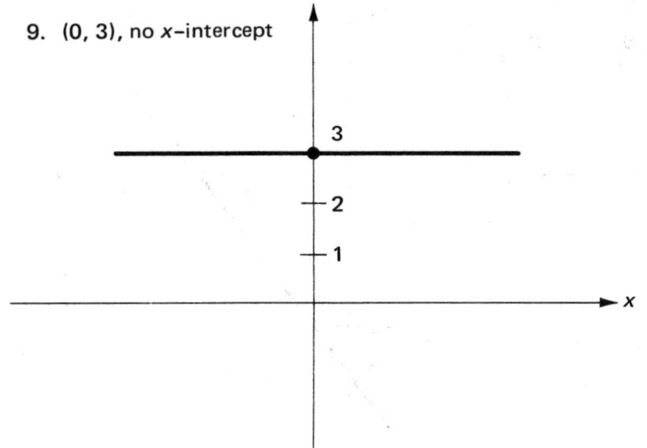

10. $(0, -\frac{7}{2})$, no x-intercept

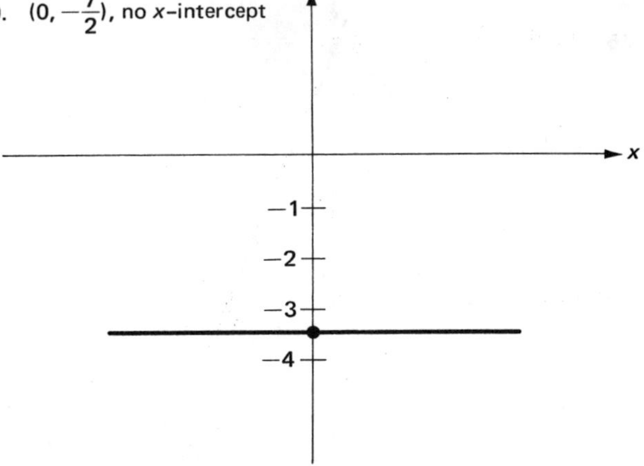

11. no *y*-intercept, $(\frac{4}{3}, 0)$

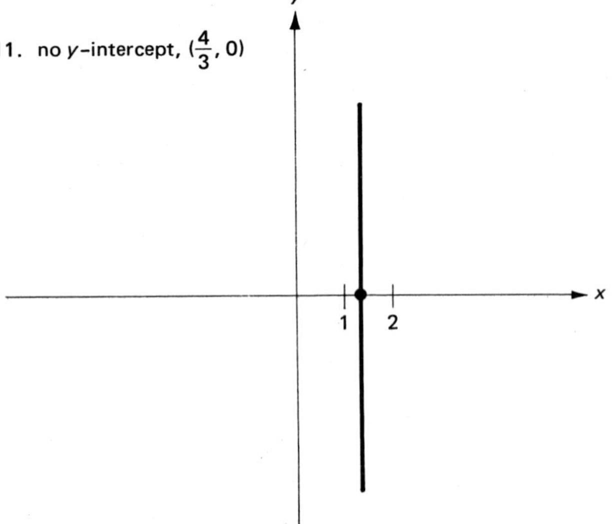

12. no *y*-intercept, $(-2, 0)$

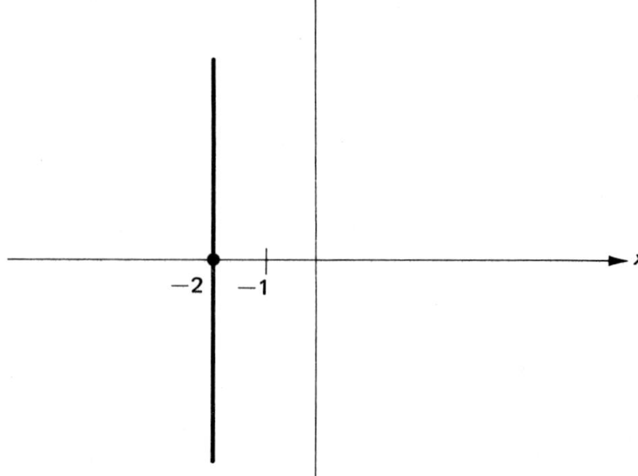

13. $(0, -6)$, 4

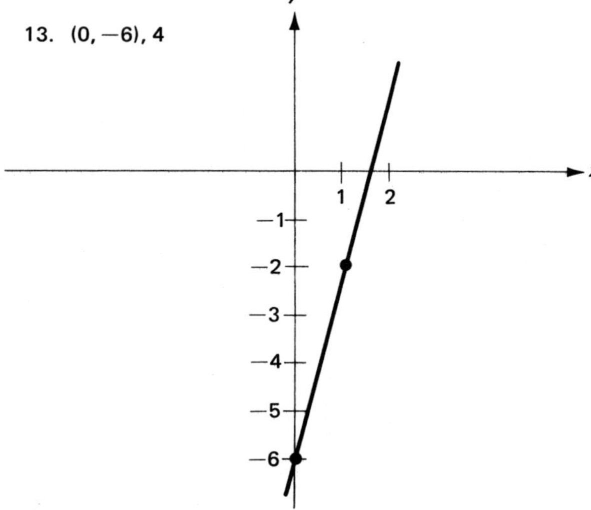

14. $(0, 2)$, -2

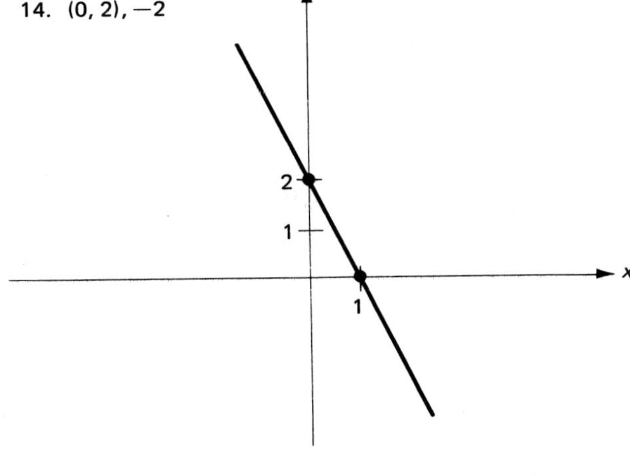

15. $(0, 5)$, $-\frac{5}{2}$

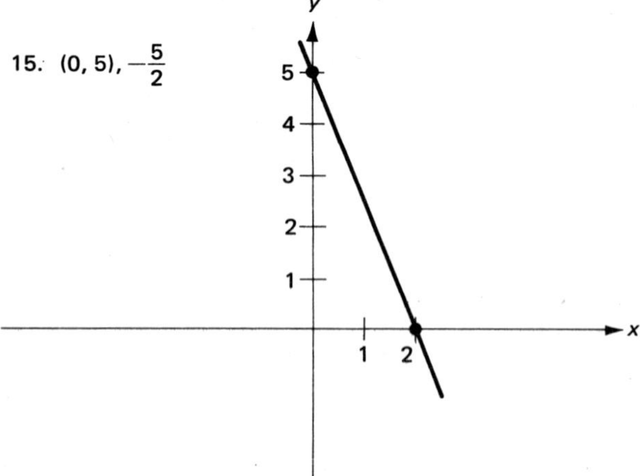

16. $(0, -4)$, $\frac{4}{3}$

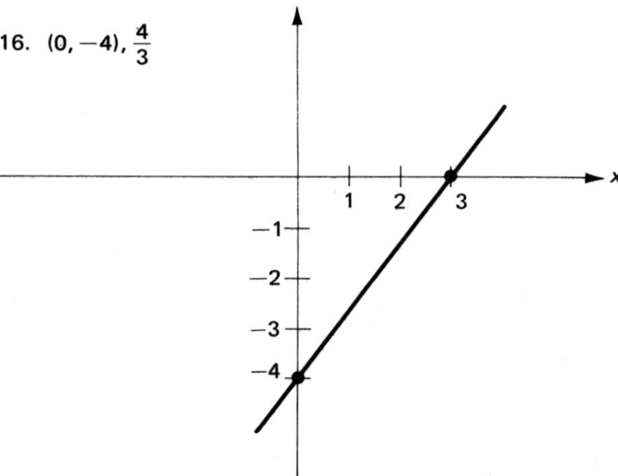

Self-test

1. $\frac{3}{4}$ (Objective 2)

2. $2\sqrt{10}$ (Objective 1)

3. $(0, -6)$, $(4, 0)$ (Objective 4)

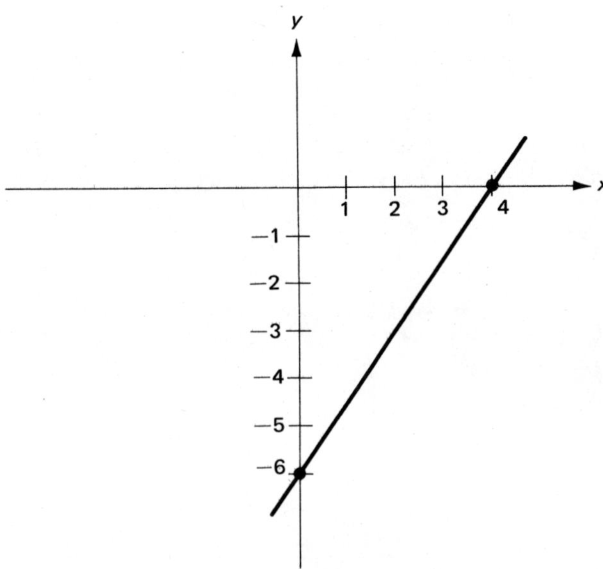

4. $3x + y + 14 = 0$ (Objective 3)

5. $2x + 5y + 4 = 0$ (Objective 3)

Unit 7

Exercise 7.1

1. $(0, -3)$, minimum

2. $(0, 4)$, maximum

3. $(\frac{1}{4}, \frac{1}{8})$, maximum

4. $(-\frac{3}{2}, -\frac{9}{4})$, minimum

5. $(2, 1)$, minimum

6. $(-1, 0)$, minimum

7. $(\frac{1}{3}, -\frac{7}{3})$, minimum

8. $(-\frac{3}{4}, -\frac{17}{8})$, minimum

9. $(-\frac{3}{2}, \frac{25}{4})$, maximum

10. $(\frac{1}{3}, \frac{4}{3})$, maximum

Exercise 7.2

1. (0, −4)
    (2, 0), (−2, 0)
    (0, −4)

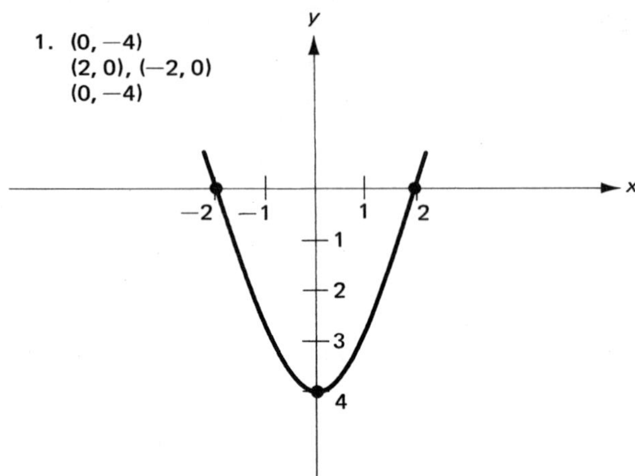

2. (0, 3)
    no x-intercepts
    (0, 3)

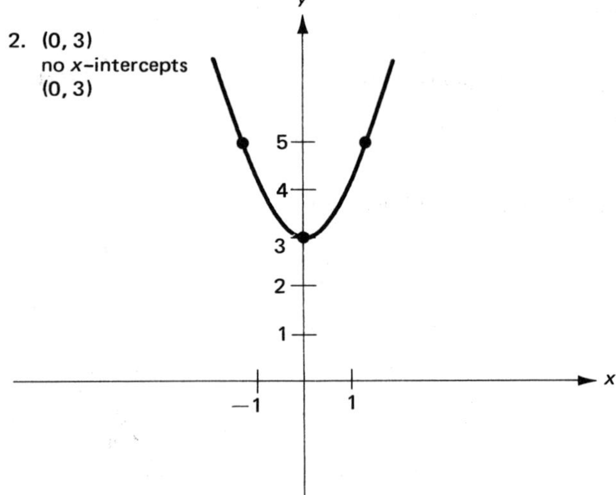

3. (0, 1)
    (.7, 0) , (−.7, 0)
    (0, 1)

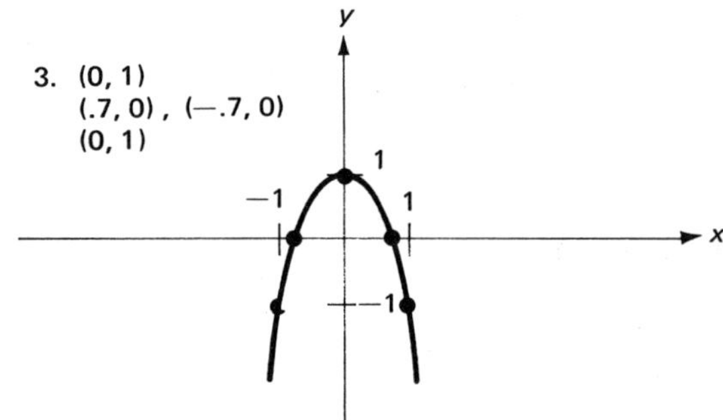

4. (0, −2)
    no x-intercepts
    (0, −2)

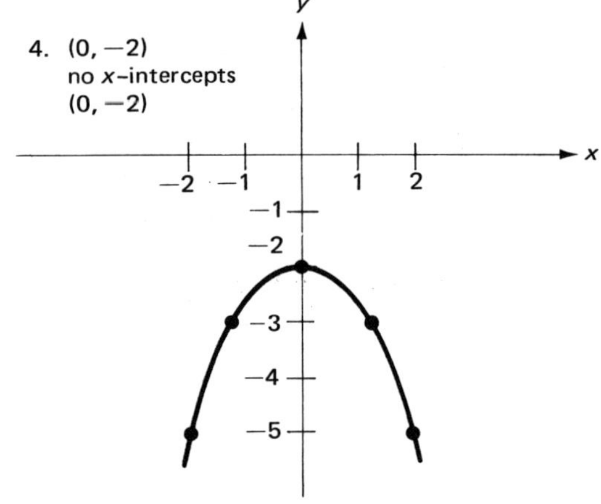

5. (0, 1)
    (−1, 0)
    (−1, 0)

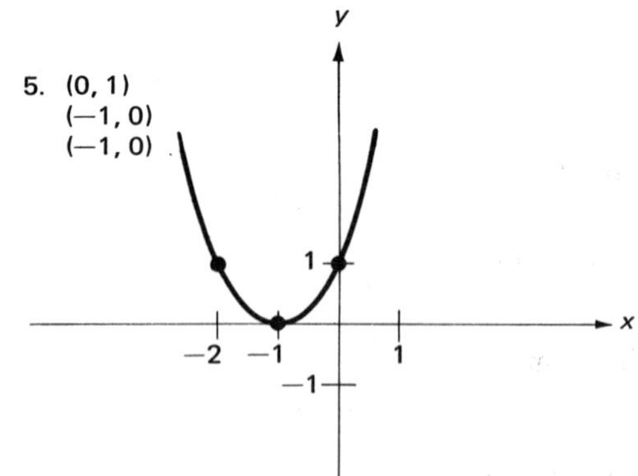

6. (0, 9)
    (3, 0)
    (3, 0)

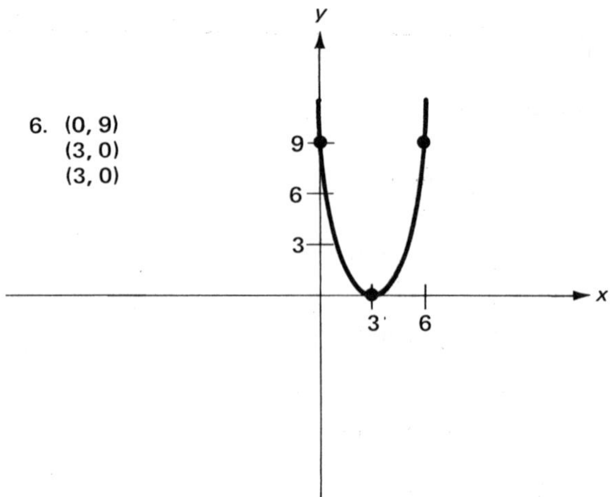

7. (0, 8)
    (2, 0), (4, 0)
    (3, −1)

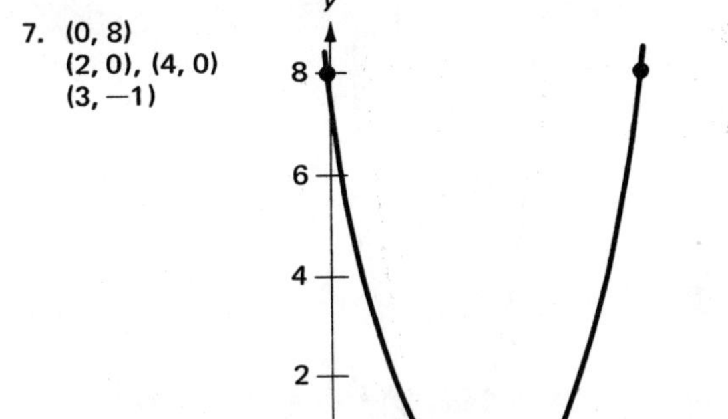

8. (0, −3)
    (−1, 0) , (3, 0)
    (1, −4)

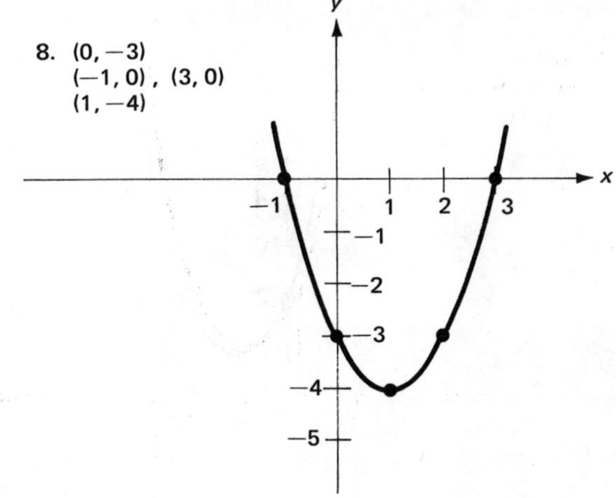

9. (0, 15)
    (−3, 0) , (5, 0)
    (1, 16)

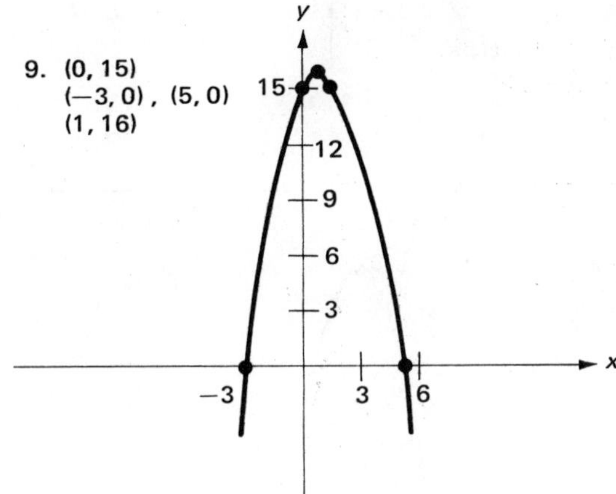

10. (0, 5)
     (−5, 0) , (1, 0)
     (−2, 9)

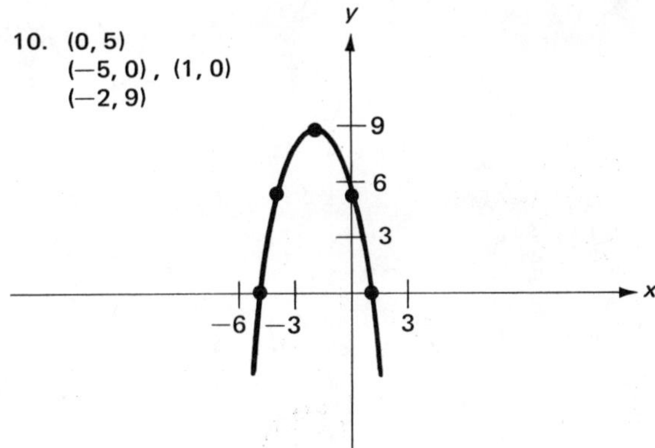

11. (0, −2)
     (−.7, 0) , (2.7, 0)
     (1, −3)

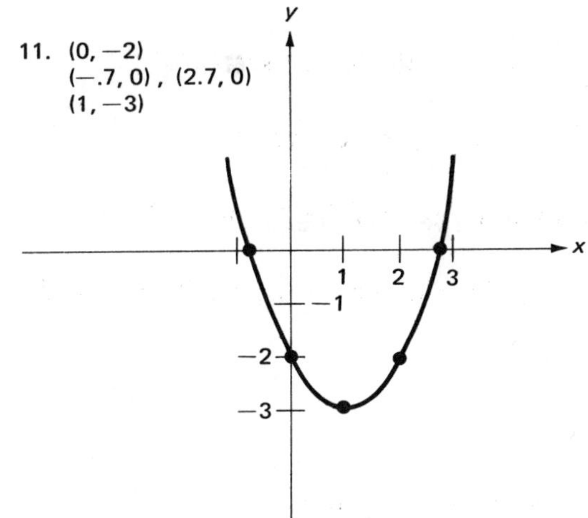

12. (0, 1)
     (−.3, 0) , (1.8, 0)
     $(\frac{3}{4}, \frac{17}{8})$

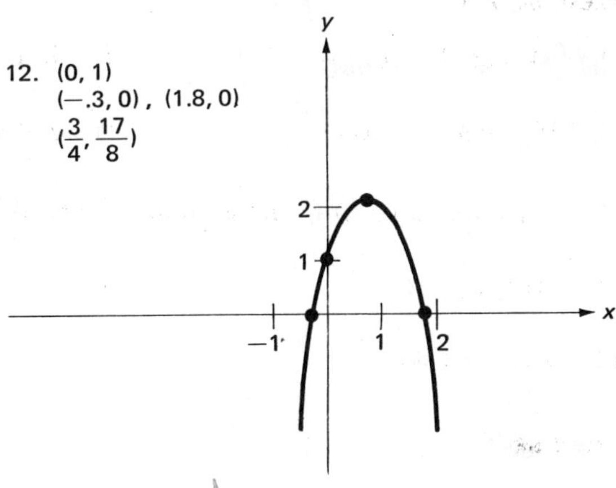

13. (0, 2)
    no *x*-intercepts
    (1, 1)

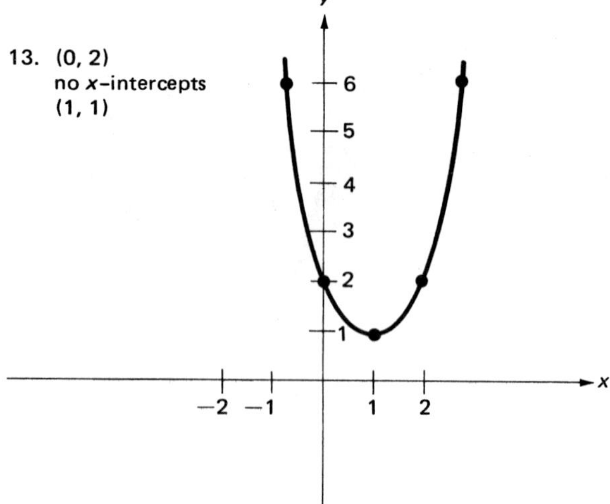

14. (0, 3)
    no *x*-intercepts
    $(-\frac{3}{4}, \frac{15}{8})$

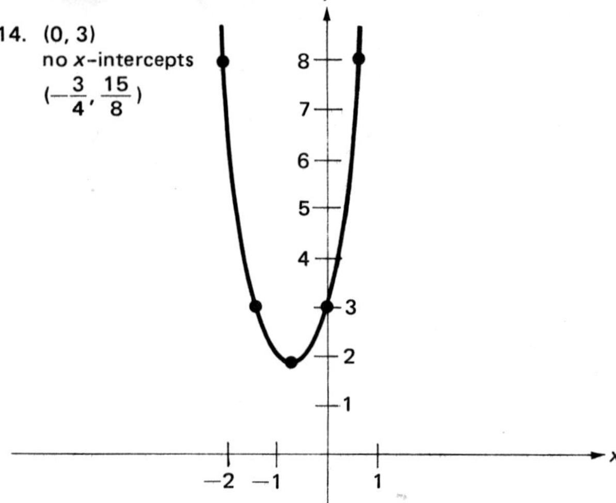

15. (0, −1)
    no *x*-intercepts
    $(\frac{1}{2}, -\frac{3}{4})$

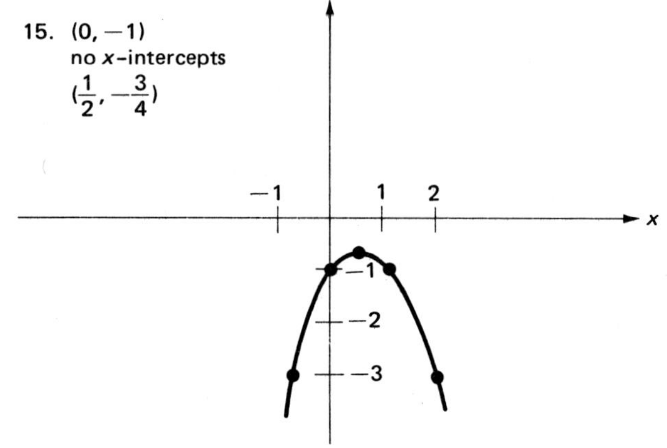

16. (0, −2)
    no *x*-intercepts
    $(-\frac{1}{2}, -\frac{5}{4})$

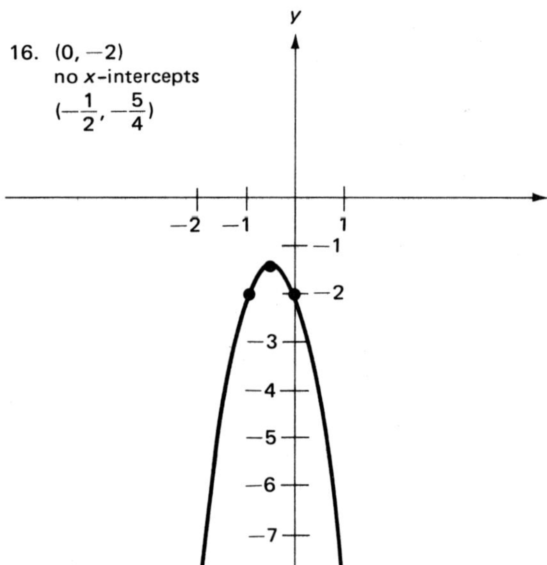

Exercise 7.3

1. 64 feet, 2 seconds
2. 32 feet, 1 second
3. 150 feet, $2\frac{1}{2}$ seconds
4. $16\frac{1}{4}$ feet, $\frac{7}{8}$ second
5. 0, 4 seconds
6. 0, $2\frac{1}{4}$ seconds
7. does not leave from the ground, $2\frac{1}{2}$ seconds
8. does not leave from the ground, 4.21 seconds
9. .414 second
10. 1.55 seconds
11. 13 feet, $\frac{3}{4}$ second, 1.65 seconds
12. 9.35 seconds

Exercise 7.4

1. function
2. not a function
3. function
4. function
5. not a function
6. function
7. not a function
8. not a function

9. not a function   10. function   11. function   12. function

13. function   14. not a function   15. not a function   16. function

Self-test

1. b, d   (Objective 4)

2. $(-\frac{3}{4}, \frac{41}{8})$, maximum   (Objective 1)

3. (0, −3)
    (−3, 0), (1, 0)
    (−1, −4)

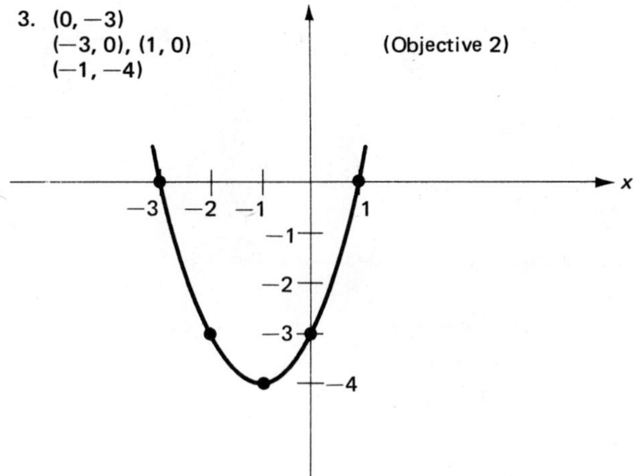

4. (0, −2)
    no x–intercepts
    (1, −1)

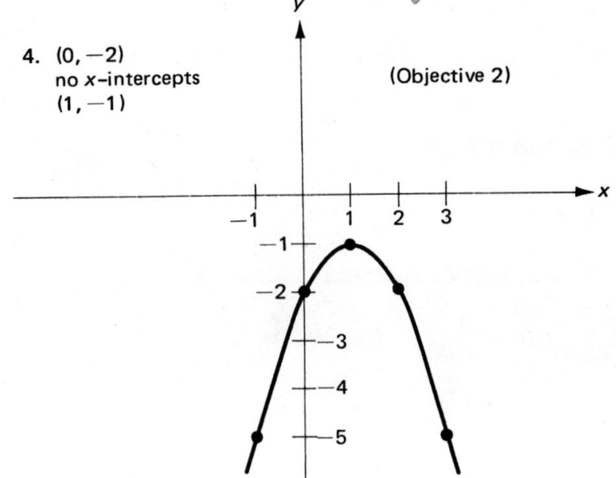

5. 4.65 seconds   (Objective 3)

Unit 8

Exercise 8.1

1. $(x+2)^2$

2. $(x-6)^2$

3. Not a perfect square

4. Not a perfect square

5. $(x-\frac{5}{2})^2$

6. Not a perfect square

7. $(x-\frac{3}{4})^2$

8. $(x+\sqrt{3})^2$

9. Not a perfect square

10. $(5x-3)^2$

11. $(4x-\frac{3}{2})^2$

12. $(\sqrt{2}\,x+\frac{1}{2})^2$

13. $(x+8y)^2$

14. Not a perfect square

15. $(2x-\frac{3}{4}y)^2$

16. $(\frac{1}{2}xy-3z)^2$

17. $-(2x-5)^2$

18. $-(x-\frac{7}{2})^2$

19. Not a perfect square

20. $-(5x+4y)^2$

Exercise 8.2

1. $(x+1)(x^2-x+1)$

2. $(x-1)(x^2+x+1)$

3. $(x-3)(x^2+3x+9)$

4. $(x+2)(x^2-2x+4)$

5. $(2x+1)(4x^2-2x+1)$

6. $(2x-3)(4x^2+6x+9)$

7. $(4x+5y)(16x^2-20xy+25y^2)$

8. $(10x-3y)(100x^2+30xy+9y^2)$

9. $(\frac{1}{2}x-y)(\frac{1}{4}x^2+\frac{1}{2}xy+y^2)$

10. $(x+\frac{1}{3}y)(x^2-\frac{1}{3}xy+\frac{1}{9}y^2)$

11. $(x+1)(x^4-x^3+x^2-x+1)$

12. $(x-1)(x^4+x^3+x^2+x+1)$

13. $(x-2y)(x^4+2x^3y+4x^2y^2+8xy^3+16y^4)$

14. $(2x+y)(16x^4-8x^3y+4x^2y^2-2xy^3+y^4)$

15. $(x+1)(x^6-x^5+x^4-x^3+x^2-x+1)$

16. $(x-2)(x^6+2x^5+4x^4+8x^3+16x^2+32x+64)$

Exercise 8.3

1. $(x+1)(x-1)$

2. Cannot be factored

3. Cannot be factored

4. $(x+6)(x-6)$

5. $(2x+3y)(2x-3y)$

6. $(7x+4y)(7x-4y)$

7. Cannot be factored

8. $(x+12y)(x-12y)$

9. $(\frac{1}{10}x+1)(\frac{1}{10}x-1)$

10. $(\frac{1}{2}x+\frac{1}{3}y)(\frac{1}{2}x-\frac{1}{3}y)$

11. Cannot be factored

12. $(x^2+y^2)(x+y)(x-y)$

13. $(x^2+9y^2)(x+3y)(x-3y)$

14. $(4x^2+1)(2x+1)(2x-1)$

15. $(x+2)(x^2-2x+4)(x-2)(x^2+2x+4)$

16. $(x^4+y^4)(x^2+y^2)(x+y)(x-y)$

Exercise 8.4

1. $(x-y)(a+b)$

2. $(a-b)(x-y)$

3. $(a-2)(x-y)$

4. $(2a-b)(x-2y)$

5. $(x-3)(x-y)$

6. $(y-2)(x-2)$

7. $(x-3)(3y+4)$

8. $(1-2y)(2x-3y)$

9. $(y-3x)(x+3)$

10. $(2y-x)(2x-1)$

11. $(x-1)(x^2+1)$

12. $(x+1)(2x^2+1)$

13. $(x-y)(x^2+1)$

14. $(x+y)(1-xy)$

15. $(x-1)^2(x+1)$

16. $(x+1)(x-1)(x+y)(x-y)$

Self-test

1. $(2x-y)(x-2)$ (Objective 4)

2. Cannot be factored   (Objective 3)

3. $(8x+3y)(8x-3y)$ (Objective 3)

4. $(x-2y)(x^2+2xy+4y^2)$ (Objective 2)

5. $(2x-\frac{9}{2}y)^2$ (Objective 1)

Unit 9

Exercise 9.1

1. $\dfrac{a}{2}$

2. $\dfrac{x}{2}$

3. $\dfrac{3}{x+3}$

4. $\dfrac{4}{x-3}$

5. $\dfrac{x}{x-4}$

6. $\dfrac{2x}{x+2}$

7. $\dfrac{x-y}{x+y}$

8. $\dfrac{1}{x^2-y^2}$

9. $\dfrac{x+y}{x-y}$

10. $2(x^2+y^2)$

11. $\dfrac{2-x}{2+x}$

12. $-(x+1)$

13. $\dfrac{x-2}{x-4}$

14. $\dfrac{x-3}{x+3}$

15. $\dfrac{2x+1}{2x+3}$

16. $\dfrac{2(x-1)}{2x-1}$

17. $\dfrac{x+y}{2}$

18. $\dfrac{x-2}{2x-1}$

19. $\dfrac{x+3}{x^2+3x+9}$

20. $\dfrac{x^2+4x+16}{4-x}$

21. $1-x$

22. $\dfrac{1}{x-5}$

23. $\dfrac{x-y}{x-3}$

24. $\dfrac{x+2}{x-2}$

25. $\dfrac{b+1}{b-1}$

26. $\dfrac{b-1}{b^2-b+1}$

27. $x+a$

28. $\dfrac{x^2+xy+y^2}{(x+y)(x^2+y^2)}$

29. $\dfrac{x^2-xy+y^2}{x-y}$

30. x^2+1

Exercise 9.2

1. $\dfrac{x(x+3)}{3(x+2)}$

2. $\dfrac{x-4}{2(x-2)}$

3. $\dfrac{x}{2}$

4. $3x$

5. 1

6. $\dfrac{y+2}{y+1}$

7. $\dfrac{1}{3}$

8. $\dfrac{x}{3}$

9. $\dfrac{2x}{(x-2)^2}$

10. $\dfrac{x}{(x+1)^2}$

11. $-\dfrac{x}{x+1}$

12. $-y$

13. $\dfrac{4x}{(x+2)(x-2)}$

14. $\dfrac{2x+1}{2(3x+1)}$

15. $\dfrac{3x}{(x+3)(x-3)}$

16. $\dfrac{x+2}{x+1}$

17. $\dfrac{(x+3)(x-2)}{(x+2)(x-3)}$

18. 1

19. $\dfrac{x(x+y)}{y(x-y)}$

20. $\dfrac{x-3}{x+3}$

21. 2

22. $\dfrac{(x+3)(x^2-3x+9)}{3x}$

23. -2

24. $-\dfrac{x^2-xy+y^2}{2}$

25. $\dfrac{x+a}{x^2+ax+a^2}$

26. a

27. $\dfrac{(x-1)(x-3)}{(x-2)(x-4)}$

28. $\dfrac{x}{y}$

29. $\dfrac{(x-a)^2}{a}$

30. $\dfrac{1}{x^2+y^2}$

Exercise 9.3

1.

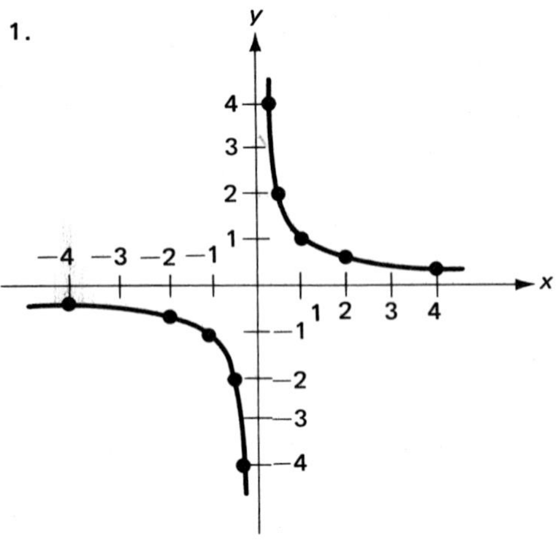

2.

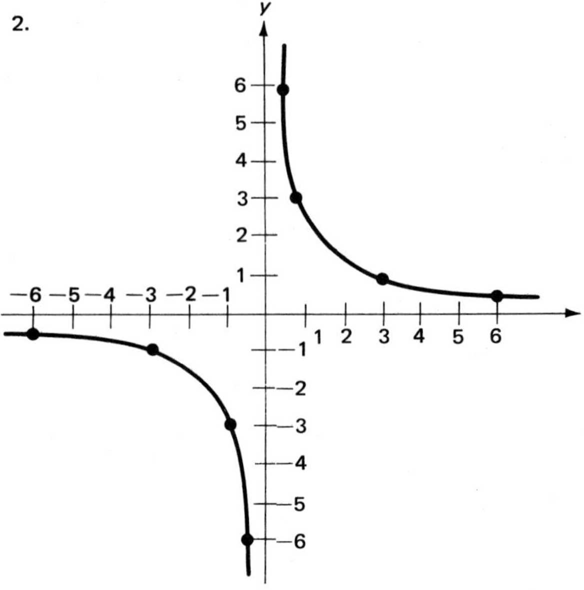

3.

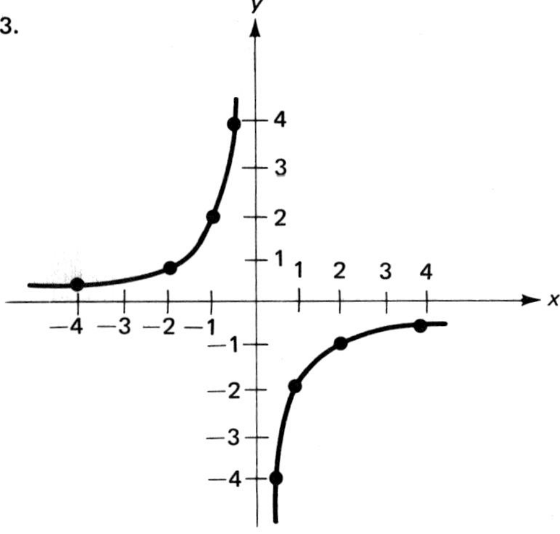

4.

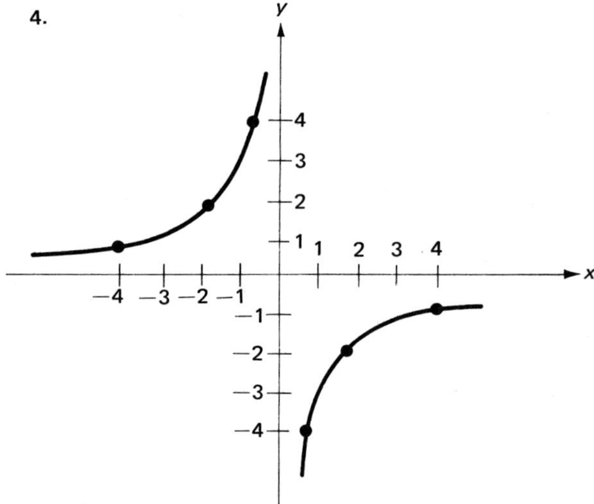

5.

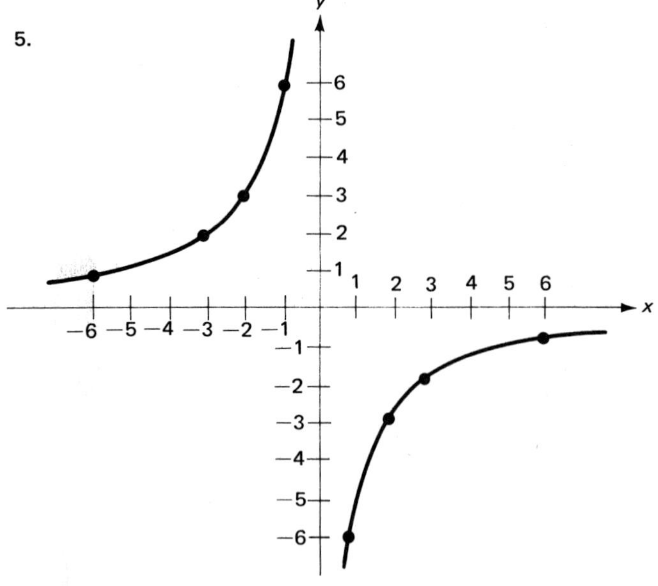

6.

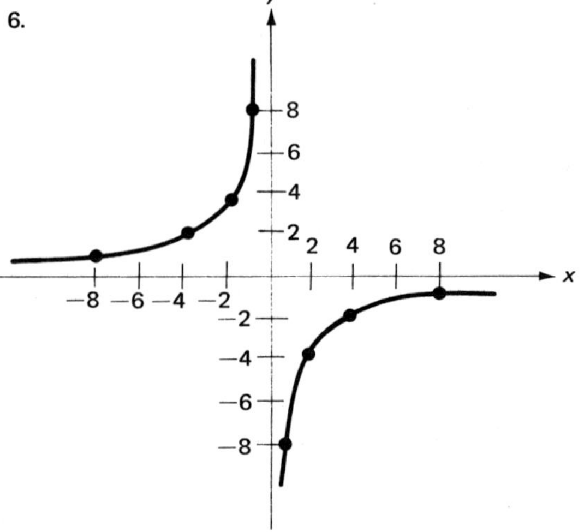

7.

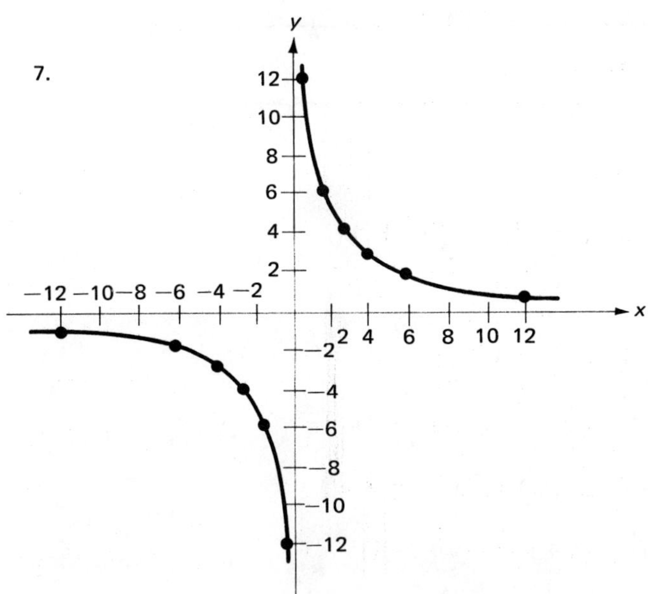

8.

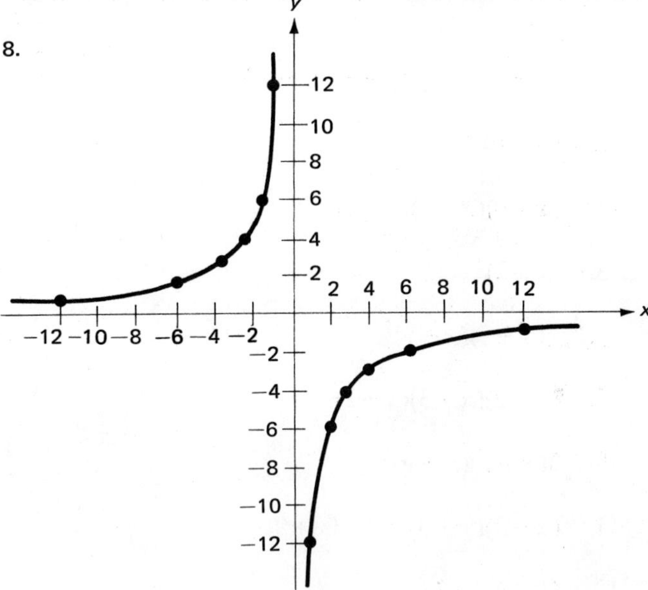

Self-test

1. $\dfrac{x + 3}{x + 5}$ (Objective 1)

2. $\dfrac{-2}{x + 2}$ (Objective 1)

3. $\dfrac{3(x + 3)}{x^2 + 3x + 9}$ (Objective 2)

4. $\dfrac{x - 2}{x + 2}$ (Objective 2)

5.

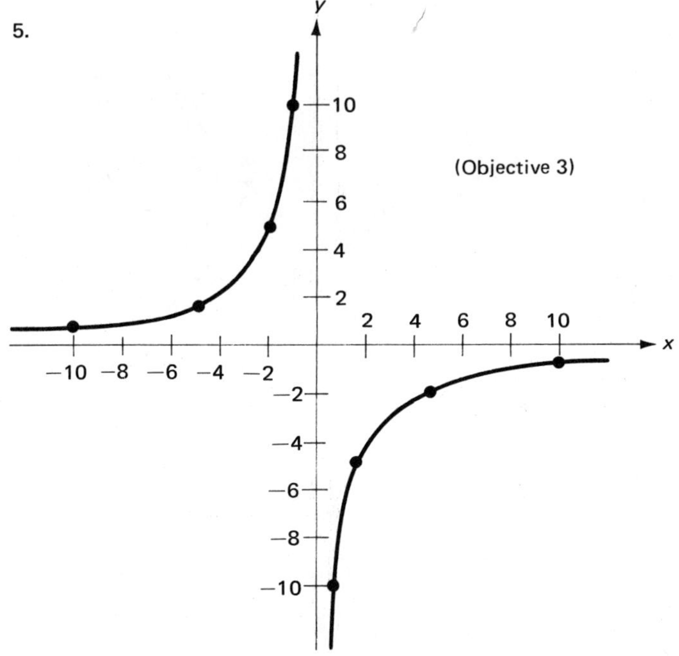

(Objective 3)

Unit 10

Exercise 10.1

1. $(x+4)(x-3)(x-2)$

2. $(2x-1)(x-1)(x+2)$

3. $(x+2)^2$

4. $2(x-3)^2$

5. $x(x+5)(x-5)^2$

6. $4x(x+4)^2$

7. $(x-2)(x+3)(x+2)$

8. $(x+3)^2(x-1)(x+1)$

9. $3(x+y)(x-y)^2$

10. $2x(x+y)^2(x^2-xy+y^2)$

11. $(x-3)(y+3)(x^2+3x+9)$

12. $(x+y)(x-y)(x-1)$

13. $(-1)(x-2)$

14. $(-2)(3x-2)$

15. $(-1)(x-4)(x+4)$

16. $(-2)(x-y)(x+y)(x^2+xy+y^2)$

Exercise 10.2

1. $\dfrac{3x}{x+3}$

2. $\dfrac{2x-4}{x+4}$

3. $\dfrac{2x}{x+1}$

4. $\dfrac{2}{x+2}$

5. $\dfrac{2-3x}{x^2}$

6. $\dfrac{x^2+4}{x^3}$

7. $\dfrac{2x^2+2}{x(x-2)}$

8. $\dfrac{x-8}{x(x+4)}$

9. $\dfrac{2x^2-x+5}{(x-1)(x+1)}$

10. $\dfrac{x-8}{(x+2)(x-3)}$

11. $\dfrac{1}{(x+1)(x+3)}$

12. $\dfrac{-1}{(x-3)(x-4)}$

13. $\dfrac{1}{(x+2)(x+1)}$

14. $\dfrac{4}{(x+5)(x+4)}$

15. $\dfrac{x+3}{x-3}$

16. 1

17. $\dfrac{x^2}{(x+4)(x-4)}$

18. $\dfrac{-x^2-1}{(x-1)(x^2+x+1)}$

19. $\dfrac{x^2y+xy^2}{(x-y)(x^2+xy+y^2)}$

20. $\dfrac{2x^2}{(x+2)(x-2)(x^2-2x+4)}$

21. $\dfrac{x^2+y}{(x+1)(x-1)(x-y)}$

22. $\dfrac{xy}{(x-3)(x-2)(y+2)}$

23. $\dfrac{-x-3}{x(x-3)^2}$

24. $\dfrac{x}{(x+2)^2}$

25. $\dfrac{x-y}{xy(x+y)}$

26. $\dfrac{2x^2+2a^2}{a(x+a)(x-a)}$

Exercise 10.3

1. $\dfrac{5x+1}{3x+1}$ 2. $\dfrac{x-2}{x-4}$ 3. $\dfrac{x-1}{2x-1}$ 4. $\dfrac{1}{3}$

5. $x-1$ 6. $\dfrac{x}{1-x}$ 7. $\dfrac{1}{x+y}$ 8. $\dfrac{y+x}{y-x}$

9. $x+y$ 10. $\dfrac{xy}{x+y}$ 11. 1 12. $\dfrac{1}{2}$

13. x 14. $\dfrac{2x}{x^2+1}$

Self-test

1. $4(x+4)(x-4)^2$ (Objective 1)   2. $\dfrac{x+8}{(x-2)(x+2)}$ (Objective 2)   3. $\dfrac{3x-15}{x(x+5)}$ (Objective 2)

4. $\dfrac{-1}{(x-4)(x-5)}$ (Objective 2)   5. x^2 (Objective 3)

Unit 11

Self-test

1. $\frac{2}{3}, -2$ (Unit 2)   2. $\dfrac{2 \pm \sqrt{2}}{2}$ (Unit 4)   3. $1\frac{1}{2}, 4\frac{1}{2}$ (Unit 3)   4. $-\frac{1}{3}$ (Unit 6)

5. $(0, -2)$
   $(\frac{1}{2}, 0)$ (Unit 6)

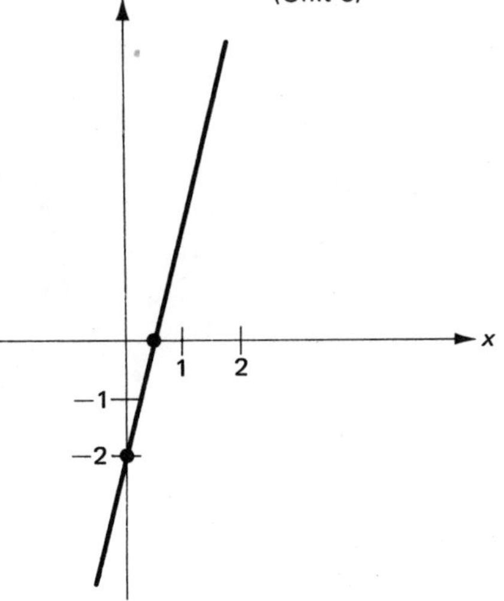

6. $(0, 12)$
   $(3, 0), (-4, 0)$
   $(-\frac{1}{2}, \frac{49}{4})$ (Unit 7)

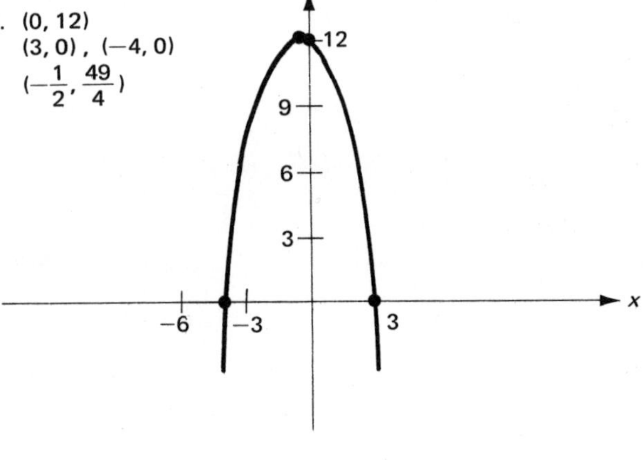

7. 80 feet   (Unit 7)   8. $-\dfrac{x^2+3x+9}{(x+3)^2}$ (Unit 9)   9. $\dfrac{x+1}{2(x-1)(x-4)}$ (Unit 10)

10. $-\dfrac{y}{x}$ (Unit 10)

Unit 12

Exercise 12.1

1. x^8　　2. x^4　　3. $\dfrac{1}{x^2}$　　4. 1　　5. x

6. x^{15}　　7. Cannot be simplified　　8. Cannot be simplified　　9. x^3y^3　　10. x^{18}

11. x^{10}　　12. $\dfrac{x^8}{y^8}$　　13. p^8q^{12}　　14. $p^{10}q^{12}$　　15. p^3q^{15}

16. $\dfrac{p^6}{q^{12}}$　　17. $\dfrac{r^2s^4}{t^6}$　　18. $\dfrac{t^6}{r^6}$　　19. rs^{11}　　20. $\dfrac{1}{rs^3t^5}$

Exercise 12.2

1. 1　　2. 1　　3. 1　　4. 1　　5. y　　6. 10

7. r^2s　　8. 1　　9. $a^4b^2c^2$　　10. $\dfrac{1}{a^6b^3}$　　11. $\dfrac{m^9}{n^6}$　　12. 1

13. p^4r^4　　14. $\dfrac{p^5}{r^{10}}$　　15. 1　　16. $x^2 + y^2$

Exercise 12.3

1. $\dfrac{1}{36}$　　2. 125　　3. 81　　4. 16

5. $\dfrac{b^2}{a^3}$　　6. $\dfrac{r^3}{p^4}$　　7. m^5n^4　　8. $\dfrac{1}{xy}$

9. $\dfrac{r}{p}$　　10. r^2s^2　　11. $\dfrac{1}{m^3n^3}$　　12. $\dfrac{k^2}{h^3}$

13. $\dfrac{a^4}{b^4}$　　14. q^{18}　　15. $\dfrac{x^{12}}{y^{12}}$　　16. $\dfrac{q^{12}r^9}{p^6}$

17. $\dfrac{p^4r^4}{q^4}$　　18. $\dfrac{1}{x^{10}y^{15}z^5}$　　19. $\dfrac{x}{y+z}$　　20. $\dfrac{z}{x+y}$

21. $\dfrac{y+x}{y-x}$　　22. $\dfrac{x+y}{xy}$　　23. $-\dfrac{x+y}{x^2y^2}$　　24. $-\dfrac{(x-y)^2}{xy}$

Exercise 12.4

1.

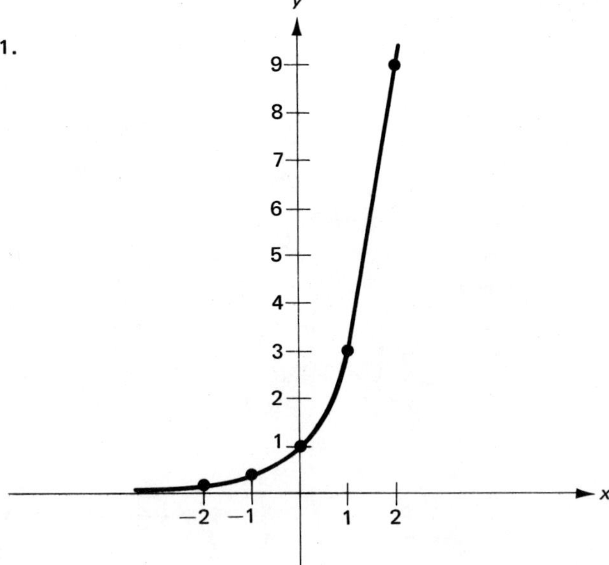

2.

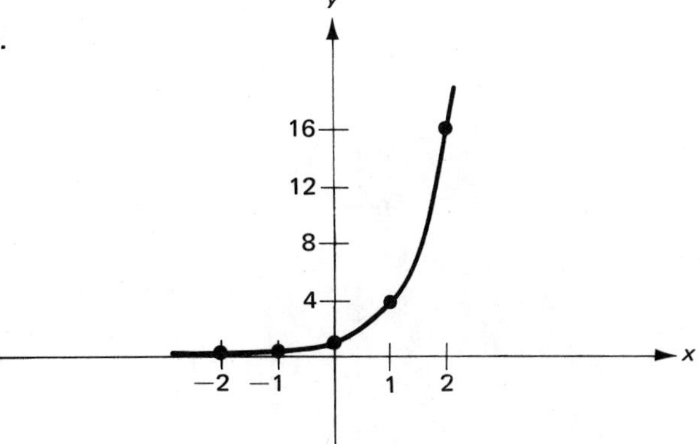

3.

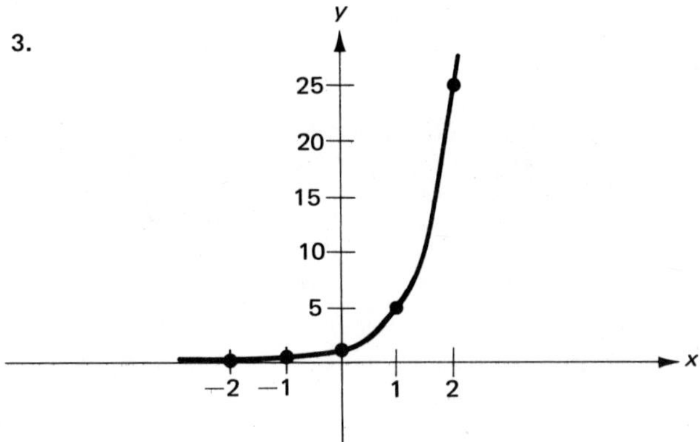

4.

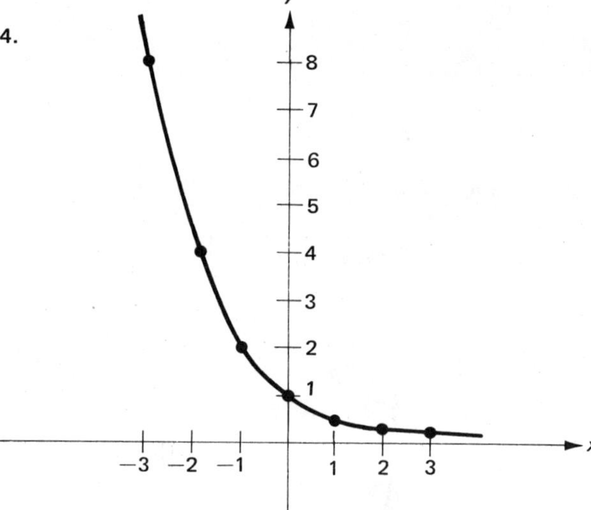

5.

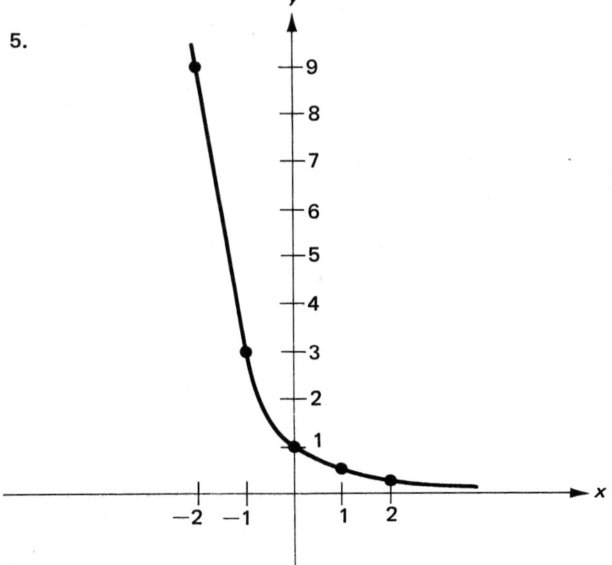

6.

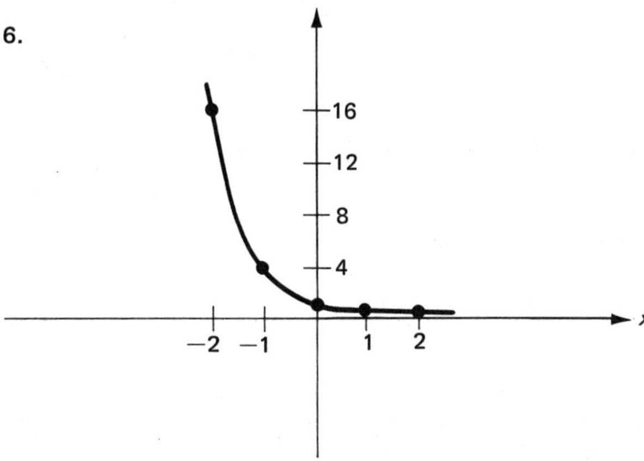

7.

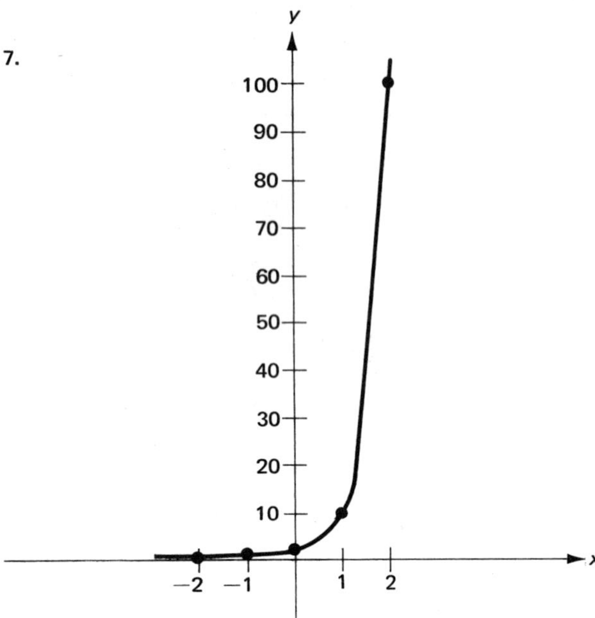

8.

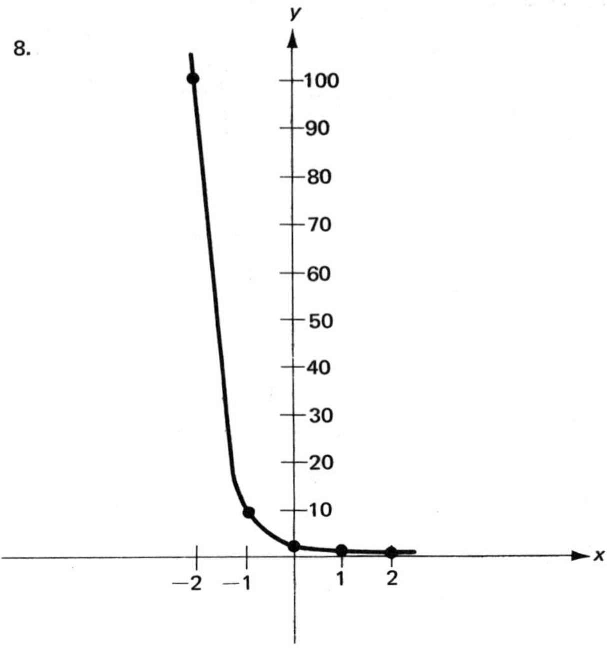

Self-test

1a. 1   (Objective 2)   1b. 25   (Objective 3)

2. a^5b^6 (Objective 1)

3. $\dfrac{r^3}{p^{12}q^{12}}$ (Objective 3)

4. $\dfrac{x^2 - xy + y^2}{x^2y^2}$ (Objective 3)

5.

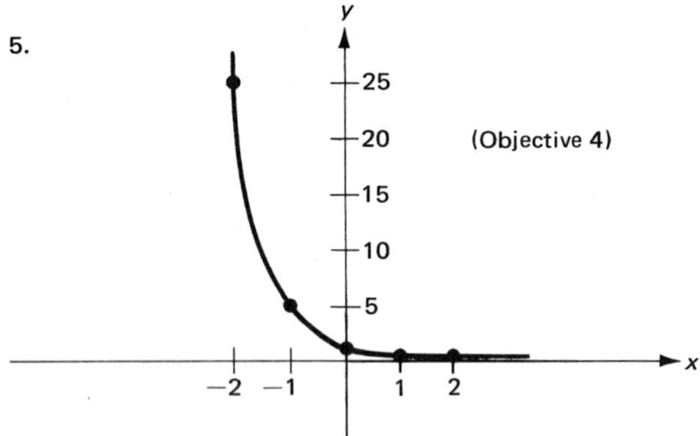

(Objective 4)

Unit 13

Exercise 13.1

1. 5   2. 2   3. -2 4. No real number solution

5. $\frac{1}{6}$ 6. $-\frac{1}{3}$ 7. No real number solution   8. $-\frac{2}{5}$

9. 64   10. 9   11. $\frac{1}{125}$ 12. No real number solution

13. $\frac{16}{25}$ 14. $-\frac{8}{27}$ 15. $-\frac{1}{32}$ 16. $\frac{625}{256}$

17. $\frac{1024}{243}$ 18. No real number solution   19. $\frac{1}{10}$ 20. $-\frac{1}{6}$

21. -27 22. 625   23. $-\frac{1024}{243}$ 24. $\frac{216}{125}$

Exercise 13.2

1. $x^{\frac{11}{12}}$ 2. $x^{\frac{5}{6}}$ 3. x^2 4. $x^{\frac{3}{2}}y^2$

5. $\dfrac{x^{\frac{1}{9}}}{y^{\frac{1}{8}}}$ 6. 1   7. $\dfrac{1}{p^{\frac{1}{10}}q^{\frac{2}{5}}}$ 8. $\dfrac{p^{\frac{8}{9}}}{q^{\frac{2}{3}}}$

9. $a^{\frac{3}{2}}b^{\frac{2}{3}}c^{\frac{1}{4}}$ 10. $\dfrac{a^{\frac{1}{9}}}{b^{\frac{1}{10}}c^{\frac{1}{2}}}$ 11. $\dfrac{q^{\frac{5}{2}}}{p^{\frac{7}{3}}}$ 12. $p^{\frac{3}{10}}q^{\frac{1}{5}}$

13. $\dfrac{q^{\frac{7}{9}}}{p^{\frac{4}{9}}r^{\frac{5}{9}}}$ 14. $\dfrac{p^6 r^5}{q^{\frac{5}{2}}}$ 15. $x^3 z^{\frac{7}{12}}$ 16. $\dfrac{1}{x^{\frac{4}{9}}y^{\frac{7}{9}}z^{\frac{1}{9}}}$

Exercise 13.3

1. $\sqrt[3]{a}$ 2. $\sqrt{a}$ 3. a^2 4. a

5. $q\sqrt{p}$ 6. $p^2\sqrt[3]{q}$ 7. $\sqrt{xy}$ 8. $x\sqrt[3]{y^2 z}$

9. $\sqrt[8]{xy}$ 10. $\sqrt[6]{xy}$ 11. $\sqrt[4]{a}$ 12. $\sqrt[6]{a}$

Self-test

1a. $\frac{1}{243}$ 1b. 4   (Objective 1)   2. $x^{\frac{1}{6}}$ (Objective 2)   3. $\dfrac{1}{pq^{\frac{1}{3}}}$ (Objective 2)

4. $\dfrac{z^{\frac{1}{3}}}{x^{\frac{1}{3}}}$ (Objective 2)   5. $b\sqrt[3]{a^2}$ (Objective 3)

Unit 14

Exercise 14.1

1. $3\sqrt[3]{2}$ 2. $2\sqrt[4]{5}$ 3. $2\sqrt[5]{4}$ 4. $5\sqrt[3]{4}$

5. $x\sqrt[3]{x^2}$ 6. $x^2\sqrt[4]{x}$ 7. $2xy\sqrt{3y}$ 8. $4x^3y^2\sqrt[3]{y^2}$

9. $2xy^2\sqrt[4]{2}$

10. $5x^2y^4$

11. $2xyz^2\sqrt[3]{5xy^2}$

12. $10x^2y^3z^4\sqrt{2xyz}$

13. $\dfrac{4x}{y}\sqrt{x}$

14. $\dfrac{xy^2}{z^2}\sqrt[4]{2xy}$

15. $\dfrac{1}{yz^2}\sqrt[3]{x^2}$

16. $\dfrac{2x}{y^2z^3}\sqrt[5]{2}$

17. $2x\sqrt[4]{x}$

18. $\dfrac{5y}{x}$

19. $\dfrac{8x}{y}\sqrt{xy}$

20. $\dfrac{3x}{y}\sqrt[3]{2xy}$

21. $\dfrac{5xz^2}{y^2}\sqrt[3]{2x^2y^2}$

22. $\dfrac{6yz^2}{x^4}\sqrt{2xyz}$

23. $\dfrac{1}{zu}\sqrt[6]{x^2y^3z^2u^3}$

24. $\dfrac{3xy^2}{z^2u^3}\sqrt[4]{xy^2z^3u^2}$

25. $\dfrac{2}{y}\sqrt[4]{xy^3}$

26. $\dfrac{xy}{z^2}\sqrt[5]{x^4z^4}$

27. $\dfrac{1}{3xy}\sqrt[3]{3xy^2}$

28. $\dfrac{1}{2xyz}\sqrt[4]{2x^3y^3z}$

29. $\dfrac{1}{y^2}\sqrt[4]{xy^2}$

30. $\dfrac{1}{xy^2}\sqrt[3]{25xy^2}$

Exercise 14.2

1. $5\sqrt{3}$

2. $3\sqrt{2}$

3. $\sqrt{xy}$

4. $3\sqrt[3]{x}$

5. $\sqrt{x}-\sqrt{xy}$

6. $2\sqrt{x}+\sqrt[3]{x}$

7. $5\sqrt[3]{x}-4\sqrt[3]{x^2}$

8. $6\sqrt{xy}-8\sqrt[3]{xy}$

9. $\sqrt{3}$

10. $24\sqrt[3]{2}$

11. $12\sqrt{2}$

12. $-3\sqrt{5}$

13. $-x\sqrt{x}$

14. $5x\sqrt[3]{x}$

15. $4xy\sqrt[3]{xy^2}$

16. $-4x^2y^2\sqrt{2xy}$

17. $7y\sqrt[3]{2x}+5xy\sqrt[3]{2}$

18. $4x\sqrt[4]{xy}-2y\sqrt[4]{xy}$

19. $\sqrt{3}$

20. $-4\sqrt{2}$

21. $\dfrac{8}{x}\sqrt{x}$

22. $\dfrac{2}{y}\sqrt{3xy}$

23. $\dfrac{10x}{y}\sqrt{xy}+\dfrac{10}{y}\sqrt{xy}$

24. $\dfrac{2x^2}{5}\sqrt{5x}-\dfrac{6x}{5}\sqrt{5x}$

Exercise 14.3

1. $12\sqrt{2}$

2. $4x\sqrt[3]{x}$

3. $3xy$

4. $x\sqrt[4]{4y^3}$

5. $10x^2y^4z^2\sqrt{2y}$

6. $5x^3y^2z^2\sqrt[3]{6xy^2z}$

7. $x^2\sqrt{3}-6x$

8. $4y\sqrt[3]{x^2}+4x\sqrt[3]{y^2}$

9. $10\sqrt[3]{x^2}+2x\sqrt[3]{75x}-10x^2\sqrt[3]{6}$

10. $12xy-8x^2y^2\sqrt[4]{2}-8x^3y\sqrt[4]{4x^2y^2}$

11. $2+2\sqrt{3}+2\sqrt{6}+6\sqrt{2}$

12. $\sqrt{30}+5\sqrt{3}-6\sqrt{2}-6\sqrt{5}$

13. $12\sqrt{5}-4$

14. $52+6\sqrt{35}$

15. 16

16. -1

17. $15x - 9x\sqrt{2}$

18. $2xy - 7x$

19. $6y\sqrt{x} - 24\sqrt{xy} + 24\sqrt{x}$

20. $\sqrt[3]{x^2y^2} + y\sqrt[3]{x^2} + x\sqrt[3]{y^2} + xy$

21. $4 + 5\sqrt{2}\sqrt[3]{4} + 4\sqrt[3]{2}$

22. $9\sqrt[3]{3} - 8\sqrt{3}\sqrt[3]{9} - 9$

23. $x - x\sqrt[5]{x}$

24. $12xy - 4\sqrt{xy}\sqrt[3]{x^2y} - 9\sqrt{xy}\sqrt[3]{xy^2}$

Exercise 14.4

1. $\dfrac{1}{y}\sqrt{xy}$

2. $\dfrac{2}{3}\sqrt{6x}$

3. $\dfrac{1}{y}\sqrt[3]{2xy^2}$

4. $\dfrac{1}{3xz}\sqrt[3]{9x^2z^2}$

5. $\dfrac{x - \sqrt{3x}}{x}$

6. $\dfrac{x\sqrt{2y} + y\sqrt{2x} - xy}{2xy}$

7. $\dfrac{x\sqrt{10} + 10\sqrt{x}}{10x}$

8. $\sqrt{x} + \sqrt{y} + \sqrt{xy}$

9. $\dfrac{\sqrt{5} + 1}{4}$

10. $6 - 4\sqrt{2}$

11. $\dfrac{x - \sqrt{x}}{x - 1}$

12. $\sqrt{5x} - \sqrt{3x}$

13. $\sqrt{15} + \sqrt{10} + 3 + \sqrt{6}$

14. $\dfrac{2\sqrt{6} - 2\sqrt{3} - 2 + \sqrt{2}}{2}$

15. $\sqrt{35} - 6$

16. $12\sqrt{2} + 17$

17. $7\sqrt{5} - 9\sqrt{3}$

18. $\dfrac{3 - 3\sqrt{6}}{5}$

19. $\dfrac{x + 2\sqrt{3x} + 3}{x - 3}$

20. $\dfrac{2x + 3\sqrt{x} - 2}{4x - 1}$

21. $\dfrac{1}{4 - \sqrt{15}}$

22. $\dfrac{-1}{4\sqrt{5} + 10}$

23. $\dfrac{1}{\sqrt{xy}}$

24. $\dfrac{1}{\sqrt{x} + \sqrt{a}}$

Exercise 14.5

1.

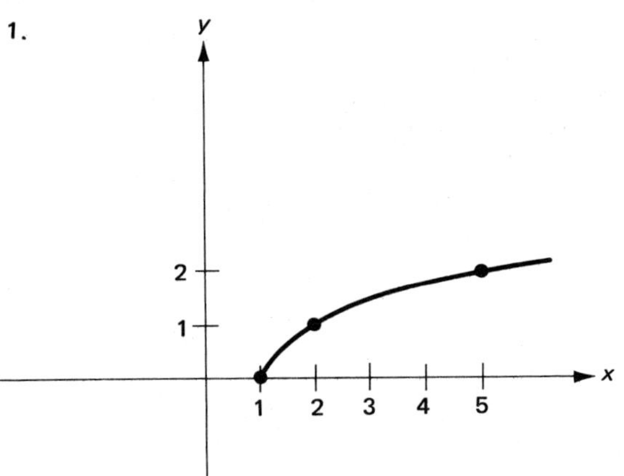

2.

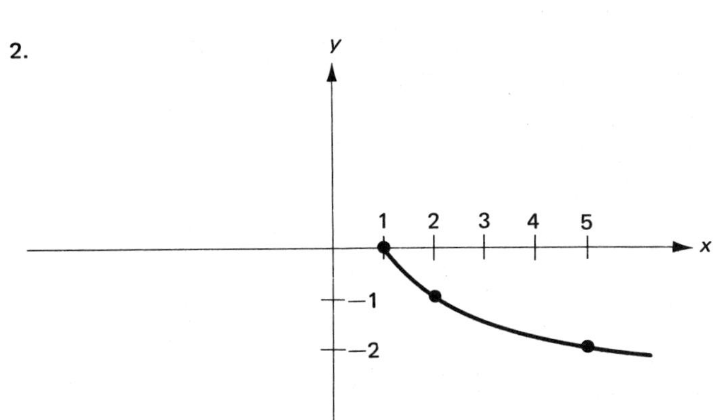

3.

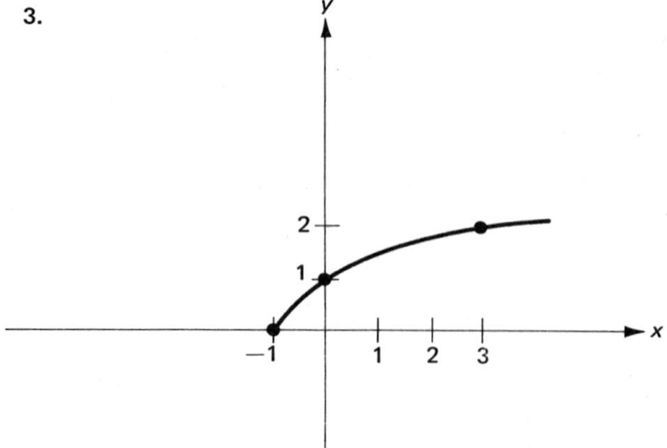

4.

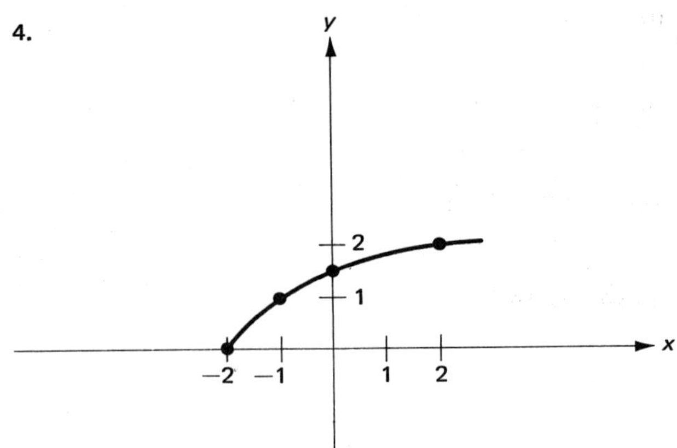

5.

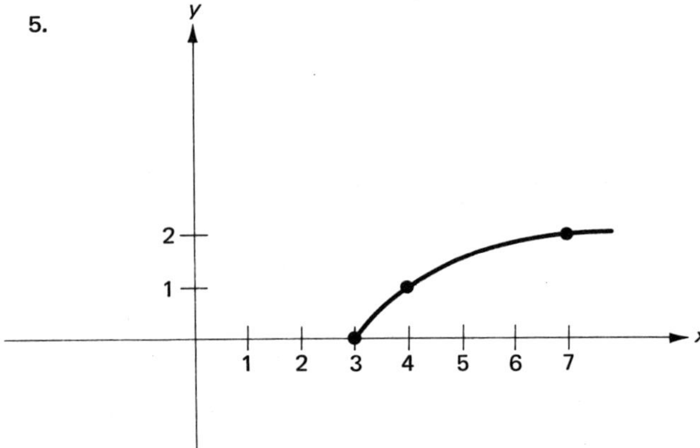

6.

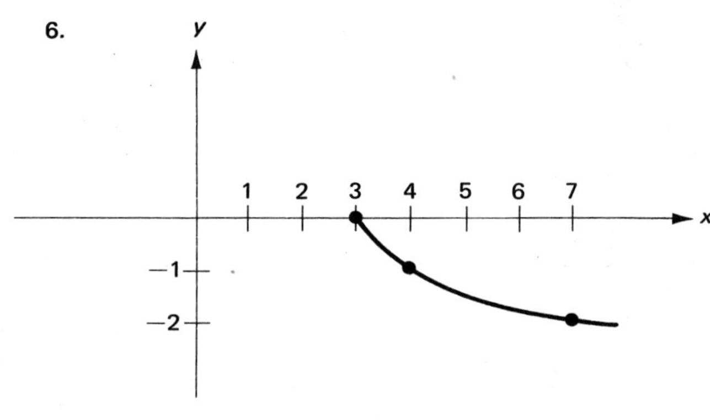

7.

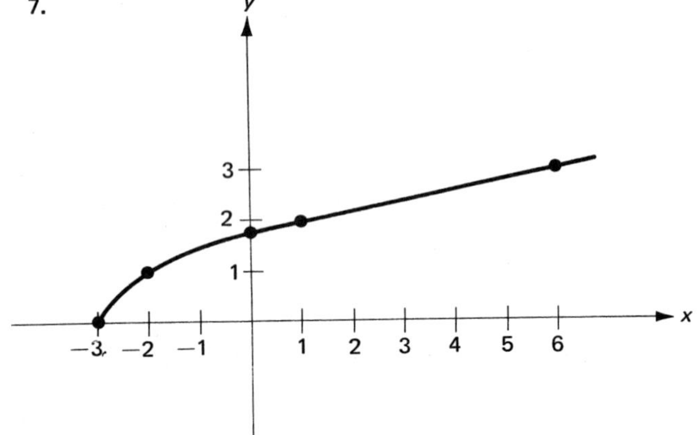

8.

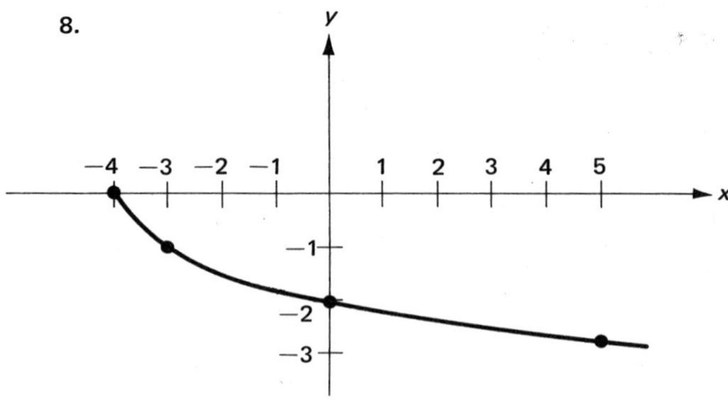

Self-test

1. $\dfrac{2x}{yz}\sqrt[4]{x^2y^3z}$ (Objective 1)

2. $4x^2\sqrt{3} - 60x$ (Objective 3)

3. $5x^2\sqrt{5x}$ (Objective 2)

4. $\sqrt{3} + 2$ (Objective 4)

5.   (Objective 5)

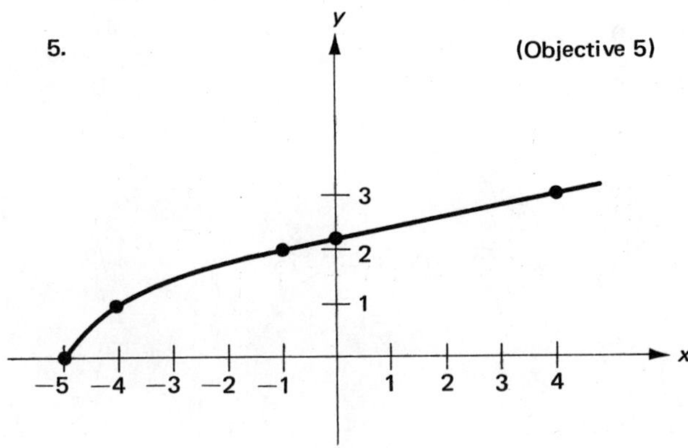

Unit 15

Exercise 15.1

1. $6i$

2. $-3i$

3. $i\sqrt{7}$

4. $-i\sqrt{11}$

5. $3i\sqrt{2}$

6. $3i\sqrt{3}$

7. $4i\sqrt{5}$

8. $7i\sqrt{2}$

9. $-5i\sqrt{2}$

10. $-3i\sqrt{5}$

11. $\frac{1}{2}i$

12. $\frac{3}{5}i$

13. $-\frac{4}{3}i$

14. $-\frac{1}{4}i$

15. $\dfrac{i\sqrt{5}}{2}$

16. $-\dfrac{2i\sqrt{2}}{3}$

17. $9+9i$

18. $4-8i$

19. $10-2i\sqrt{5}$

20. $\sqrt{3}+6i\sqrt{2}$

21. $\frac{3}{4}-\frac{3}{2}i$

22. $\frac{2}{3}+\dfrac{i\sqrt{2}}{3}$

Exercise 15.2

1. $5i$

2. $3i$

3. $3+6i$

4. $7-3i$

5. $3+4i$

6. $-1+2i$

7. $6-12i$

8. $5-3i$

9. $-2+4i$

10. $-5-10i$

11. $4i$

12. $-i$

13. 4

14. 8

15. $5-3i$

16. $-7i$

17. $8-14i$

18. $9+i$

19. $2\sqrt{2}+5i\sqrt{2}$

20. $3\sqrt{3}+(2\sqrt{5}-2\sqrt{3})i$

Exercise 15.3

1. -14

2. -4

3. -4

4. -6

5. -4

6. $-3\sqrt{2}$

7. $24i - 16$

8. $8 + 12i$

9. $1 - 4i$

10. $-10 - 25i$

11. $5 + 14i$

12. $19i - 9$

13. $26 + 18i$

14. $-3 - 36i$

15. 41

16. 5

17. $13i$

18. $-20i$

19. $4 + 5i\sqrt{2}$

20. $8 + 3i\sqrt{3}$

21. $3 + \sqrt{5} + (\sqrt{3} - \sqrt{15})i$

22. 8

23. $6 - \sqrt{15} + (2\sqrt{5} + 3\sqrt{3})i$

24. $9i$

Exercise 15.4

1. $-\frac{5}{6}i$

2. $\frac{2}{3}i$

3. $-\frac{3}{4}i$

4. $-5i$

5. $-i$

6. $-\dfrac{i\sqrt{2}}{4}$

7. $\dfrac{8 - 5i}{2}$

8. $-\dfrac{4 + 3i}{9}$

9. $\dfrac{8 + 2i}{5}$

10. $1 - i$

11. $\dfrac{2 + i}{2}$

12. $1 - 2i$

13. $\dfrac{9 + 12i}{5}$

14. $\dfrac{22i - 7}{13}$

15. $\dfrac{14 + 5i}{17}$

16. $\dfrac{9 - 13i}{25}$

17. $\dfrac{5 - 5i}{4}$

18. $\dfrac{1 - 6i}{3}$

19. $\dfrac{4 + 3i}{5}$

20. $-\dfrac{3 + 4i}{5}$

21. i

22. $-i$

23. $\dfrac{\sqrt{5} - 3i}{2}$

24. $1 + i\sqrt{2}$

Exercise 15.5

1. $\pm i$

2. $\pm 5i$

3. $\pm 3i\sqrt{5}$

4. $\pm 2i\sqrt{3}$

5. $\dfrac{-1 \pm i\sqrt{3}}{2}$

6. $\dfrac{5 \pm i\sqrt{15}}{4}$

7. $1 \pm i$

8. $-1 \pm 5i$

9. $\dfrac{1 \pm i\sqrt{17}}{2}$

10. $\dfrac{3 \pm i\sqrt{39}}{6}$

Self-test

1. $-3i\sqrt{10}$ (Objective 1)

2. $14 - 34i$ (Objective 3)

3. $2 + 6i$ (Objective 2)

4. $\dfrac{3 - 4i}{5}$ (Objective 4)

5. $-1 \pm i\sqrt{14}$ (Objective 5)

Unit 16

Self-test

1. $-\frac{2}{3}, 4$ (Unit 2)

2. $3 \pm 2i$ (Unit 15)

3. $\frac{1}{4}$ hour   (Unit 3)

4. $-\frac{4}{3}, 4x + 3y - 12 = 0$ (Unit 6)

5. (0, 3)
   no x–intercepts
   $(-1, 2)$

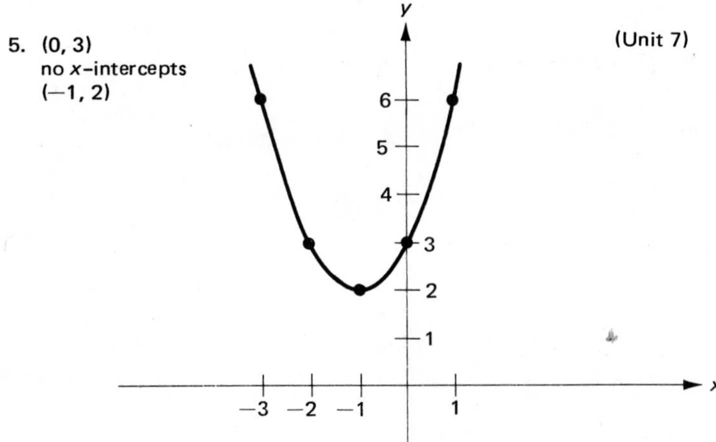

(Unit 7)

6. $\dfrac{xz^2}{y^2}$ (Unit 13)

7. $\dfrac{-2x}{x + 5}$ (Unit 9)

8. $2x\sqrt[3]{3x^2}$ (Unit 14)

9. $\dfrac{-1 + i}{2}$ (Unit 15)

10.

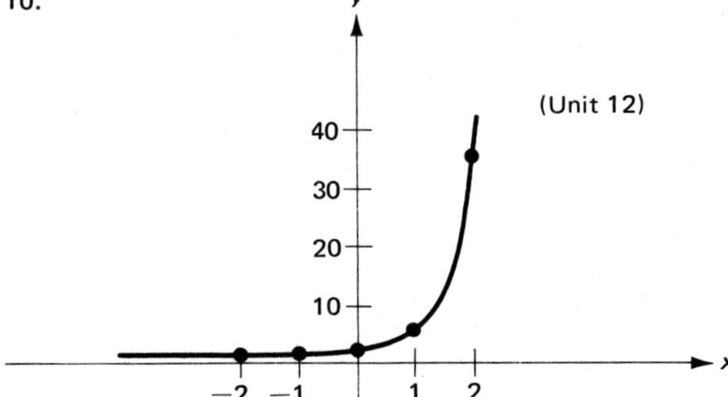

(Unit 12)

Unit 17

Exercise 17.1

1. -1

2. $\frac{3}{2}$

3. $-\frac{1}{2}$

4. 0

5. No solution

6. No solution

7. $4, 9$

8. $-1, 3$

9. $2, \frac{2}{3}$

10. $1, \frac{3}{2}$

11. -2

12. $0, 5$

13. $\pm\frac{\sqrt{10}}{2}$

14. $\frac{-3 \pm \sqrt{41}}{4}$

15. $\frac{2}{3}$

16. -2

17. 4

18. 10

19. No solution

20. 4

21. $2 \pm \sqrt{10}$

22. $\frac{1 \pm \sqrt{17}}{2}$

23. $\frac{1}{2}$

24. No solution

Exercise 17.2

1. $\frac{3}{2}, -\frac{2}{3}$

2. $3, -\frac{1}{3}$

3. $-1, 2$

4. $-2, \frac{3}{2}$

5. $2, -\frac{1}{4}$

6. $\frac{3}{4}, -\frac{1}{3}$

7. $1 \pm \sqrt{2}$

8. $\frac{-3 \pm \sqrt{17}}{4}$

9. 12 minutes

10. 2 minutes

11. $1\frac{7}{8}$ hours

12. $2\frac{2}{3}$ hours

13. 30 minutes

14. 3 hours

15. 12 hours

16. $3\frac{1}{3}$ hours

17. 120 minutes

18. 18 minutes

19. 20 miles per hour, 10 miles per hour

20. 6 miles per hour, $1\frac{1}{2}$ miles per hour

21. 4 miles per hour, 8 miles per hour

22. 6 miles per hour, 16 miles per hour

Self-test

1. No solution   (Objective 1)

2. $-4, 3$ (Objective 1)

3. $-\frac{2}{3}$ (Objective 1)

4. $\frac{1 \pm \sqrt{13}}{6}$ (Objective 2)

5. $\frac{14}{15}$ hour   (Objective 2)

Unit 18

Exercise 18.1

1. 7

2. $\frac{7}{2}$

3. $\frac{3}{2}$

4. $-\frac{1}{2}$

5. $-\frac{4}{3}$

6. -7

7. $-\frac{1}{3}$

8. 5

9. No solution

10. No solution

11. 9

12. No solution

13. 5

14. 3

15. -2

16. No solution

17. $-1+\sqrt{2}$

18. $2-\sqrt{5}$

19. No solution

20. 25

21. 1, 5

22. 11

23. -3

24. No solution

Exercise 18.2

1. $y=\frac{1}{4}x^2$

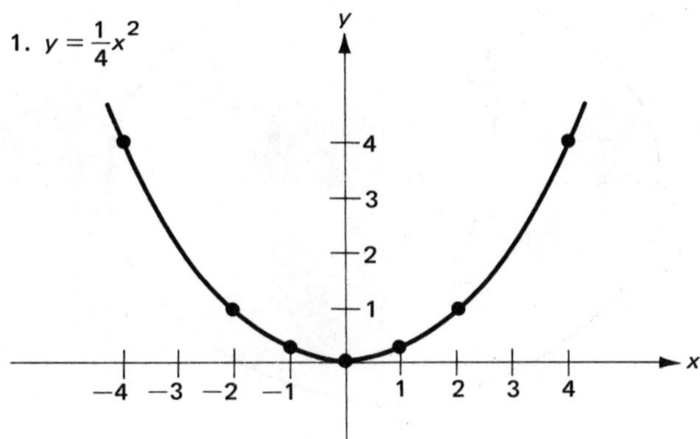

2. $y=x^2$

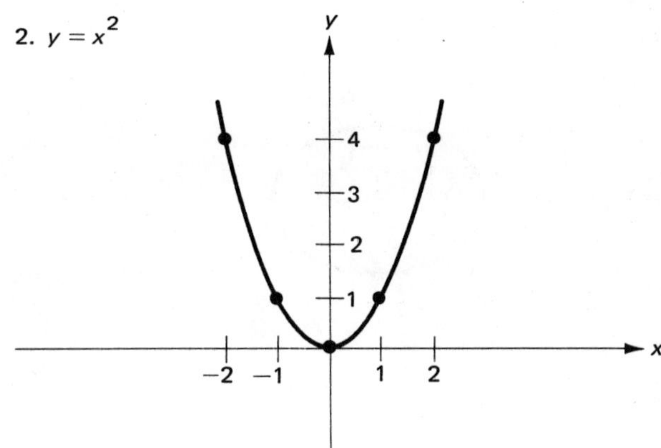

3. $y^2=8x$

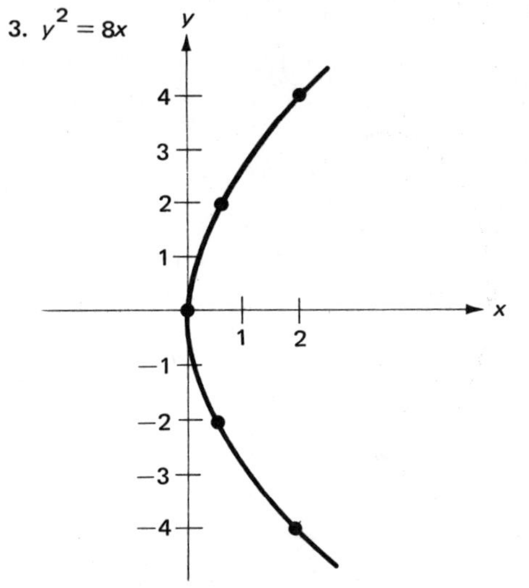

4. $y^2=\frac{1}{2}x$

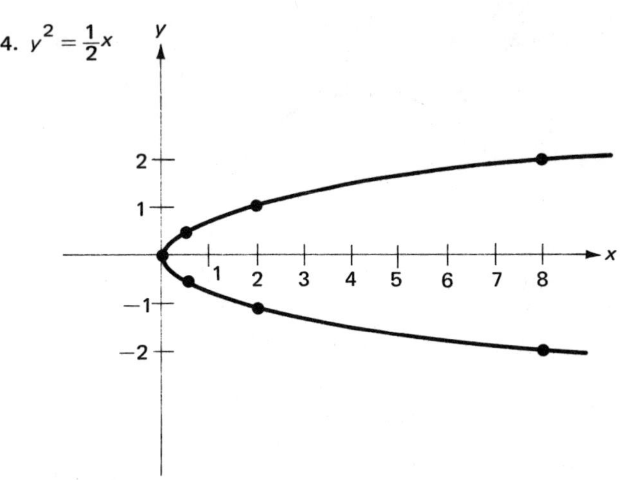

5. $y^2 = -2x$

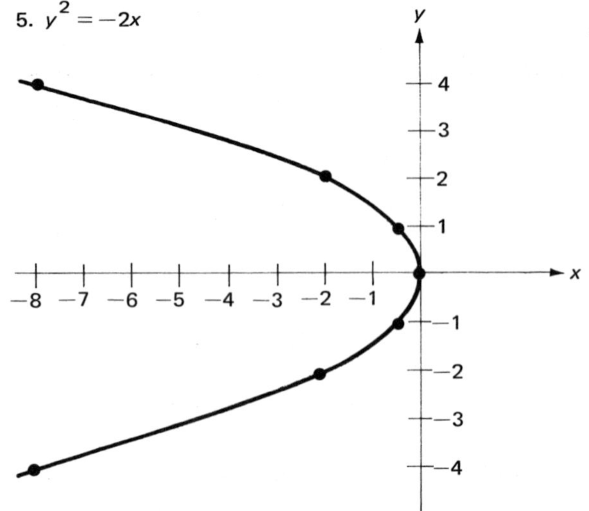

6. $y = -\frac{1}{4}x^2$

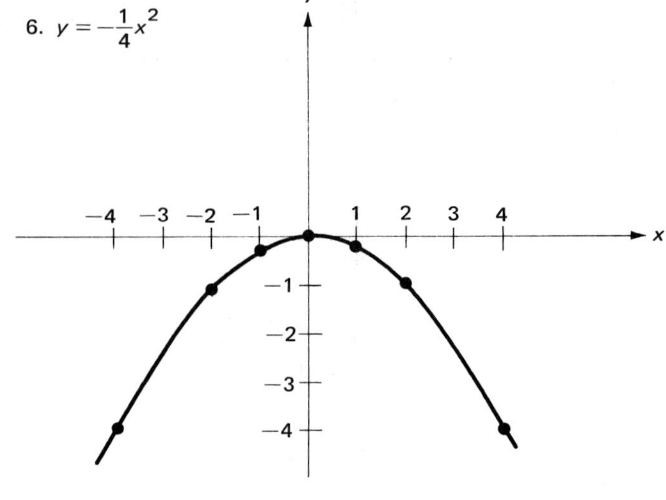

7. $3x^2 + 4y^2 = 12$

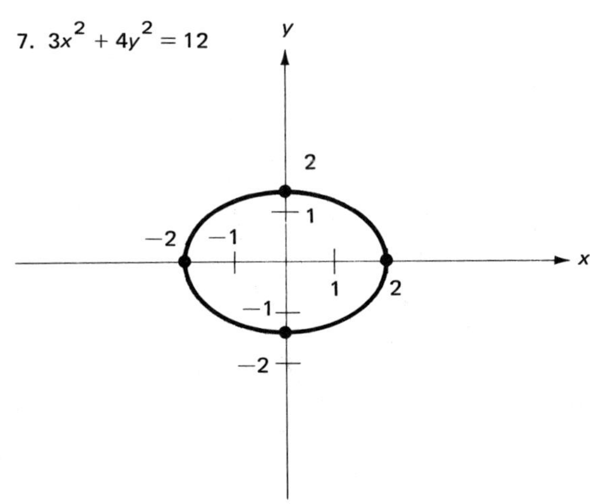

8. $16x^2 + 25y^2 = 400$

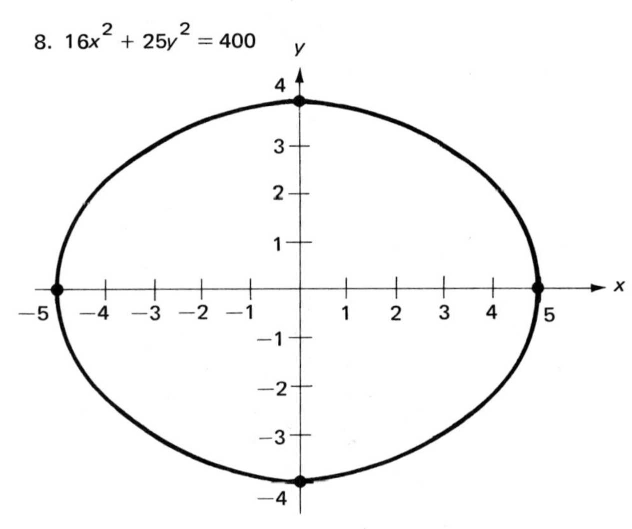

9. $4x^2 + 3y^2 = 48$

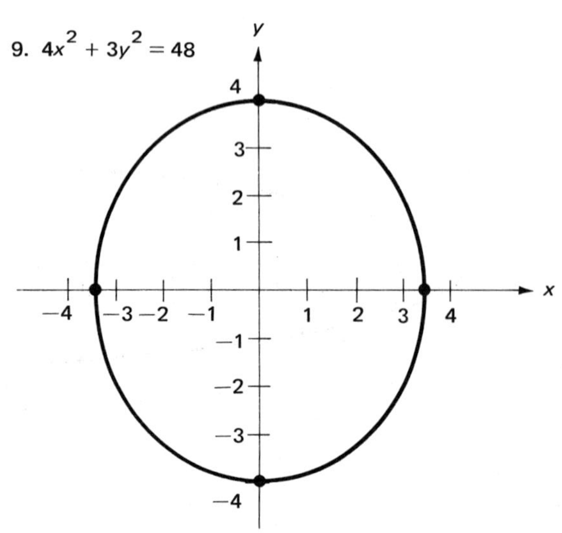

10. $9x^2 + 4y^2 = 36$

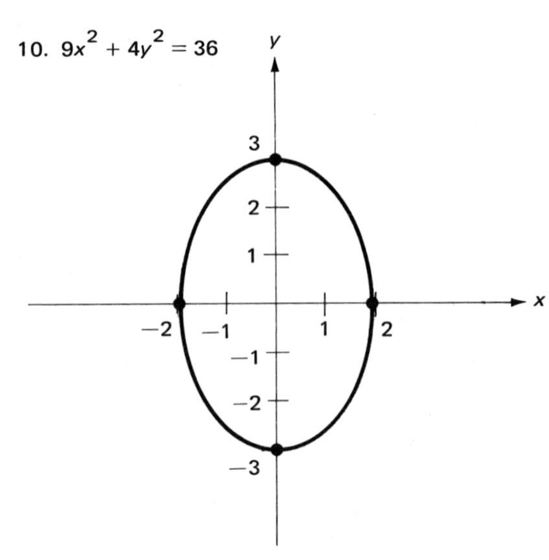

11. $x^2 + y^2 = 16$

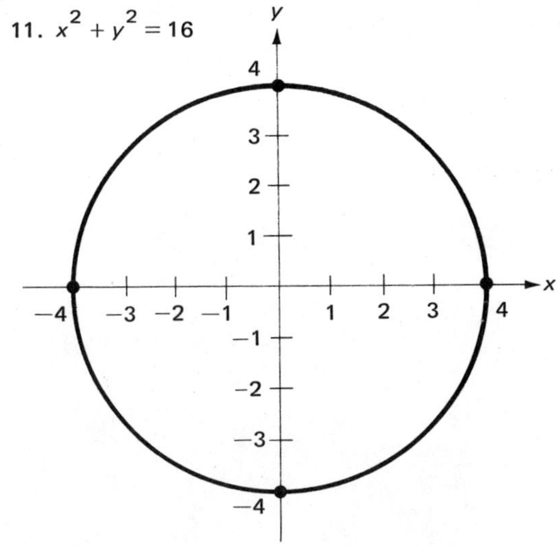

12. $x^2 + y^2 = 3$

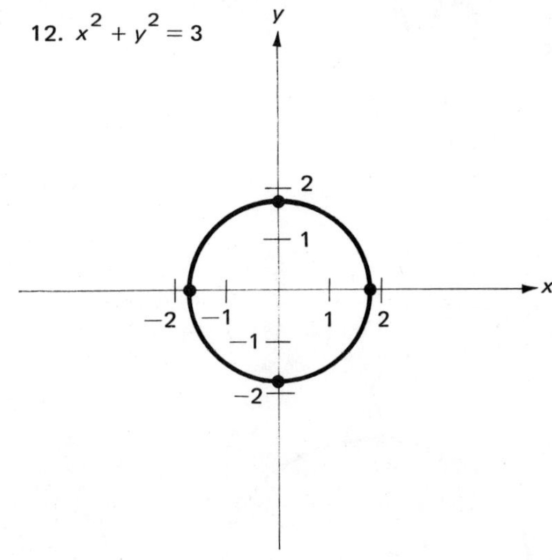

13. $3x^2 - y^2 = 3$

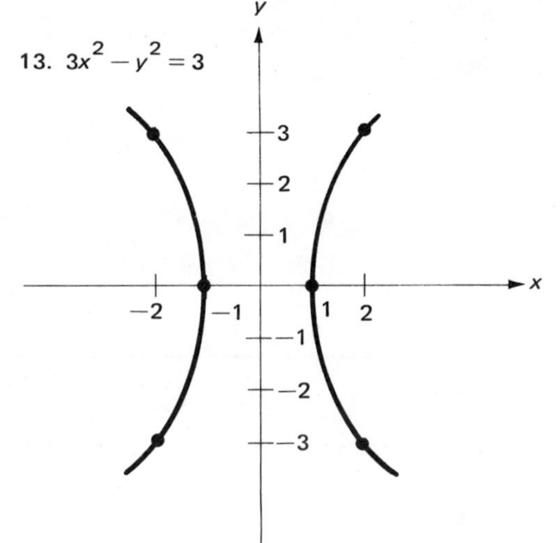

14. $x^2 - y^2 = 1$

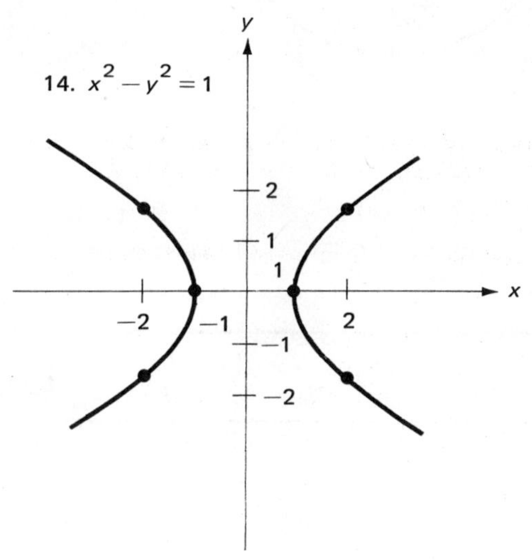

15. $16y^2 - 9x^2 = 144$

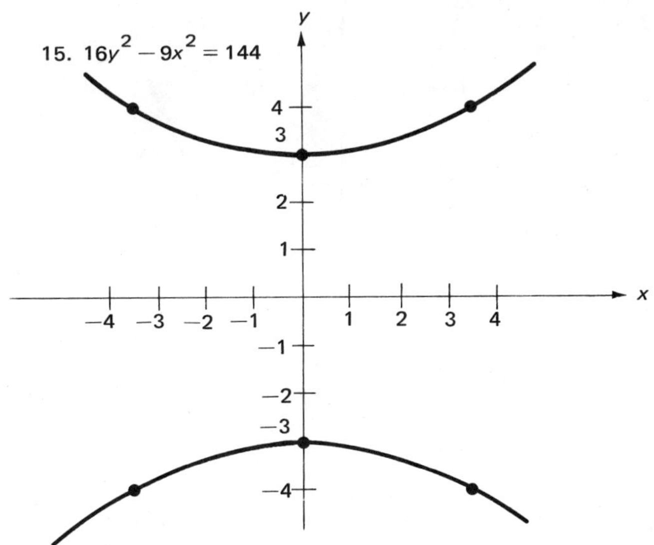

16. $9y^2 - 4x^2 = 36$

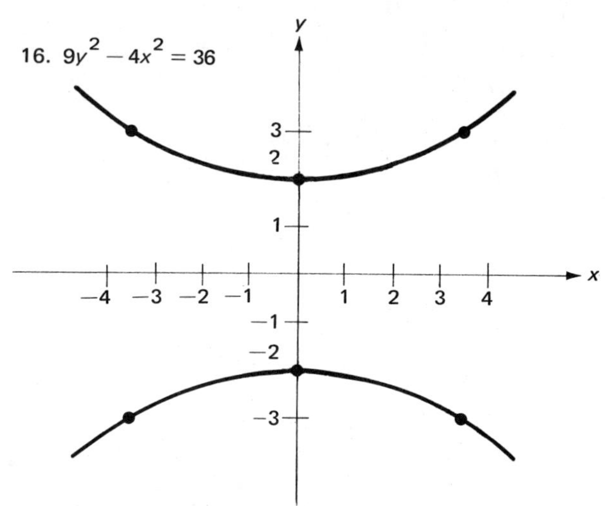

Self-test

1. No solution   (Objective 1)

2. $4 + \sqrt{5}$ (Objective 1)

3. 16 (Objective 1)

4. 2   (Objective 1)

5. $8x^2 + 9y^2 = 72$ (Objective 2)

Unit 19

Exercise 19.1

1. (1, 1)

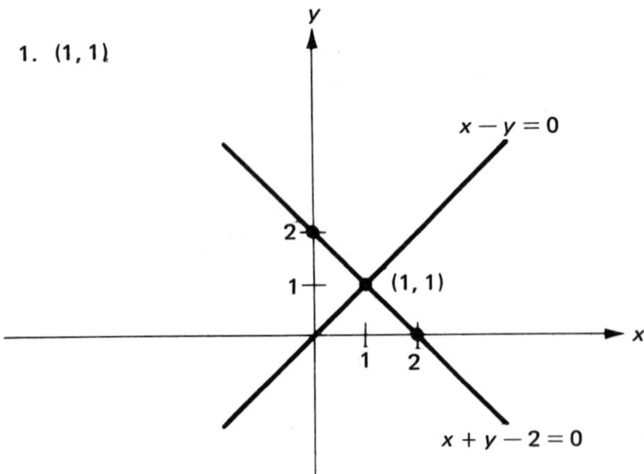

2. (2, −2)

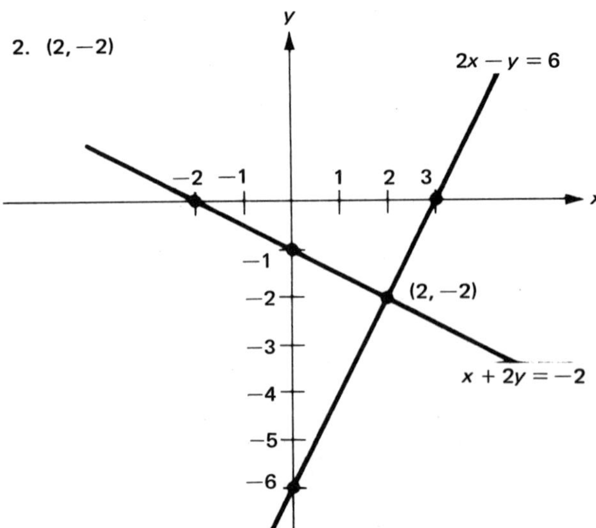

3. $(-3, -2)$

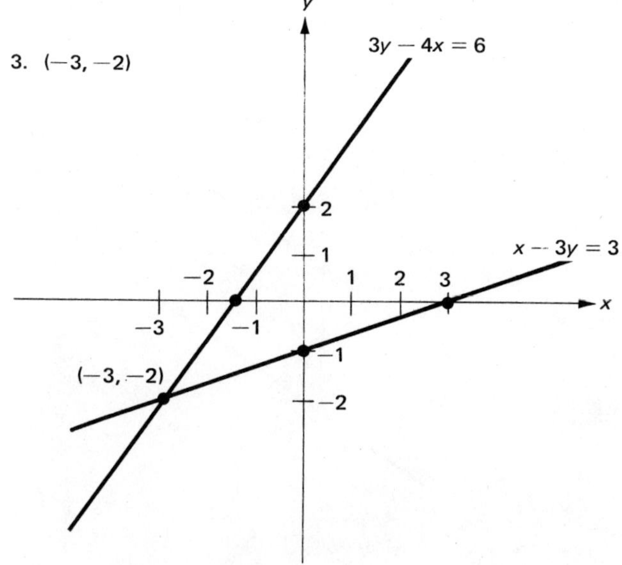

4. $(1, -2)$

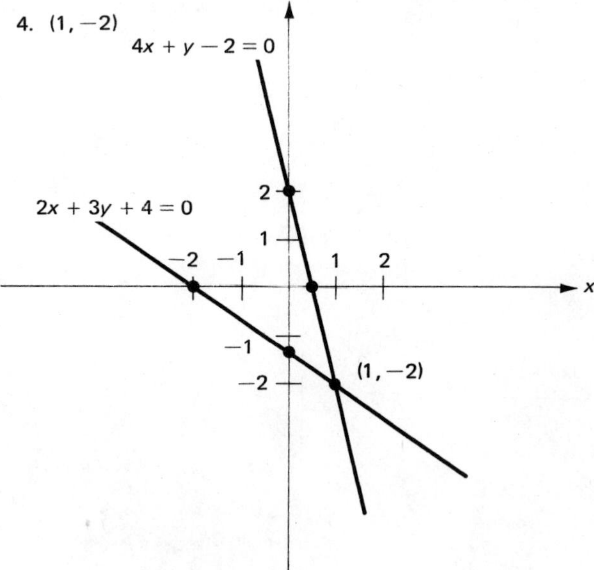

5. $(4, 0)$

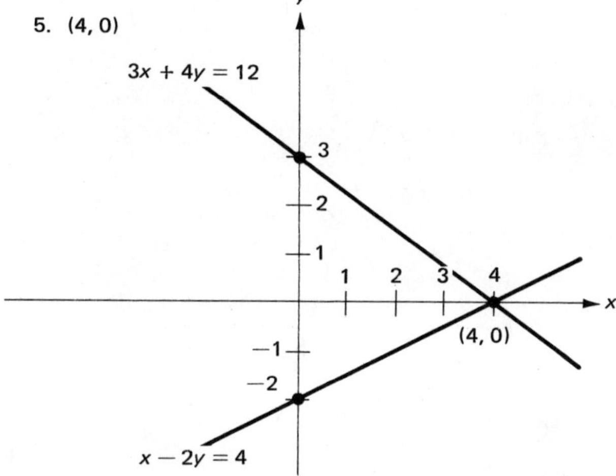

6. $\left(0, -\dfrac{5}{2}\right)$

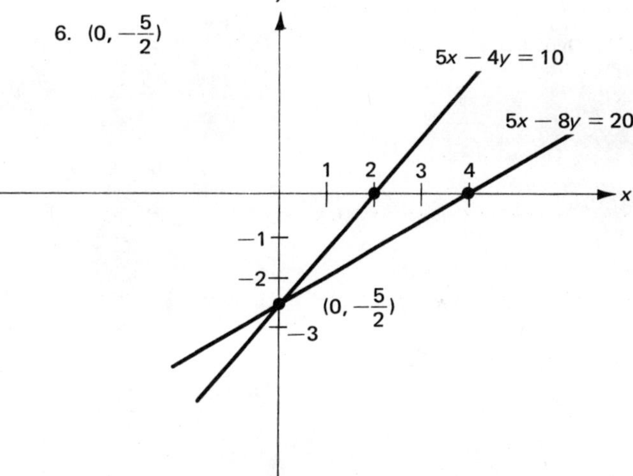

7. $(-3, -1)$

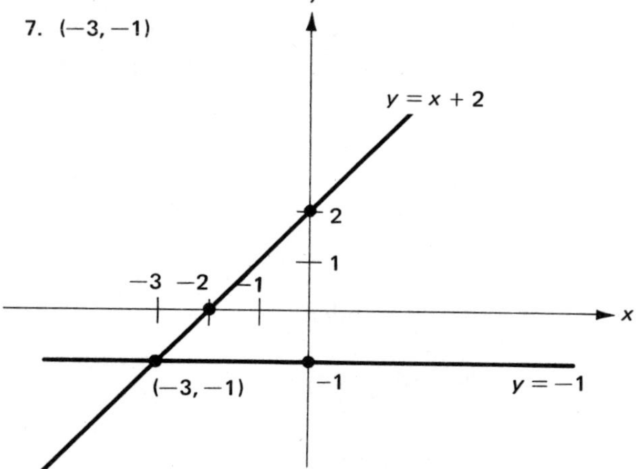

8. $(-1, 2)$

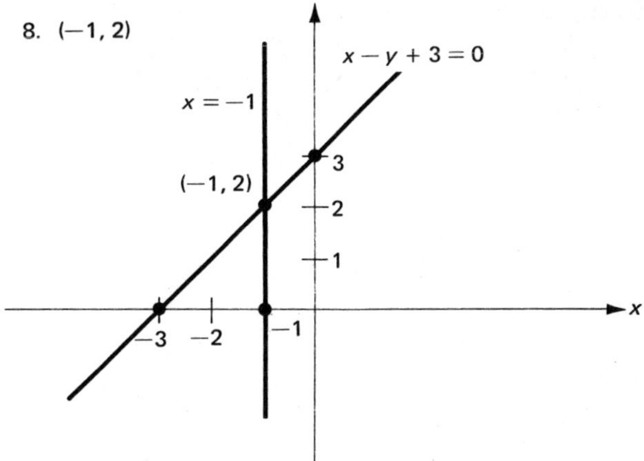

9. Inconsistent

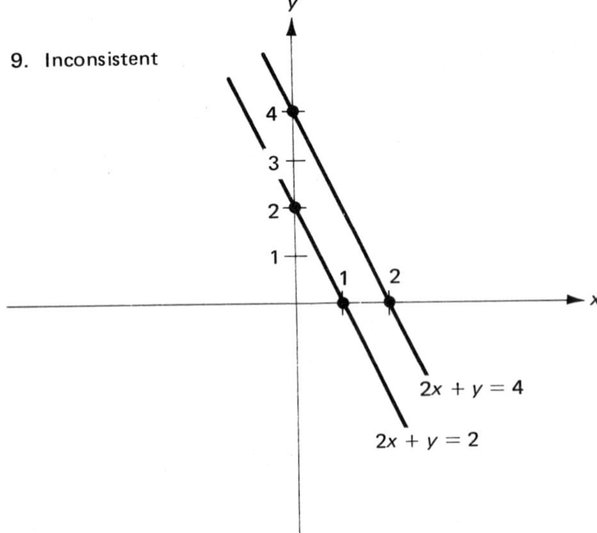

10. Inconsistent

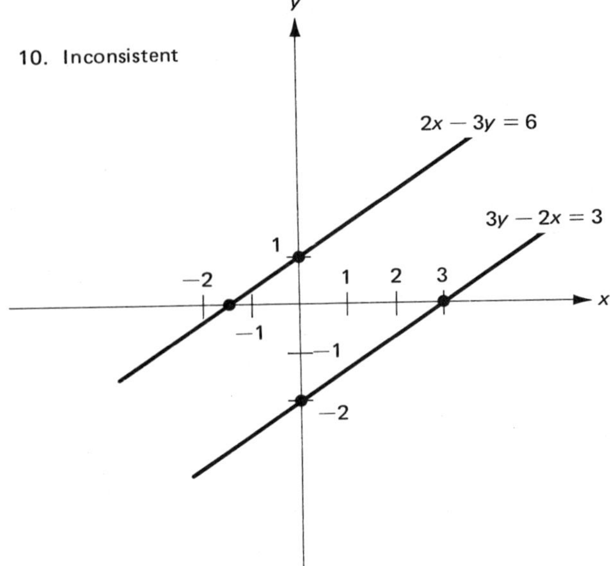

11. Dependent

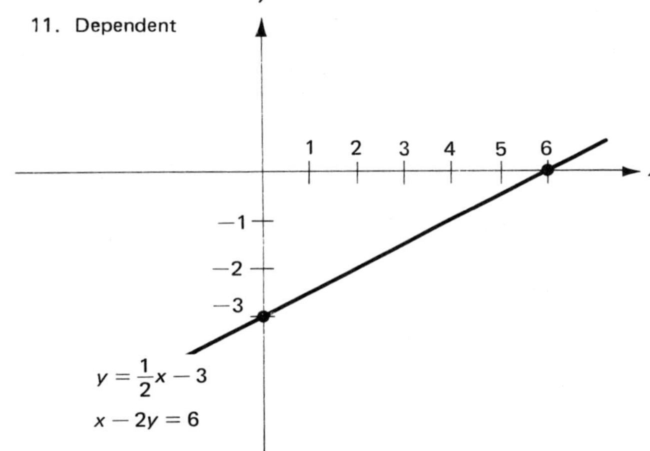

12. Dependent

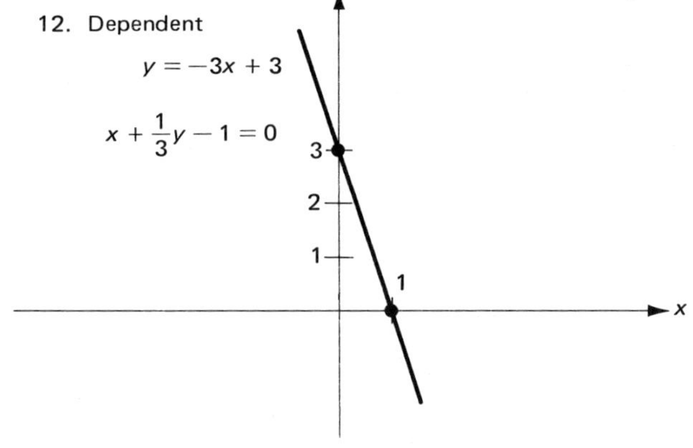

Exercise 19.2

1. $(5, 3)$

2. $(2, 4)$

3. $\left(\frac{9}{2}, \frac{1}{2}\right)$

4. $(3, -1)$

5. $(-2, 3)$

6. $(-1, -2)$

7. $(3, 1)$

8. $(-4, -3)$

9. $(2, 5)$

10. $(0, 0)$

11. $\left(\frac{1}{2}, \frac{3}{2}\right)$

12. $\left(-\frac{1}{3}, 3\right)$

13. Inconsistent

14. Inconsistent

15. Dependent

16. Dependent

17. $(12, 4)$

18. $\left(-2, \frac{5}{2}\right)$

19. $\left(\frac{1}{3}, 2\right)$

20. $(-7, -2)$

21. $(-3, -4)$

22. $\left(-\frac{1}{2}, \frac{7}{2}\right)$

23. Dependent

24. Inconsistent

Exercise 19.3

1. $(2, 2, -2)$
2. $(\frac{2}{3}, \frac{3}{2}, -1)$
3. $(2, 1, -1)$
4. $(-3, -1, 2)$

5. $(2, 1, 4)$
6. $(1, -2, 1)$
7. $(-1, -2, -3)$
8. $(\frac{3}{2}, -\frac{3}{2}, \frac{1}{2})$

9. Dependent
10. Inconsistent

Exercise 19.4

1. 46, 18
2. $16\frac{1}{2}, 8\frac{1}{2}$

3. $10\frac{1}{2}$ ounces corn, $5\frac{1}{2}$ ounces lima beans

4. 35 pounds of bottles, 15 pounds of cans

5. 24 ounces oil, 4 ounces wine vinegar, 4 ounces cider vinegar

6. 5 quarts potting soil, $2\frac{1}{2}$ quarts peat moss, $2\frac{1}{2}$ quarts sand

7. 3 miles per hour, 1 mile per hour

8. 6 miles per hour, 4 miles per hour

9. 440 miles per hour, 40 miles per hour

10. 16 kilometers per hour, 4 kilometers per hour

11. $y = 3x + 2$
12. $y = \frac{1}{2}x - \frac{3}{4}$

13. $y = 2x^2 - x - 1$
14. $y = -x^2 + 4x - 4$

15. $x = y^2 - 4$
16. $x = \frac{1}{4}y^2$

Exercise 19.5

1. $(2, 1), (2, -1), (-2, 1), (-2, -1)$

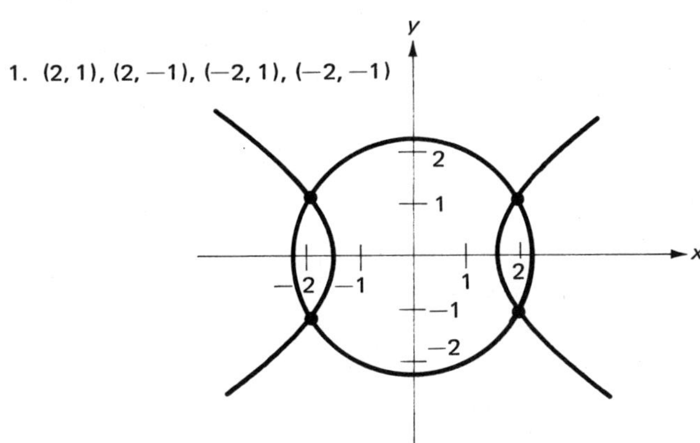

2. $(1, 0), (-1, 0)$

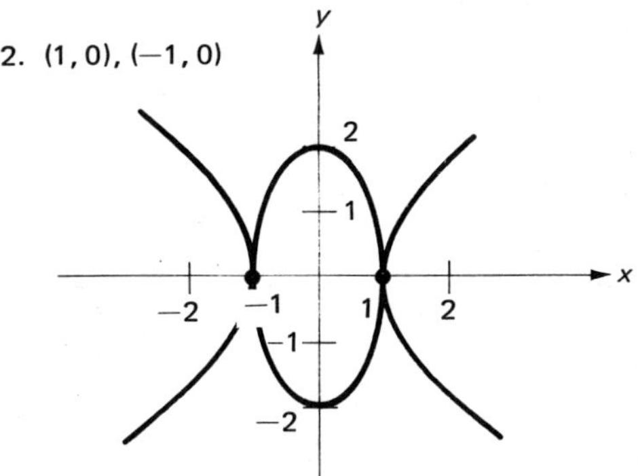

3. $(1, 1)$, $(1, -1)$, $(-1, 1)$, $(-1, -1)$

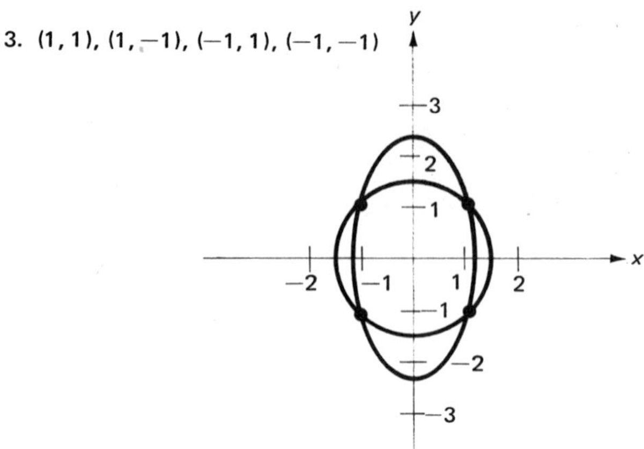

4. $(0, 2)$, $(0, -2)$

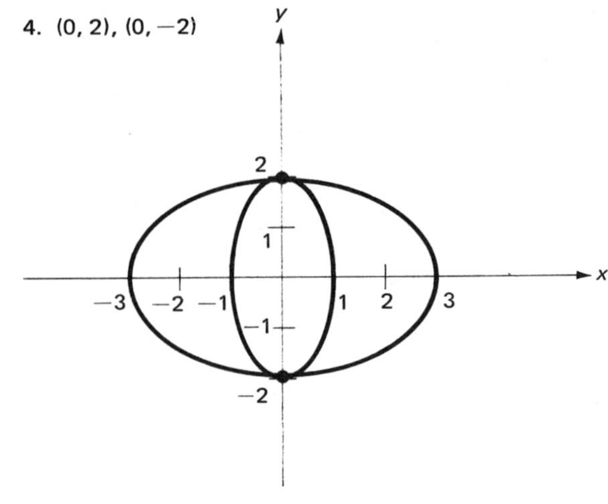

6. $(\sqrt{3}, \sqrt{3})$, $(\sqrt{3}, -\sqrt{3})$, $(-\sqrt{3}, \sqrt{3})$, $(-\sqrt{3}, -\sqrt{3})$

5. $(2, \sqrt{2})$, $(2, -\sqrt{2})$, $(-2, \sqrt{2})$, $(-2, -\sqrt{2})$

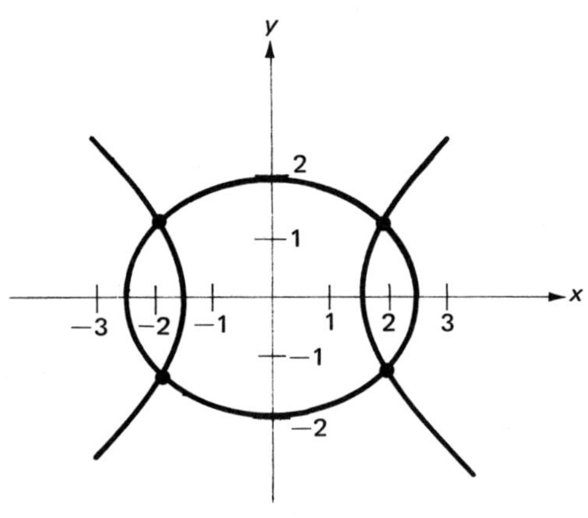

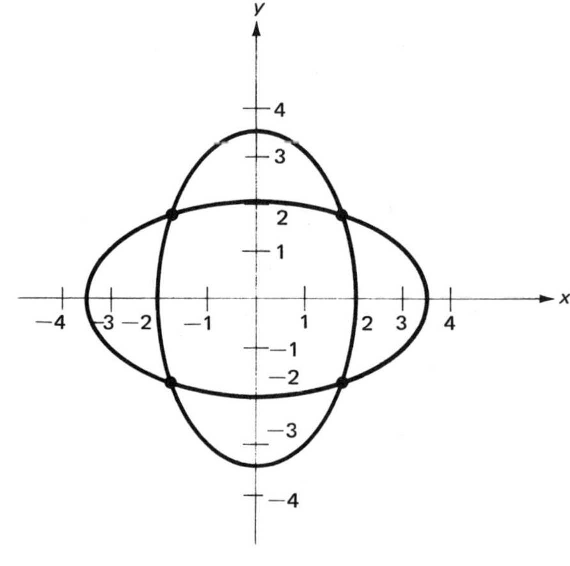

7. No real number solution

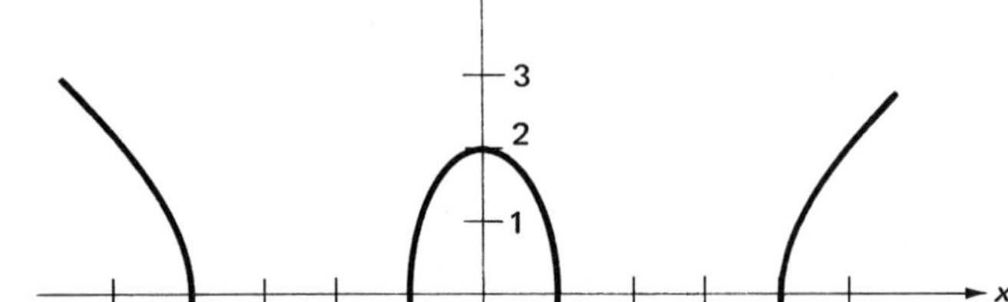

8. No real number solution

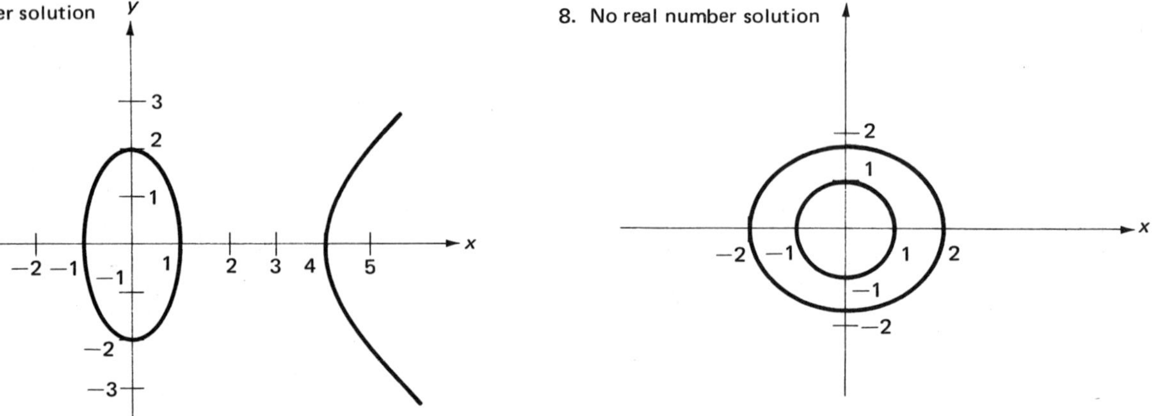

9. $(1, 1), (-1, 1)$

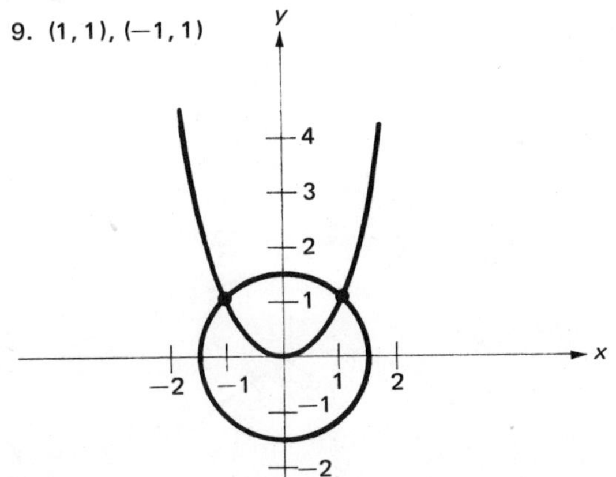

10. $(\sqrt{2}, 1), (-\sqrt{2}, 1)$

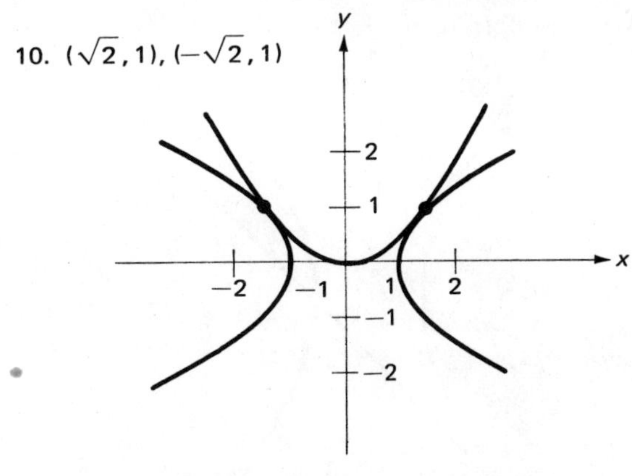

11. $(-2, 2), (4, 8)$

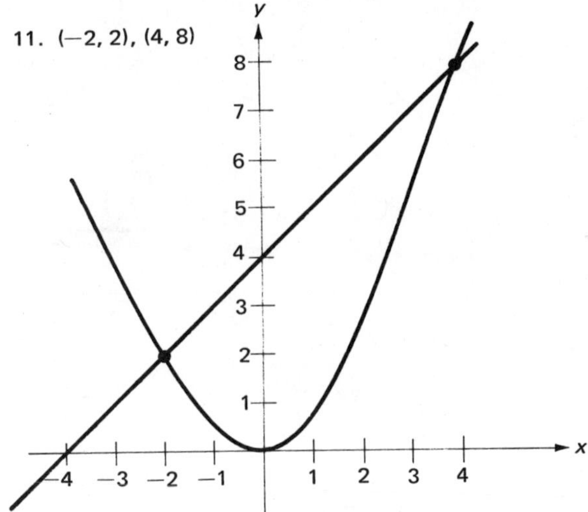

12. $(1, -2)$

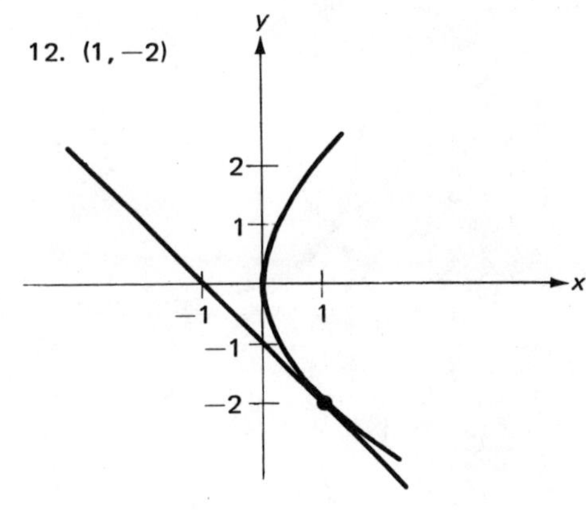

13. $(2, 3), (-3, -2)$

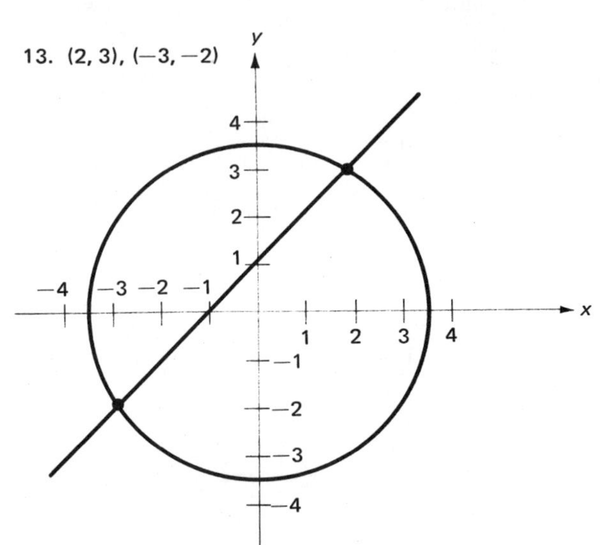

14. No real number solution

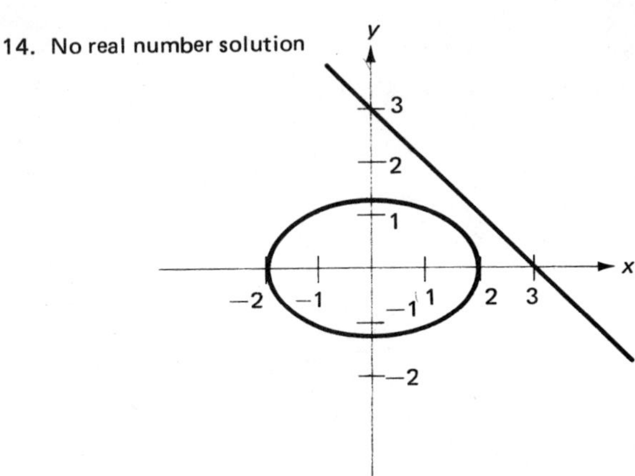

15. $(-3, -4)$, $(1, 0)$

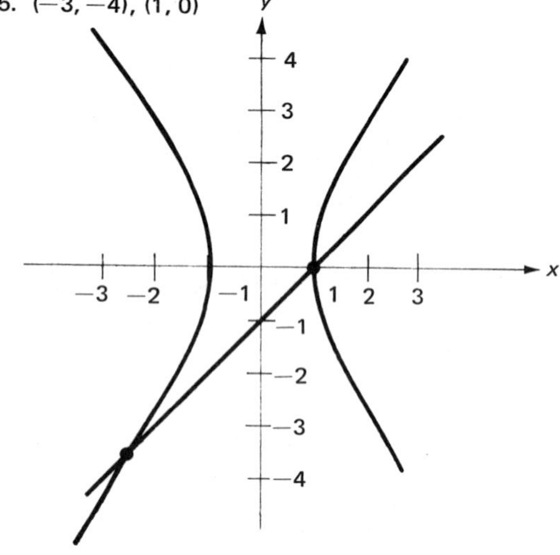

16. No real number solution

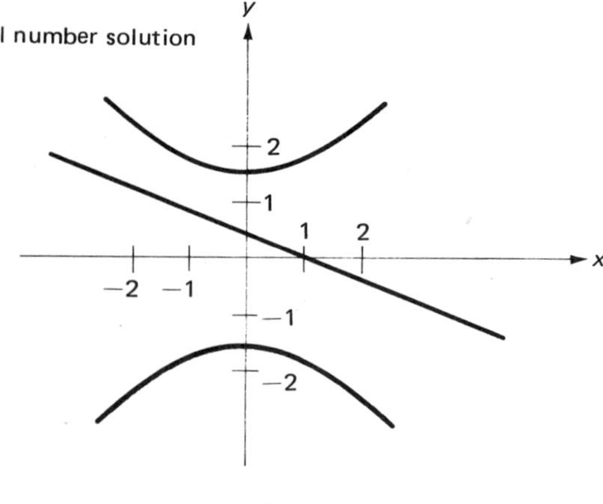

17. $(2, 2)$

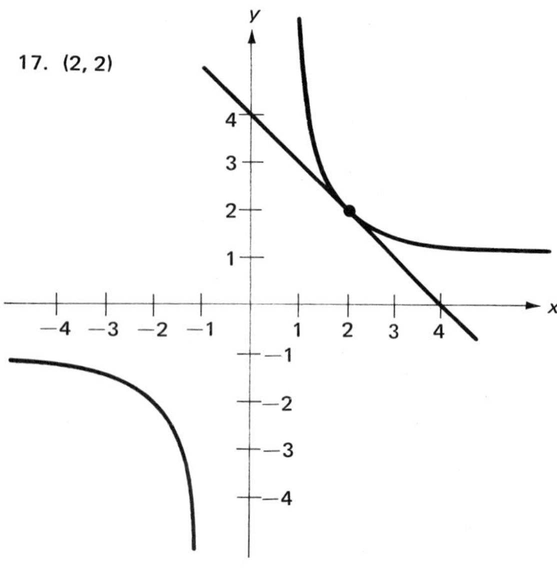

18. $(1, 6)$, $(-6, -1)$

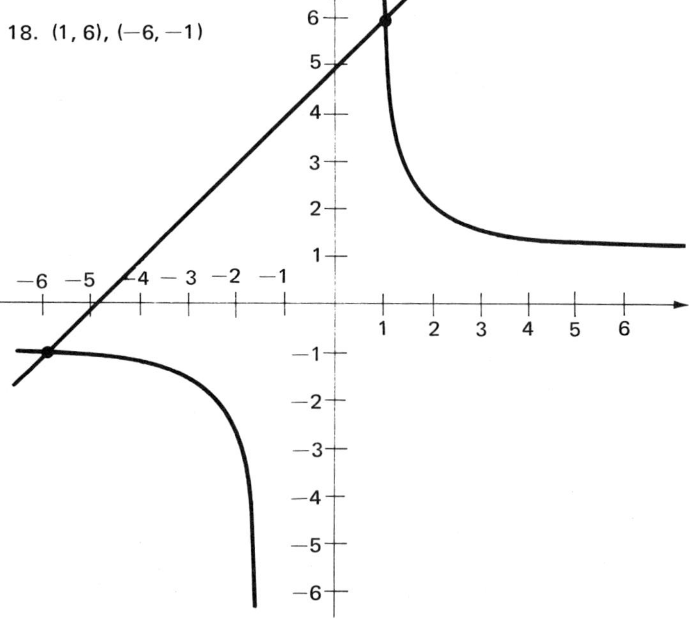

19. $(1, 2)$, $(-1, -2)$, $(2, 1)$, $(-2, -1)$

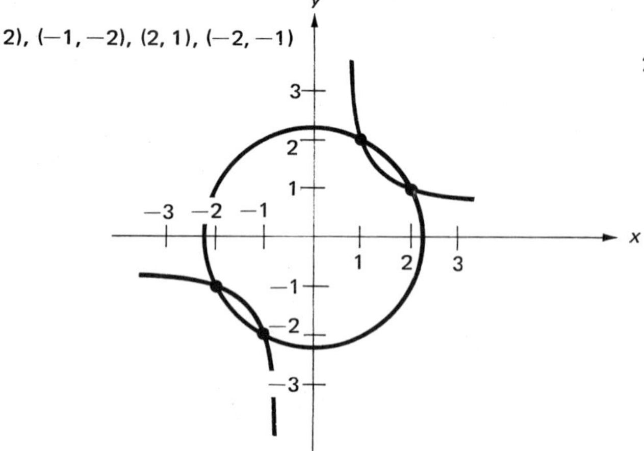

20. $(1, 3)$, $(-1, -3)$

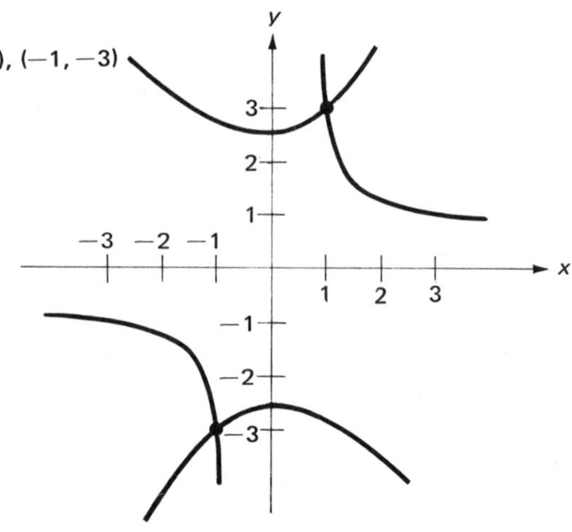

Self-test

1. (6, 3)   (Objective 1)

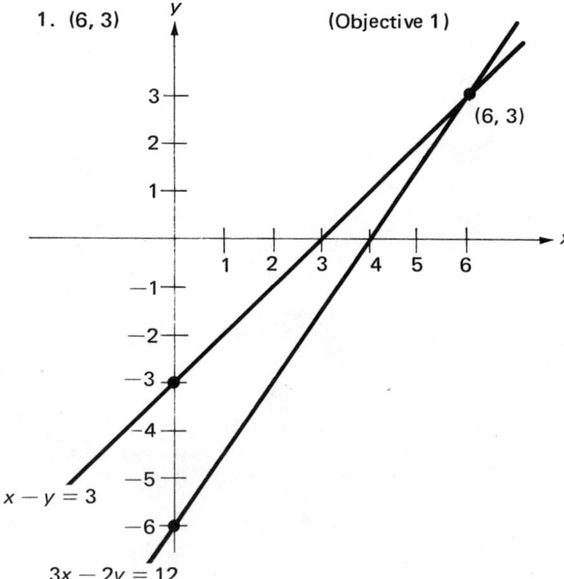

2. $\left(-\frac{1}{2}, -4\right)$ (Objective 2)

3. $\left(2, 2, -\frac{1}{2}\right)$ (Objective 3)

4. $(2, 1), (2, -1), (-2, 1), (-2, -1)$ (Objective 5)

5. 270 miles per hour, 30 miles per hour   (Objective 4)

Unit 20

Exercise 20.1

1. $x > 1$

2. $x < \frac{5}{2}$

3. $x < 3$

4. $x > -3$

5. $x \leqslant -\frac{1}{3}$

6. $x \geqslant \frac{2}{3}$

7. $x < 1$

8. $x < -1$

9. $x < \frac{1}{2}$

10. $x > -2$

11. $x > \frac{1}{2}$

12. $x < -4$

13. $x \geqslant 1$

14. $x \geqslant 0$

15. $x < 1$ or $x > 5$

16. $x \leqslant -1$ or $x \geqslant 2$

17. $-5 < x < 2$

18. $-2 \leqslant x \leqslant -\frac{1}{2}$

19. $1 - \sqrt{5} < x < 1 + \sqrt{5}$

20. $x \geqslant 2 + \sqrt{2}$ or $x \leqslant 2 - \sqrt{2}$

21. All real numbers

22. No solution

23. $x < 1$ or $x > 2$

24. $-1 < x < 1$

25. $\frac{1}{2} < x \leqslant 3$

26. $x \geqslant 0$ or $x < -\frac{3}{2}$

Exercise 20.2

1.

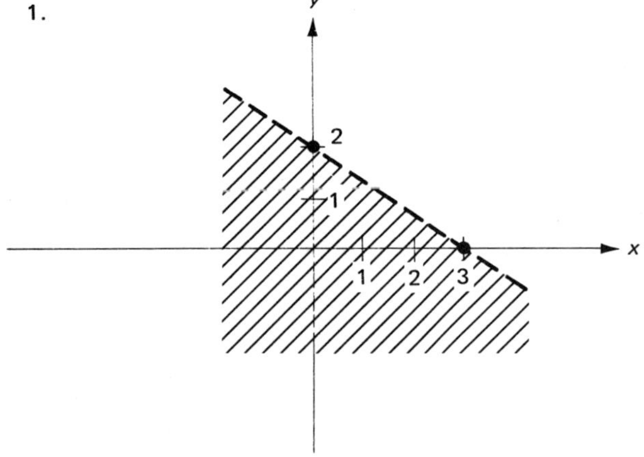

2.

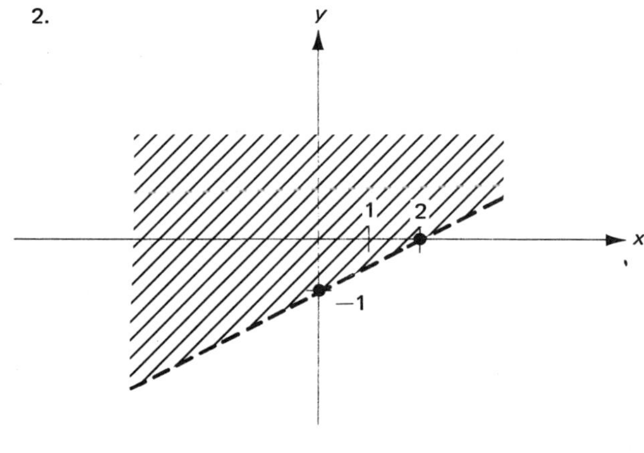

3.

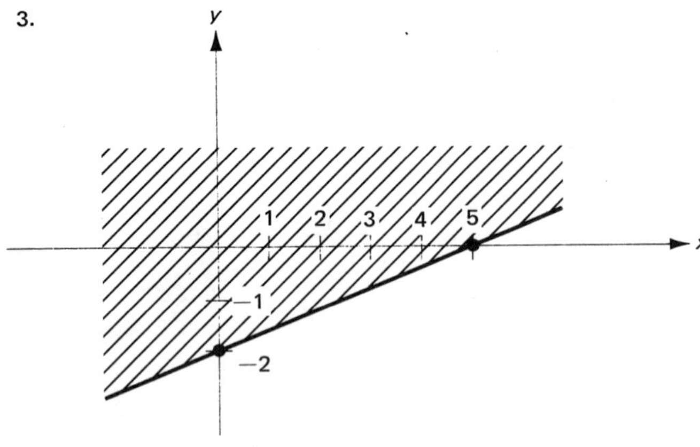

4.

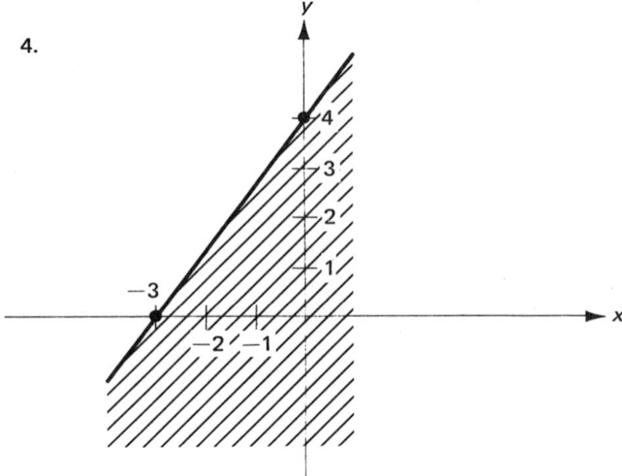

5.

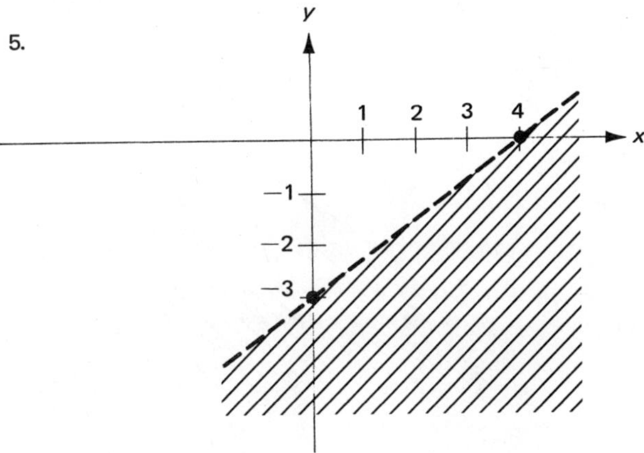

6.

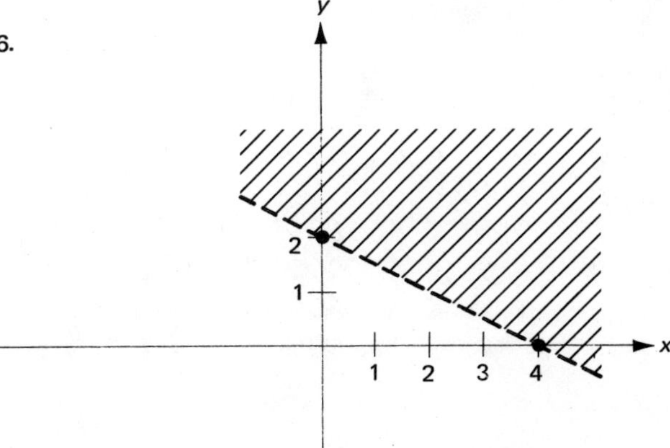

7.

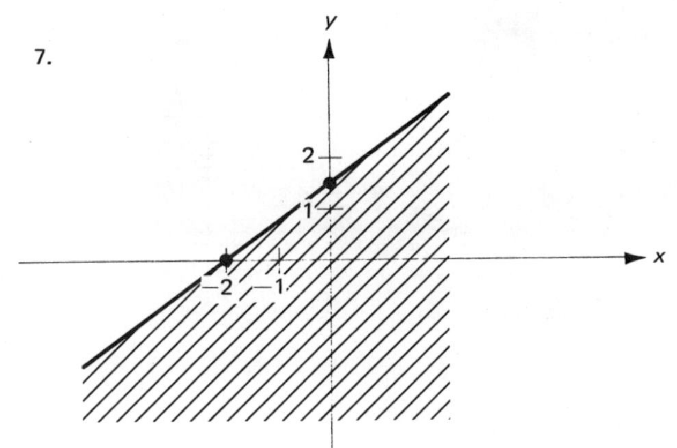

8.

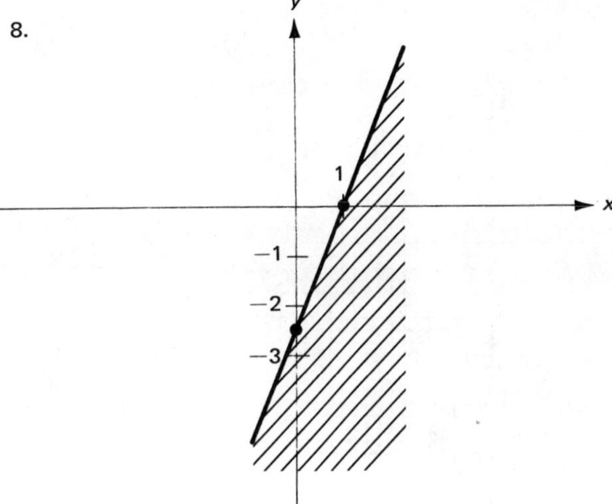

9.

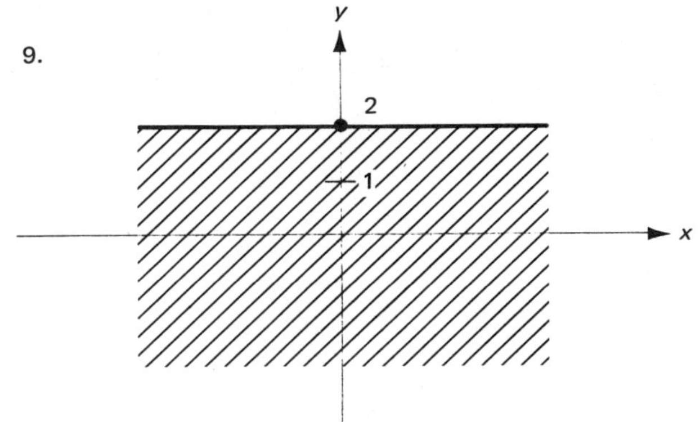

10.

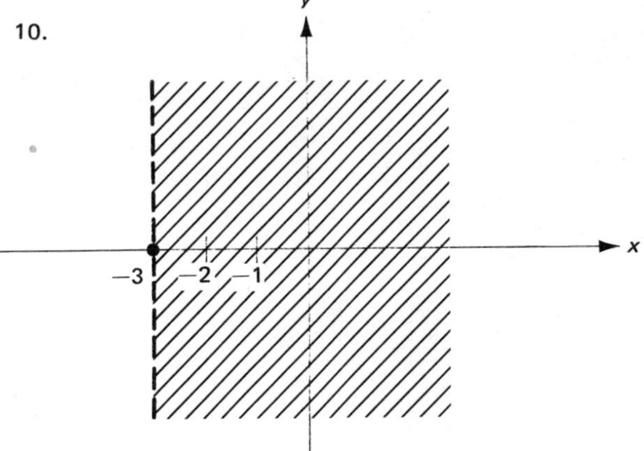

11.

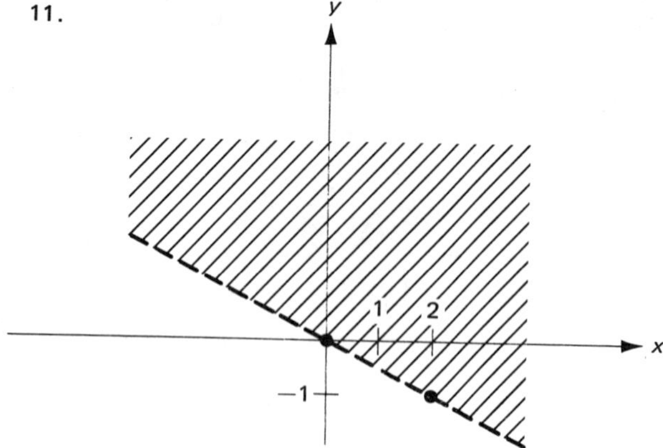

12.

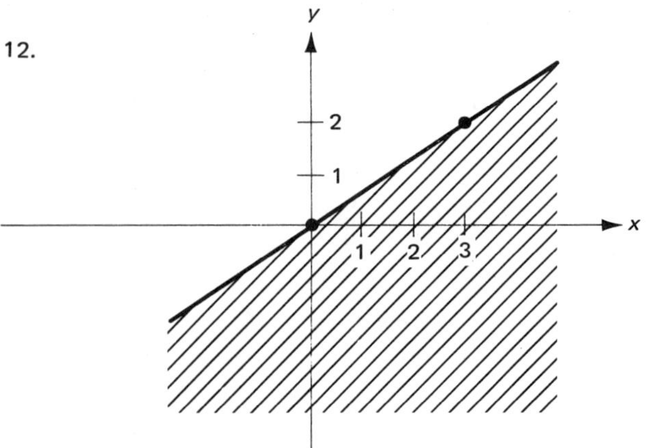

Exercise 20.3

1.

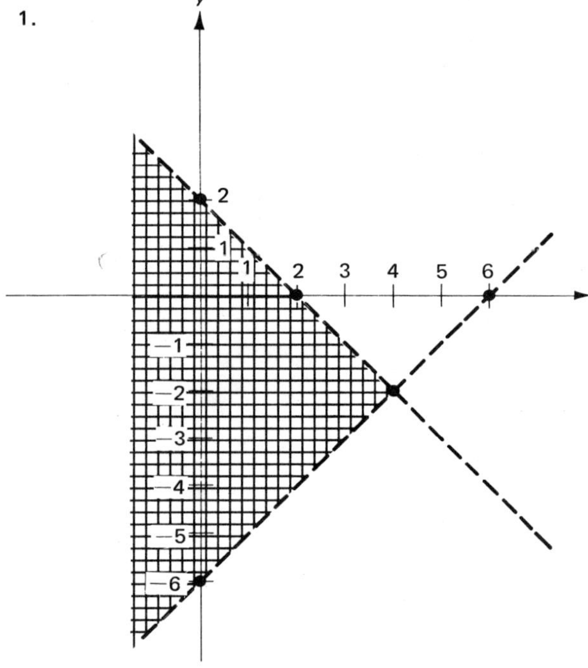

2.

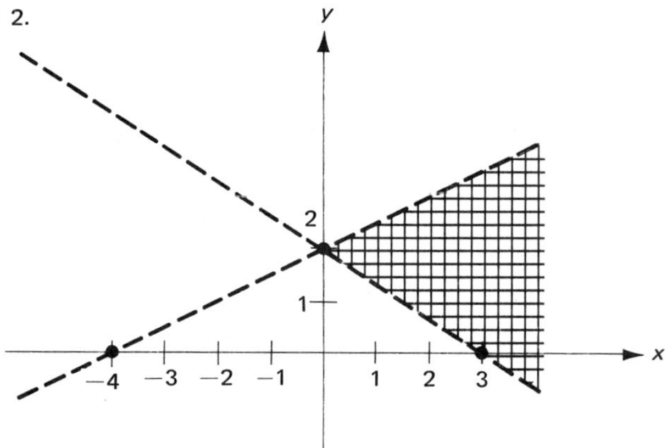

3.

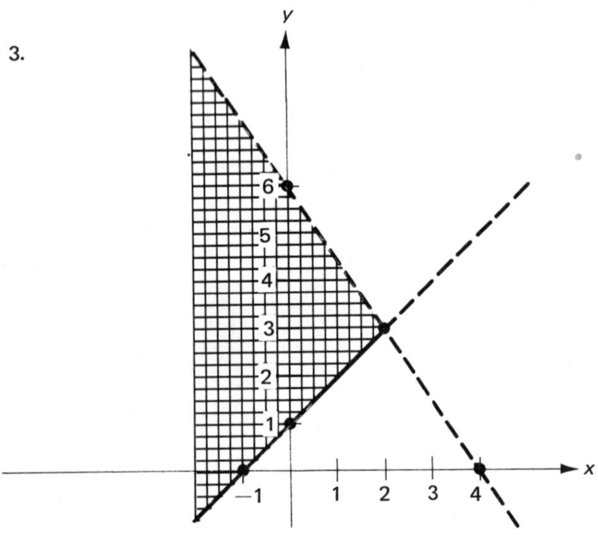

4.

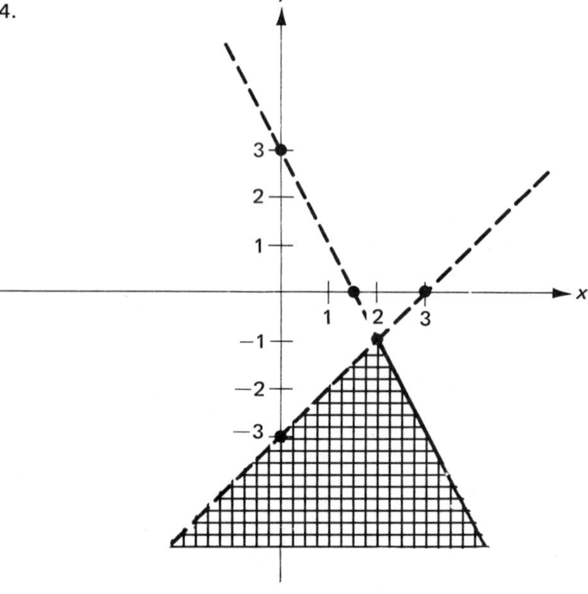

5.

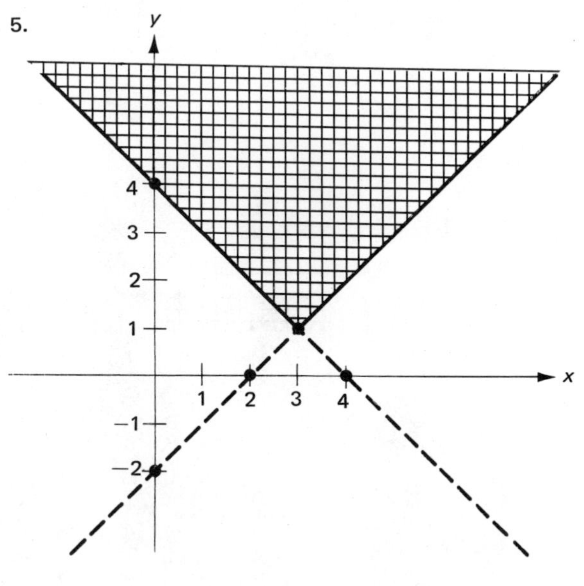

6.

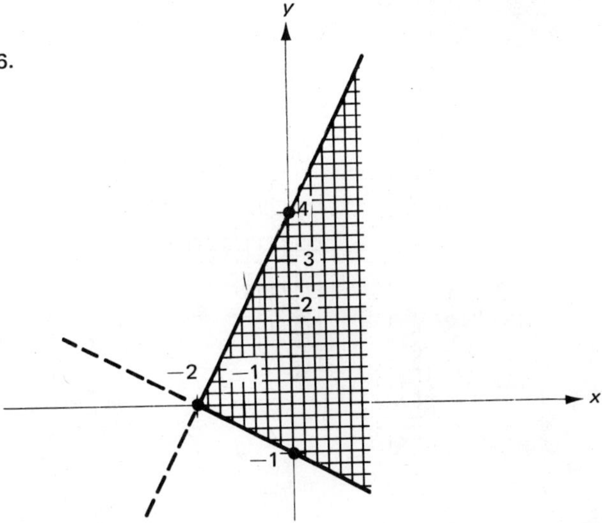

7.

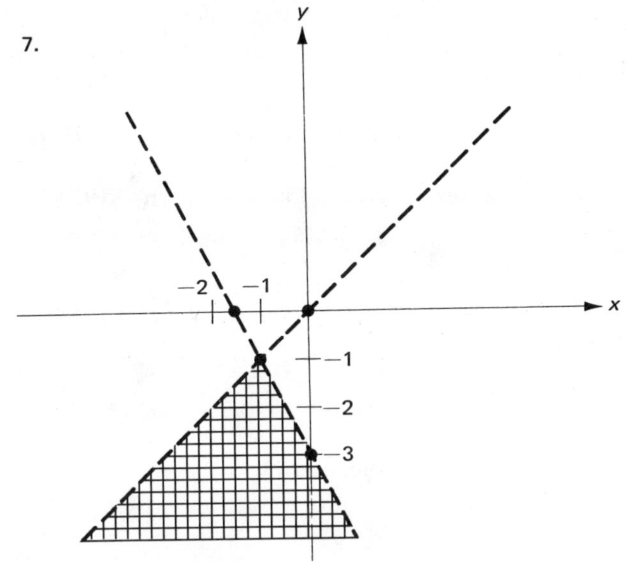

8.

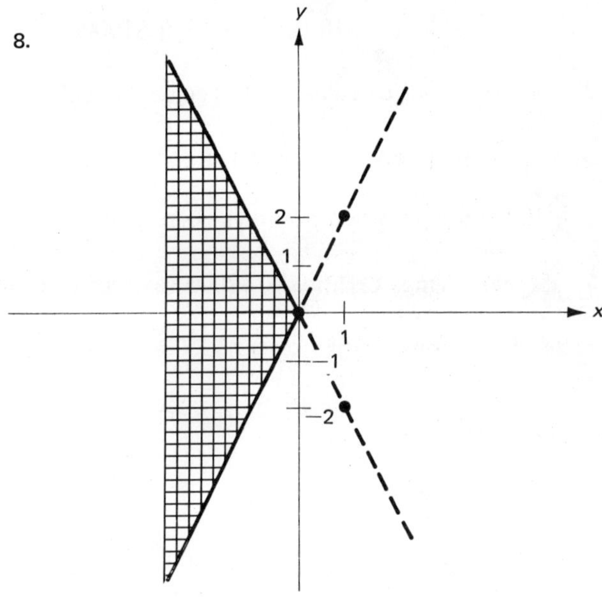

9.

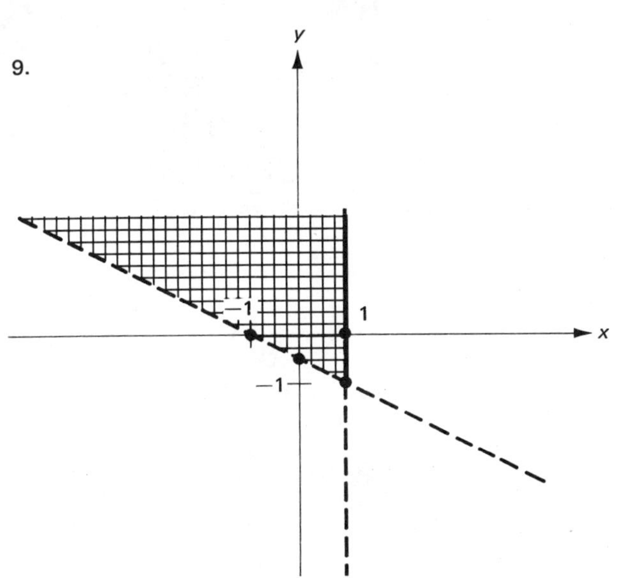

10.

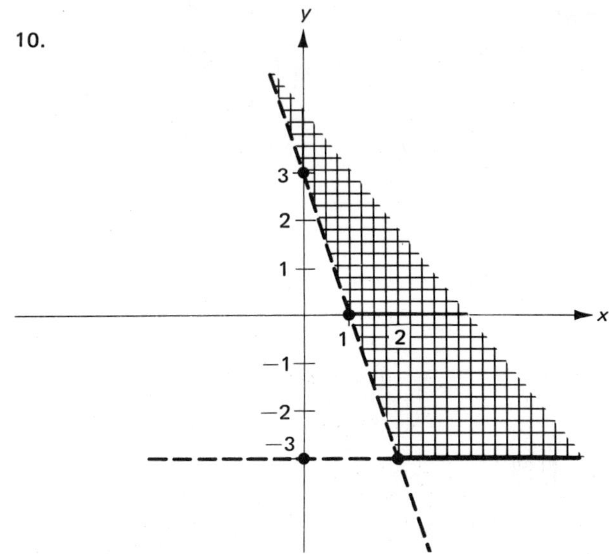

11.

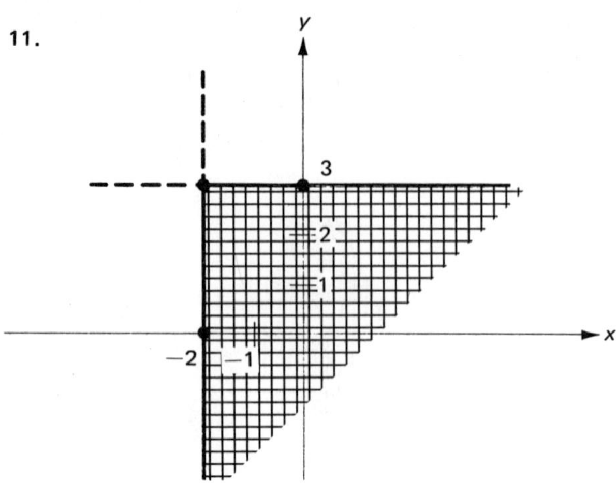

12.

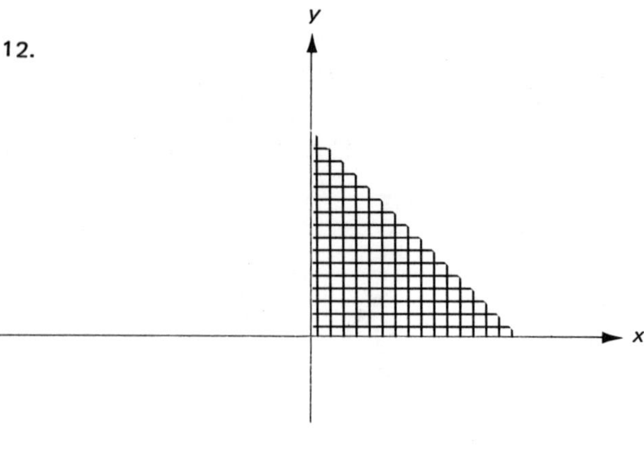

Exercise 20.4

1. 20 portable TVs, 10 console TVs; $1600

2. 500 personal calculators, 150 scientific calculators; $550

3. 20 plain spoons, 20 decorated spoons; $220

4. 60 plain spoons, no decorated spoons; $300

5. 300 acres potatoes, no acres corn; $24,000

6. 150 acres potatoes, 150 acres corn; $19,500

7. $8000 savings certificate, $2000 savings account; $760

8. $15,000 conservative stock, $5000 speculative stock; $1550

Self-test

1. $x > 2$ (Objective 1)

2. $-3 < x < 5$ (Objective 1)

3.

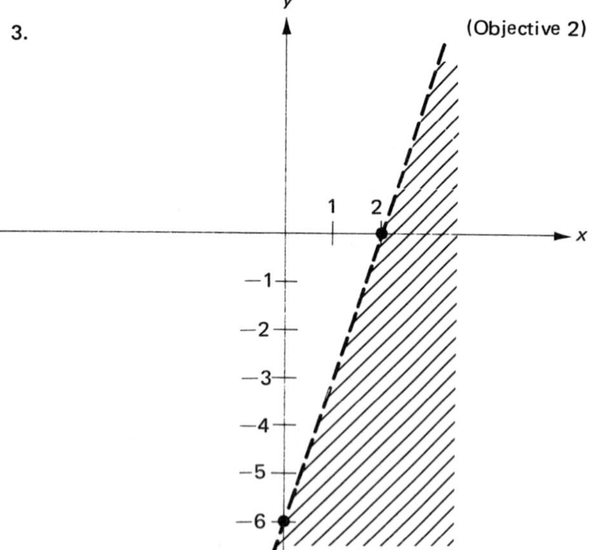

(Objective 2)

4.

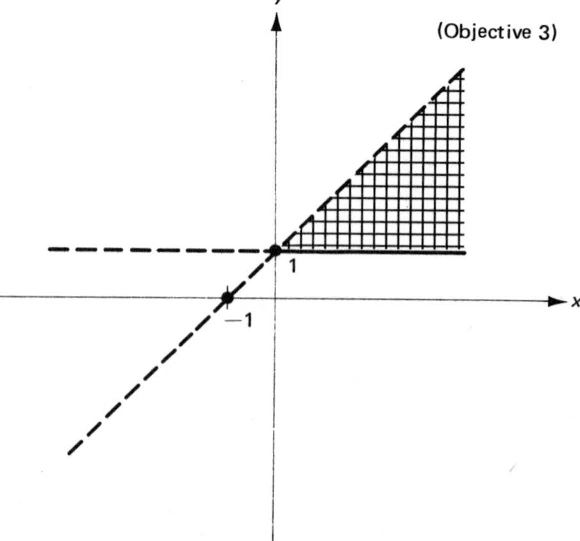

(Objective 3)

5. 50 unlined drapes, no lined drapes; $250   (Objective 4)

Unit 21

Exercise 21.1

1. 10
2. 0
3. 10
4. $\frac{2}{5}$
5. 15

6. 24
7. $\frac{4}{3}$
8. $\frac{2}{3}$
9. 5
10. 7

11. 25
12. $\frac{3}{4}$
13. 12
14. 0
15. 7

16. 3
17. $4|x+2|$
18. $6|2x-1|$
19. $\dfrac{|x-3|}{6}$
20. $\dfrac{|x+2|}{2}$

21. $3|x+5|$
22. $|x+1|$

Exercise 21.2

1. $4, -4$
2. No solution
3. $3, -3$
4. $6, -6$

5. $1, -1$
6. No solution
7. $8, 2$
8. $-2, -4$

9. $-\frac{1}{3}, -1$
10. $6, -1$
11. 2
12. $-\frac{1}{4}$

13. No solution
14. No solution
15. $6, -2$
16. $\frac{7}{2}, -\frac{9}{2}$

17. $5, 1$
18. $0, -2$
19. $1, -9$
20. $3, -4$

21. $4, -2$
22. $3, \frac{1}{2}$
23. 10
24. $\frac{5}{2}$

25. No solution
26. No solution

Exercise 21.3

1.

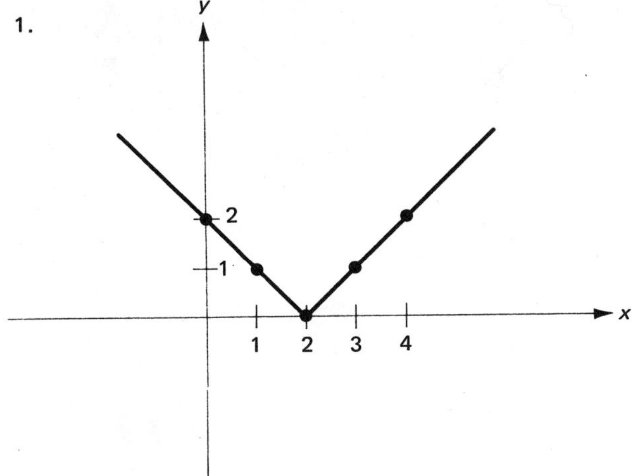

2.

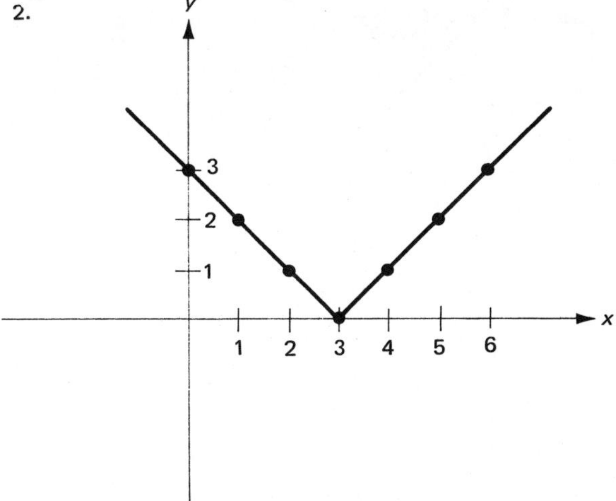

3.

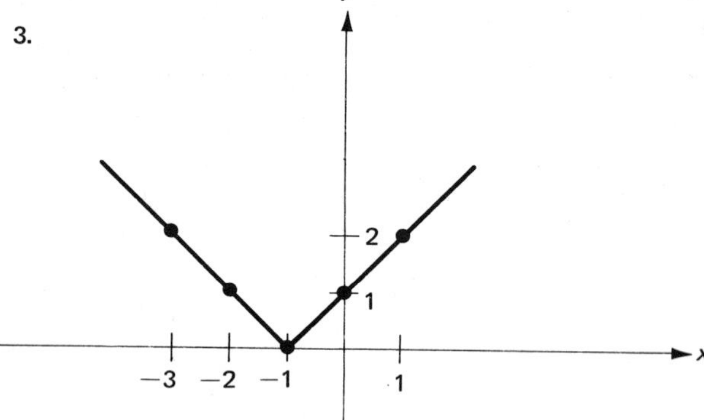

4.

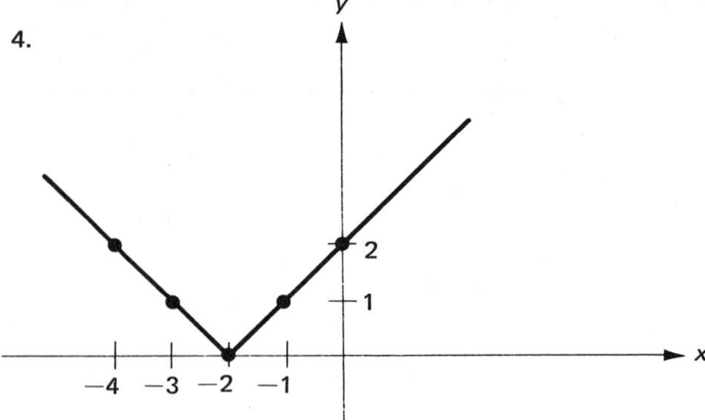

5.

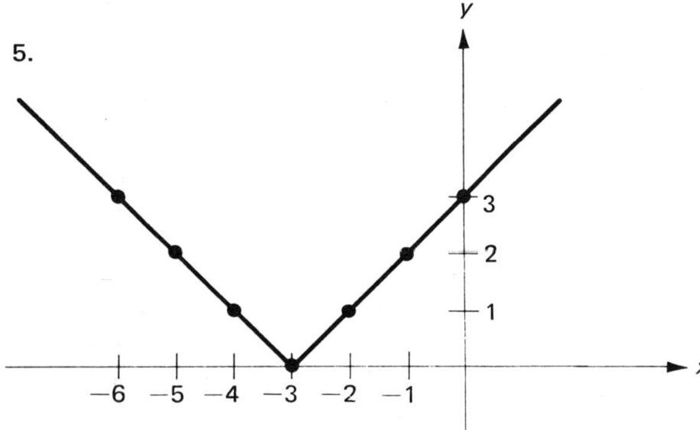

6.

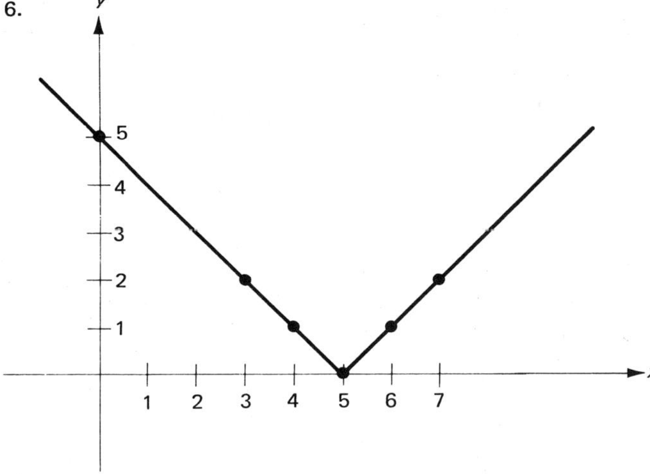

7.

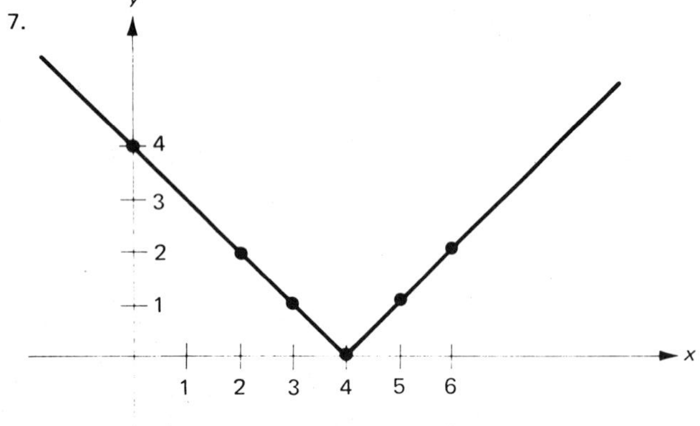

8.

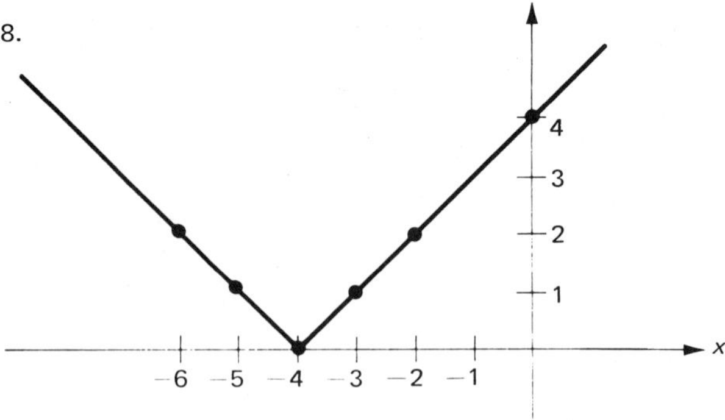

Self-test

1a. 5   1b. 13   (Objective 1)   2.   No solution   (Objective 2)

3. $2, -\frac{1}{2}$ (Objective 2)   4.   6   (Objective 2)

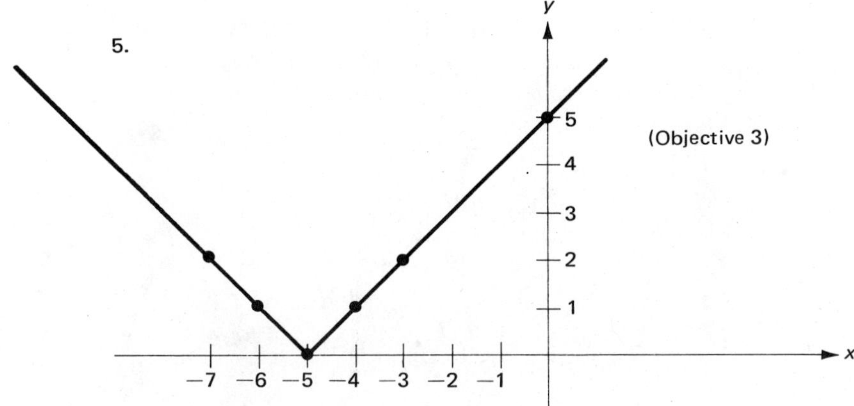

(Objective 3)

Unit 22

Self-test

1a. $\frac{1}{16}$ (Unit 13)   1b. $2i$ (Unit 15)   2. $\dfrac{x+3}{x^2-1}$ (Unit 10)

3. $-3+2\sqrt{6}$ (Unit 14)   4. $-2\pm\sqrt{7}$ (Unit 4)

5.   10   (Unit 18)   6.   8   (Unit 21)

7. $x>-6$ (Unit 20)   8.   10 miles per hour, 8 miles per hour (Unit 17)

9. $(\frac{20}{3}, -\frac{4}{3})$ (Unit 19)

10a. $y=\dfrac{-4}{x}$

10b. $y=(\frac{1}{4})^x$

10c. None of these

10d. $y=\sqrt{x+4}$

10e. $y=x^2-4$

10f. $y=|x-4|$
     (Units 7, 9, 12, 14, 21)

Unit 23

Exercise 23.1

1. $\log_4 64 = 3$

2. $\log_{10} .01 = -2$

3. $\log_{64} 8 = \frac{1}{2}$

4. $\log_{64} 4 = \frac{1}{3}$

5. $\log_4 \frac{1}{64} = -3$

6. $\log_4 2 = \frac{1}{2}$

7. $\log_{10} 10,000 = 4$

8. $\log_{10} .0001 = -4$

9. $2^6 = 64$

10. $27^{\frac{1}{3}} = 3$

11. $10^2 = 100$

12. $10^0 = 1$

13. $36^{\frac{1}{2}} = 6$

14. $2^{-6} = \frac{1}{64}$

15. $10^{-1} = .1$

16. $10^{-4} = .0001$

Exercise 23.2

1. 3

2. 3

3. 2

4. -2

5. $\frac{1}{2}$

6. $\frac{1}{4}$

7. 3

8. $\frac{1}{3}$

9. -3

10. -4

11. -3

12. $\frac{1}{3}$

13. 3

14. 4

15. 5

16. -3

17. -4

18. -5

19. 1

20. 0

Exercise 23.3

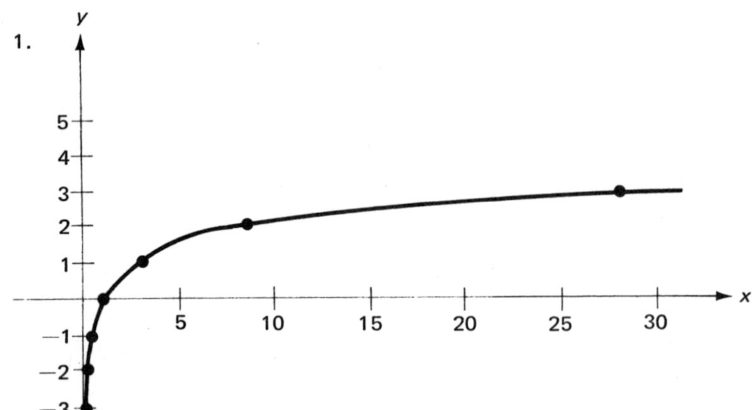

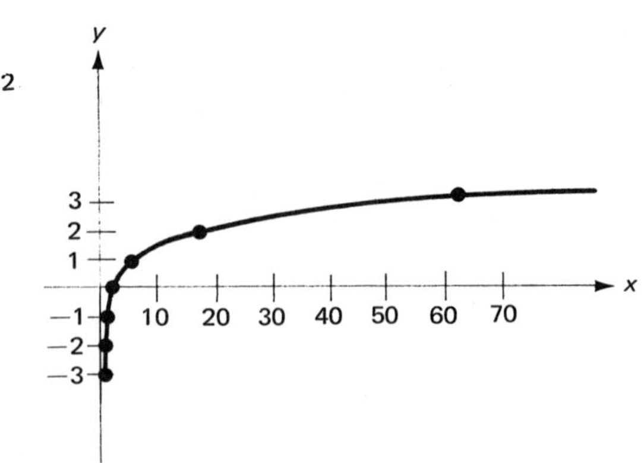

3.

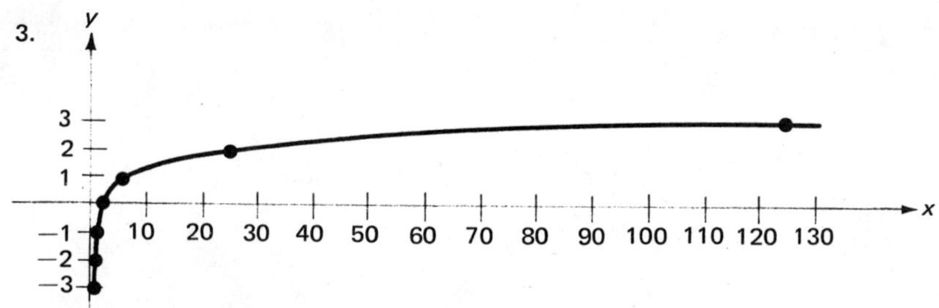

4.

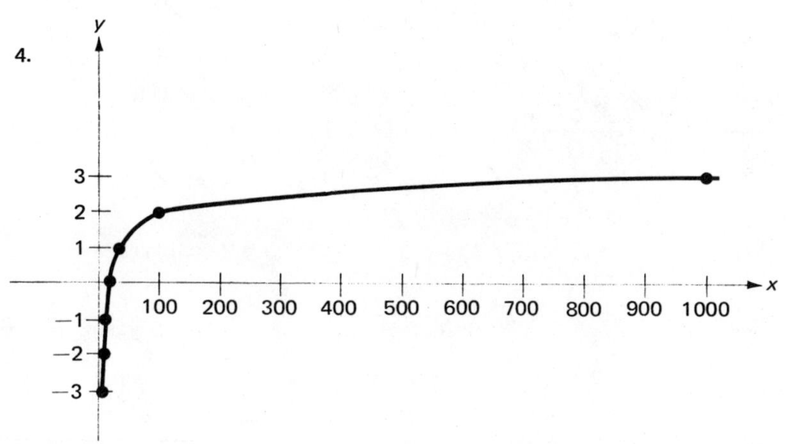

Exercise 23.4

1. $\log_b x + \log_b y + \log_b z$ 2. $2\log_b x + 2\log_b y$

3. $\log_b x + \log_b y - \log_b z$ 4. $\log_b x - \log_b y - \log_b z$

5. $2\log_b k + \log_b m + 3\log_b n$ 6. $2\log_b k - \log_b m - 3\log_b n$

7. $\frac{1}{2}\log_b s - 3\log_b t$ 8. $2\log_b u - \frac{1}{2}\log_b v - \frac{1}{2}\log_b w$

9. $\frac{1}{3}\log_b p + \frac{2}{3}\log_b q - \frac{1}{3}\log_b r$ 10. $\frac{1}{2}\log_b p - \frac{1}{4}\log_b q - \frac{3}{4}\log_b r$

Exercise 23.5

1. .3404   2. .9191   3. 1.9274   4. $.5092 - 1$

5. $.7404 - 3$ 6. 4.6693   7. 3.9685   8. 6.6263

9. $.8096 - 3$ 10. $.6010 - 5$ 11. 3.44   12. 1.60

13. .678   14. 545   15. 23,600   16. 590,000

17. .00519   18. .00089   19. .0000827   20. 9,620,000,000

Self-test

1. $2\log_b x + \frac{1}{2}\log_b y - \log_b z$ (Objective 4)

2a. -4 2b. $\frac{1}{4}$ (Objective 2)

3. $\log_3 9 = 2$ (Objective 1)

4a. $.8021 - 1$ 4b. 505 (Objective 5)

5.

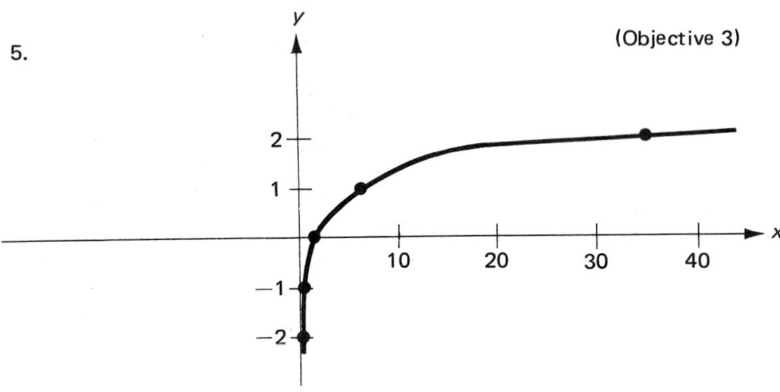

Unit 24

Note: These answers were found using a scientific calculator. If you are using Tables II and III, the third significant digit of your answer to some problems may differ by one.

Exercise 24.1

1. 1.63   2. 1.96   3. 2.73   4. .699

5. 1.23   6. −.468   7. .753   8. 1.84

Exercise 24.2

1. $4620   2. $650

3. 11.7, or about 12 years   4. 3.63, or about $3\frac{3}{4}$ years

5. 5.79 years   6. $21,200

7. 3310 bacteria   8. 34.7 hours

9. 6.77 grams   10. 576 hours

11. 5570 years   12. 12,900 years

Exercise 24.3

1. 1 hour

2. 9.96 seconds

3. $2110

4. 10.7 years

5. 5.76 milligrams

6. Superior: 431 years; Michigan: 71.4 years; Erie: 6.05 years; Ontario: 17.6 years

Self-test

1. .861   (Objective 1)

2. $743   (Objective 2)

3. 1.73 years   (Objective 2 or 3)

4. 485 bacteria   (Objective 2 or 3)

5. 4.61 seconds   (Objective 2 or 3)

Appendix

Squares and Square Roots

N	N^2	$\sqrt{N}$	$\sqrt{10N}$	N	N^2	$\sqrt{N}$	$\sqrt{10N}$
1	1	1.0 00	3.1 62	51	2 601	7.1 41	22.5 83
2	4	1.4 14	4.4 72	52	2 704	7.2 11	22.8 04
3	9	1.7 32	5.4 77	53	2 809	7.2 80	23.0 22
4	16	2.0 00	6.3 25	54	2 916	7.3 48	23.2 38
5	25	2.2 36	7.0 71	55	3 025	7.4 16	23.4 52
6	36	2.4 49	7.7 46	56	3 136	7.4 83	23.6 64
7	49	2.6 46	8.3 67	57	3 249	7.5 50	23.8 75
8	64	2.8 28	8.9 44	58	3 364	7.6 16	24.0 83
9	81	3.0 00	9.4 87	59	3 481	7.6 81	24.2 90
10	100	3.1 62	10.0 00	60	3 600	7.7 46	24.4 95
11	121	3.3 17	10.4 88	61	3 721	7.8 10	24.6 98
12	144	3.4 64	10.9 54	62	3 844	7.8 74	24.9 00
13	169	3.6 06	11.4 02	63	3 969	7.9 37	25.1 00
14	196	3.7 42	11.8 32	64	4 096	8.0 00	25.2 98
15	225	3.8 73	12.2 47	65	4 225	8.0 62	25.4 95
16	256	4.0 00	12.6 49	66	4 356	8.1 24	25.6 90
17	289	4.1 23	13.0 38	67	4 489	8.1 85	25.8 84
18	324	4.2 43	13.4 16	68	4 624	8.2 46	26.0 77
19	361	4.3 59	13.7 84	69	4 761	8.3 07	26.2 68
20	400	4.4 72	14.1 42	70	4 900	8.3 67	26.4 58
21	441	4.5 83	14.4 91	71	5 041	8.4 26	26.6 46
22	484	4.6 90	14.8 32	72	5 184	8.4 85	26.8 33
23	529	4.7 96	15.1 66	73	5 329	8.5 44	27.0 19
24	576	4.8 99	15.4 92	74	5 476	8.6 02	27.2 03
25	625	5.0 00	15.8 11	75	5 625	8.6 60	27.3 86
26	676	5.0 99	16.1 25	76	5 776	8.7 18	27.5 68
27	729	5.1 96	16.4 32	77	5 929	8.7 75	27.7 49
28	784	5.2 92	16.7 33	78	6 084	8.8 32	27.9 28
29	841	5.3 85	17.0 29	79	6 241	8.8 88	28.1 07
30	900	5.4 77	17.3 21	80	6 400	8.9 44	28.2 84
31	961	5.5 68	17.6 07	81	6 561	9.0 00	28.4 60
32	1 024	5.6 57	17.8 89	82	6 724	9.0 55	28.6 36
33	1 089	5.7 45	18.1 66	83	6 889	9.1 10	28.8 10
34	1 156	5.8 31	18.4 39	84	7 056	9.1 65	28.9 83

TABLE I (Continued)

N	N^2	$\sqrt{N}$	$\sqrt{10N}$	N	N^2	$\sqrt{N}$	$\sqrt{10N}$
35	1 225	5.9 16	18.7 08	85	7 225	9.2 20	29.1 55
36	1 296	6.0 00	18.9 74	86	7 396	9.2 74	29.3 26
37	1 369	6.0 83	19.2 35	87	7 569	9.3 27	29.4 96
38	1 444	6.1 64	19.4 94	88	7 744	9.3 81	29.6 65
39	1 521	6.2 45	19.7 48	89	7 921	9.4 34	29.8 33
40	1 600	6.3 25	20.0 00	90	8 100	9.4 87	30.0 00
41	1 681	6.4 03	20.2 48	91	8 281	9.5 39	30.1 66
42	1 764	6.4 81	20.4 94	92	8 464	9.5 92	30.3 32
43	1 849	6.5 57	20.7 36	93	8 649	9.6 44	30.4 96
44	1 936	6.6 33	20.9 76	94	8 836	9.6 95	30.6 59
45	2 025	6.7 08	21.2 13	95	9 025	9.7 47	30.8 22
46	2 116	6.7 82	21.4 48	96	9 216	9.7 98	30.9 84
47	2 209	6.8 56	21.6 79	97	9 409	9.8 49	31.1 45
48	2 304	6.9 28	21.9 09	98	9 604	9.8 99	31.3 05
49	2 401	7.0 00	22.1 36	99	9 801	9.9 50	31.4 64
50	2 500	7.0 71	22.3 61	100	10 000	10.0 00	31.6 23
N	N^2	$\sqrt{N}$	$\sqrt{10N}$	N	N^2	$\sqrt{N}$	$\sqrt{10N}$

Common Logarithms

N	0	1	2	3	4	5	6	7	8	9
1.0	.0000	.0043	.0086	.0128	.0170	.0212	.0253	.0294	.0334	.0374
1.1	.0414	.0453	.0492	.0531	.0569	.0607	.0645	.0682	.0719	.0755
1.2	.0792	.0828	.0864	.0899	.0934	.0969	.1004	.1038	.1072	.1106
1.3	.1139	.1173	.1206	.1239	.1271	.1303	.1335	.1367	.1399	.1430
1.4	.1461	.1492	.1523	.1553	.1584	.1614	.1644	.1673	.1703	.1732
1.5	.1761	.1790	.1818	.1847	.1875	.1903	.1931	.1959	.1987	.2014
1.6	.2041	.2068	.2095	.2122	.2148	.2175	.2201	.2227	.2253	.2279
1.7	.2304	.2330	.2355	.2380	.2405	.2430	.2455	.2480	.2504	.2529
1.8	.2553	.2577	.2601	.2625	.2648	.2672	.2695	.2718	.2742	.2765
1.9	.2788	.2810	.2833	.2856	.2878	.2900	.2923	.2945	.2967	.2989
2.0	.3010	.3032	.3054	.3075	.3096	.3118	.3139	.3160	.3181	.3201
2.1	.3222	.3243	.3263	.3284	.3304	.3324	.3345	.3365	.3385	.3404
2.2	.3424	.3444	.3464	.3483	.3502	.3522	.3541	.3560	.3579	.3598
2.3	.3617	.3636	.3655	.3674	.3692	.3711	.3729	.3747	.3766	.3784
2.4	.3802	.3820	.3838	.3856	.3874	.3892	.3909	.3927	.3945	.3962
2.5	.3979	.3997	.4014	.4031	.4048	.4065	.4082	.4099	.4116	.4133
2.6	.4150	.4166	.4183	.4200	.4216	.4232	.4249	.4265	.4281	.4298
2.7	.4314	.4330	.4346	.4362	.4378	.4393	.4409	.4425	.4440	.4456
2.8	.4472	.4487	.4502	.4518	.4533	.4548	.4564	.4579	.4594	.4609
2.9	.4624	.4639	.4654	.4669	.4683	.4698	.4713	.4728	.4742	.4757
3.0	.4771	.4786	.4800	.4814	.4829	.4843	.4857	.4871	.4886	.4900
3.1	.4914	.4928	.4942	.4955	.4969	.4983	.4997	.5011	.5024	.5038
3.2	.5051	.5065	.5079	.5092	.5105	.5119	.5132	.5145	.5159	.5172
3.3	.5185	.5198	.5211	.5224	.5237	.5250	.5263	.5276	.5289	.5302
3.4	.5315	.5328	.5340	.5353	.5366	.5378	.5391	.5403	.5416	.5428
3.5	.5441	.5453	.5465	.5478	.5490	.5502	.5514	.5527	.5539	.5551
3.6	.5563	.5575	.5587	.5599	.5611	.5623	.5635	.5647	.5658	.5670
3.7	.5682	.5694	.5705	.5717	.5729	.5740	.5752	.5763	.5775	.5786
3.8	.5798	.5809	.5821	.5832	.5843	.5855	.5866	.5877	.5888	.5899
3.9	.5911	.5922	.5933	.5944	.5955	.5966	.5977	.5988	.5999	.6010
4.0	.6021	.6031	.6042	.6053	.6064	.6075	.6085	.6096	.6107	.6117
4.1	.6128	.6138	.6149	.6160	.6170	.6180	.6191	.6201	.6212	.6222
4.2	.6232	.6243	.6253	.6263	.6274	.6284	.6294	.6304	.6314	.6325
4.3	.6335	.6345	.6355	.6365	.6375	.6385	.6395	.6405	.6415	.6425
4.4	.6435	.6444	.6454	.6464	.6474	.6484	.6493	.6503	.6513	.6522
4.5	.6532	.6542	.6551	.6561	.6571	.6580	.6590	.6599	.6609	.6618
4.6	.6628	.6637	.6646	.6656	.6665	.6675	.6684	.6693	.6702	.6712
4.7	.6721	.6730	.6739	.6749	.6758	.6767	.6776	.6785	.6794	.6803
4.8	.6812	.6821	.6830	.6839	.6848	.6857	.6866	.6875	.6884	.6893
4.9	.6902	.6911	.6920	.6928	.6937	.6946	.6955	.6964	.6972	.6981
5.0	.6990	.6998	.7007	.7016	.7024	.7033	.7042	.7050	.7059	.7067
5.1	.7076	.7084	.7093	.7101	.7110	.7118	.7126	.7135	.7143	.7152
5.2	.7160	.7168	.7177	.7185	.7193	.7202	.7210	.7218	.7226	.7235
5.3	.7243	.7251	.7259	.7267	.7275	.7284	.7292	.7300	.7308	.7316
5.4	.7324	.7332	.7340	.7348	.7356	.7364	.7372	.7380	.7388	.7396
N	0	1	2	3	4	5	6	7	8	9

TABLE II   (Continued)

N	0	1	2	3	4	5	6	7	8	9
5.5	.7404	.7412	.7419	.7427	.7435	.7443	.7451	.7459	.7466	.7474
5.6	.7482	.7490	.7497	.7505	.7513	.7520	.7528	.7536	.7543	.7551
5.7	.7559	.7566	.7574	.7582	.7589	.7597	.7604	.7612	.7619	.7627
5.8	.7634	.7642	.7649	.7657	.7664	.7672	.7679	.7686	.7694	.7701
5.9	.7709	.7716	.7723	.7731	.7738	.7745	.7752	.7760	.7767	.7774
6.0	.7782	.7789	.7796	.7803	.7810	.7818	.7825	.7832	.7839	.7846
6.1	.7853	.7860	.7868	.7875	.7882	.7889	.7896	.7903	.7910	.7917
6.2	.7924	.7931	.7938	.7945	.7952	.7959	.7966	.7973	.7980	.7987
6.3	.7993	.8000	.8007	.8014	.8021	.8028	.8035	.8041	.8048	.8055
6.4	.8062	.8069	.8075	.8082	.8089	.8096	.8102	.8109	.8116	.8122
6.5	.8129	.8136	.8142	.8149	.8156	.8162	.8169	.8176	.8182	.8189
6.6	.8195	.8202	.8209	.8215	.8222	.8228	.8235	.8241	.8248	.8254
6.7	.8261	.8267	.8274	.8280	.8287	.8293	.8299	.8306	.8312	.8319
6.8	.8325	.8331	.8338	.8344	.8351	.8357	.8363	.8370	.8376	.8382
6.9	.8388	.8395	.8401	.8407	.8414	.8420	.8426	.8432	.8439	.8445
7.0	.8451	.8457	.8463	.8470	.8476	.8482	.8488	.8494	.8500	.8506
7.1	.8513	.8519	.8525	.8531	.8537	.8543	.8549	.8555	.8561	.8567
7.2	.8573	.8579	.8585	.8591	.8597	.8603	.8609	.8615	.8621	.8627
7.3	.8633	.8639	.8645	.8651	.8657	.8663	.8669	.8675	.8681	.8686
7.4	.8692	.8698	.8704	.8710	.8716	.8722	.8727	.8733	.8739	.8745
7.5	.8751	.8756	.8762	.8768	.8774	.8779	.8785	.8791	.8797	.8802
7.6	.8808	.8814	.8820	.8825	.8831	.8837	.8842	.8848	.8854	.8859
7.7	.8865	.8871	.8876	.8882	.8887	.8893	.8899	.8904	.8910	.8915
7.8	.8921	.8927	.8932	.8938	.8943	.8949	.8954	.8960	.8965	.8971
7.9	.8976	.8982	.8987	.8993	.8998	.9004	.9009	.9015	.9020	.9025
8.0	.9031	.9036	.9042	.9047	.9053	.9058	.9063	.9069	.9074	.9079
8.1	.9085	.9090	.9096	.9101	.9106	.9112	.9117	.9122	.9128	.9133
8.2	.9138	.9143	.9149	.9154	.9159	.9165	.9170	.9175	.9180	.9186
8.3	.9191	.9196	.9201	.9206	.9212	.9217	.9222	.9227	.9232	.9238
8.4	.9243	.9248	.9253	.9258	.9263	.9269	.9274	.9279	.9284	.9289
8.5	.9294	.9299	.9304	.9309	.9315	.9320	.9325	.9330	.9335	.9340
8.6	.9345	.9350	.9355	.9360	.9365	.9370	.9375	.9380	.9385	.9390
8.7	.9395	.9400	.9405	.9410	.9415	.9420	.9425	.9430	.9435	.9440
8.8	.9445	.9450	.9455	.9460	.9465	.9469	.9474	.9479	.9484	.9489
8.9	.9494	.9499	.9504	.9509	.9513	.9518	.9523	.9528	.9533	.9538
9.0	.9542	.9547	.9552	.9557	.9562	.9566	.9571	.9576	.9581	.9586
9.1	.9590	.9595	.9600	.9605	.9609	.9614	.9619	.9624	.9628	.9633
9.2	.9638	.9643	.9647	.9652	.9657	.9661	.9666	.9671	.9675	.9680
9.3	.9685	.9689	.9694	.9699	.9703	.9708	.9713	.9717	.9722	.9727
9.4	.9731	.9736	.9741	.9745	.9750	.9754	.9759	.9763	.9768	.9773
9.5	.9777	.9782	.9786	.9791	.9795	.9800	.9805	.9809	.9814	.9818
9.6	.9823	.9827	.9832	.9836	.9841	.9845	.9850	.9854	.9859	.9863
9.7	.9868	.9872	.9877	.9881	.9886	.9890	.9894	.9899	.9903	.9908
9.8	.9912	.9917	.9921	.9926	.9930	.9934	.9939	.9943	.9948	.9952
9.9	.9956	.9961	.9965	.9969	.9974	.9978	.9983	.9987	.9991	.9996
N	0	1	2	3	4	5	6	7	8	9

Table III

Natural Logarithms

N	0	1	2	3	4	5	6	7	8	9
1.0	0000	0100	0198	0296	0392	0488	0583	0677	0770	0862
1.1	0953	1044	1133	1222	1310	1398	1484	1570	1655	1740
1.2	1823	1906	1989	2070	2151	2231	2311	2390	2469	2546
1.3	2624	2700	2776	2852	2927	3001	3075	3148	3221	3293
1.4	3365	3436	3507	3577	3646	3716	3874	3853	3920	3988
1.5	4055	4121	4187	4253	4318	4383	4447	4511	4574	4637
1.6	4700	4762	4824	4886	4947	5008	5068	5128	5188	5247
1.7	5306	5365	5423	5481	5539	5596	5653	5710	5766	5822
1.8	5878	5933	5988	6043	6098	6152	6206	6259	6313	6366
1.9	6419	6471	6523	6575	6627	6678	6729	6780	6831	6881
2.0	6931	6981	7031	7080	7129	7178	7227	7275	7324	7372
2.1	7419	7467	7514	7561	7608	7655	7701	7747	7793	7839
2.2	7885	7930	7975	8020	8065	8109	8154	8198	8242	8286
2.3	8329	8372	8416	8459	8502	8544	8587	8629	8671	8713
2.4	8755	8796	8838	8879	8920	8961	9002	9042	9083	9123
2.5	9163	9203	9243	9282	9322	9361	9400	9439	9478	9517
2.6	9555	9594	9632	9670	9708	9746	9783	9821	9858	9895
2.7	9933	9969	*0006	*0043	*0080	*0116	*0152	*0188	*0225	*0260
2.8	1.0296	0332	0367	0403	0438	0473	0508	0543	0578	0613
2.9	0647	0682	0716	0750	0784	0818	0852	0886	0919	0953
3.0	1.0986	1019	1053	1086	1119	1151	1184	1217	1249	1282
3.1	1314	1346	1378	1410	1442	1474	1506	1537	1569	1600
3.2	1632	1663	1694	1725	1756	1787	1817	1848	1878	1909
3.3	1939	1969	2000	2030	2060	2090	2119	2149	2179	2208
3.4	2238	2267	2296	2326	2355	2384	2413	2442	2470	2499
3.5	1.2528	2556	2585	2613	2641	2669	2698	2726	2754	2782
3.6	2809	2837	2865	2892	2920	2947	2975	3002	3029	3056
3.7	3083	3110	3137	3164	3191	3218	3244	3271	3297	3324
3.8	3350	3376	3403	3429	3455	3481	3507	3533	3558	3584
3.9	3610	3635	3661	3686	3712	3737	3762	3788	3813	3838
4.0	1.3863	3888	3913	3938	3962	3987	4012	4036	4061	4085
4.1	4110	4134	4159	4183	4207	4231	4255	4279	4303	4327
4.2	4351	4375	4398	4422	4446	4469	4493	4516	4540	4563
4.3	4586	4609	4633	4656	4679	4702	4725	4748	4770	4793
4.4	4816	4839	4861	4884	4907	4929	4951	4974	4996	5019
4.5	1.5041	5063	5085	5107	5129	5151	5173	5195	5217	5239
4.6	5261	5282	5304	5326	5347	5369	5390	5412	5433	5454
4.7	5476	5497	5518	5539	5560	5581	5602	5623	5644	5665
4.8	5686	5707	5728	5748	5769	5790	5810	5831	5851	5872
4.9	5892	5913	5933	5953	5974	5994	6014	6034	6054	6074
5.0	1.6094	6114	6134	6154	6174	6194	6214	6233	6253	6273
5.1	6292	6312	6332	6351	6371	6390	6409	6429	6448	6467
5.2	6487	6506	6525	6544	6563	6582	6601	6620	6639	6658
5.3	6677	6696	6715	6734	6752	6771	6790	6808	6827	6845
5.4	6864	6882	6901	6919	6938	6956	6974	6993	7011	7029
N	0	1	2	3	4	5	6	7	8	9

TABLE III   **(Continued)**

N	0	1	2	3	4	5	6	7	8	9
5.5	1.7047	7066	7084	7102	7120	7138	7156	7174	7192	7210
5.6	7228	7246	7263	7281	7299	7317	7334	7352	7370	7387
5.7	7405	7422	7440	7457	7475	7492	7509	7527	7544	7561
5.8	7579	7596	7613	7630	7647	7664	7681	7699	7716	7733
5.9	7750	7766	7783	7800	7817	7834	7851	7867	7884	7901
6.0	1.7918	7934	7951	7967	7984	8001	8017	8034	8050	8066
6.1	8083	8099	8116	8132	8148	8165	8181	8197	8213	8229
6.2	8245	8262	8278	8294	8310	8326	8342	8358	8374	8390
6.3	8405	8421	8437	8453	8469	8485	8500	8516	8532	8547
6.4	8563	8579	8594	8610	8625	8641	8656	8672	8687	8703
6.5	1.8718	8733	8749	8764	8779	8795	8810	8825	8840	8856
6.6	8871	8886	8901	8916	8931	8946	8961	8976	8991	9006
6.7	9021	9036	9051	9066	9081	9095	9110	9125	9140	9155
6.8	9169	9184	9199	9213	9228	9242	9257	9272	9286	9301
6.9	9315	9330	9344	9359	9373	9387	9402	9416	9430	9445
7.0	1.9459	9473	9488	9502	9516	9530	9544	9559	9573	9587
7.1	9601	9615	9629	9643	9657	9671	9685	9699	9713	9727
7.2	9741	9755	9769	9782	9796	9810	9824	9838	9851	9865
7.3	9879	9892	9906	9920	9933	9947	9961	9974	9988	*0001
7.4	2.0015	0028	0042	0055	0069	0082	0096	0109	0122	0136
7.5	2.0149	0162	0176	0189	0202	0215	0229	0242	0255	0268
7.6	0281	0295	0308	0321	0334	0347	0360	0373	0386	0399
7.7	0412	0425	0438	0451	0464	0477	0490	0503	0516	0528
7.8	0541	0554	0567	0580	0592	0605	0618	0630	0643	0656
7.9	0669	0681	0694	0707	0719	0732	0744	0757	0769	0782
8.0	2.0794	0807	0819	0832	0844	0857	0869	0882	0894	0906
8.1	0919	0931	0943	0956	0968	0980	0992	1005	1017	1029
8.2	1041	1054	1066	1078	1090	1102	1114	1126	1138	1150
8.3	1163	1175	1187	1199	1211	1223	1235	1247	1258	1270
8.4	1282	1294	1306	1318	1330	1342	1353	1365	1377	1389
8.5	2.1401	1412	1424	1436	1448	1459	1471	1483	1494	1506
8.6	1518	1529	1541	1552	1564	1576	1587	1599	1610	1622
8.7	1633	1645	1656	1668	1679	1691	1702	1713	1725	1736
8.8	1748	1759	1770	1782	1793	1804	1815	1827	1838	1849
8.9	1861	1872	1883	1894	1905	1917	1928	1939	1950	1961
9.0	2.1972	1983	1994	2006	2017	2028	2039	2050	2061	2072
9.1	2083	2094	2105	2116	2127	2138	2148	2159	2170	2181
9.2	2192	2203	2214	2225	2235	2246	2257	2268	2279	2289
9.3	2300	2311	2322	2332	2343	2354	2364	2375	2386	2396
9.4	2407	2418	2428	2439	2450	2460	2471	2481	2492	2502
9.5	2.2513	2523	2534	2544	2555	2565	2576	2586	2597	2607
9.6	2618	2628	2638	2649	2659	2670	2680	2690	2701	2711
9.7	2721	2732	2742	2752	2762	2773	2783	2793	2803	2814
9.8	2824	2834	2844	2854	2865	2875	2885	2895	2905	2915
9.9	2925	2935	2946	2956	2966	2976	2986	2996	3006	3016
N	0	1	2	3	4	5	6	7	8	9

Index